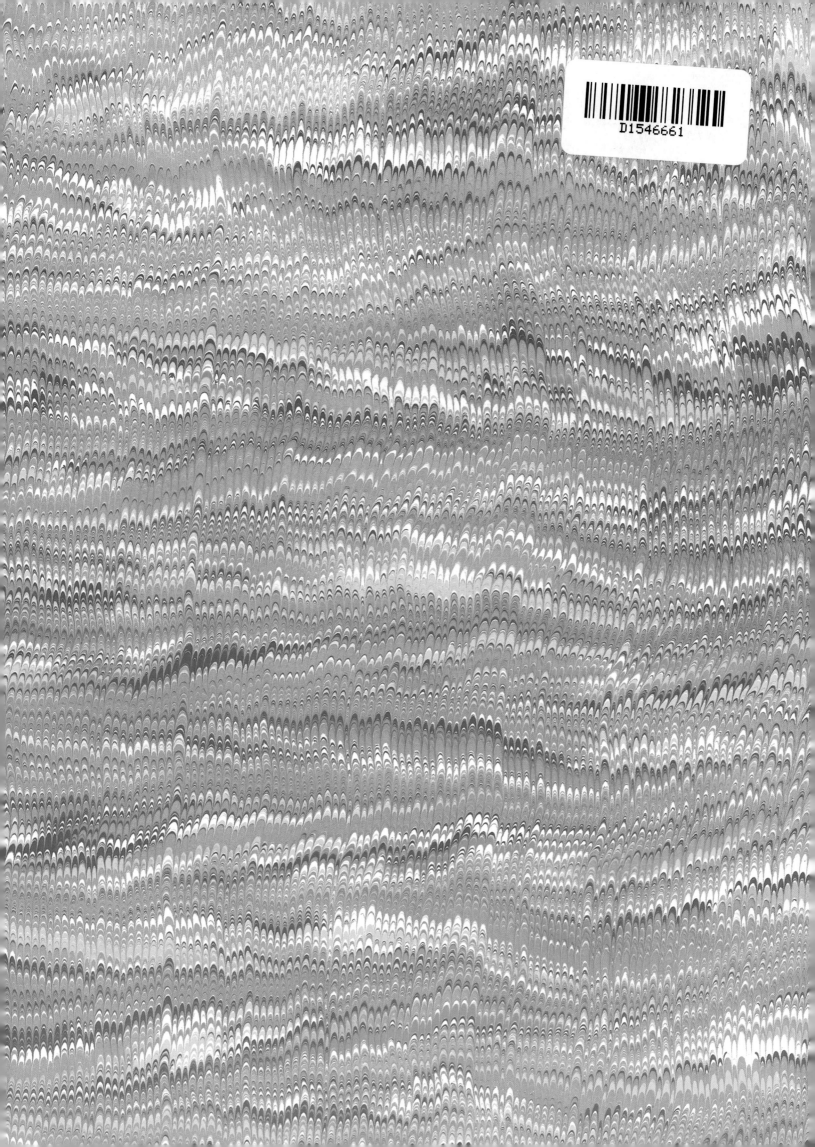

SECRET AIRCRAFT DESIGNS OF THE THIRD REICH

DAVID MYHRA

SECRET AIRCRAFT DESIGNS OF THE THIRD REICH

DAVID MYHRA

Schiffer Military/Aviation History
Atglen, PA

ACKNOWLEDGMENTS

In the time I spent researching, writing, and assembling this book, I benefitted from the generous help and assistance of a large number of individuals and organizations throughout the world. I extend my heartfelt thanks to everyone, but am especially grateful to Dipl.-Ing. Werner Wilde of the Deutsche Gesellschaft für Luft-und Raumfahrt, Köln, and to Reinhard Roeser of Langenhagen, Germany. It was these two individuals who helped me collect research material, some of it from their personal collections, for many of the chapters in this book. My thanks to Philip Edwards of the Smithsonian Institution's National Air and Space Museum Library, Washington, District of Colombia and Dieter Multhopp of Cincinnati, Ohio, for their splendid cooperation.

A substantial number of photographs appearing in this book came from private collections. I wish to thank the Deutsches Museum, München, Germany; David Goode of the Royal Aircraft Establishment in Farnborough, Hants, England; Frau Gertrude Lippisch, Cedar Rapids, Iowa; Frau Emmy Vogt, Santa Barbara, California; Dr. Peter Vogt, Port Republic, Maryland; Dipl.-Ing. Hans Wocke, Hamburg, Germany; Hans Amtmann, Rancho Santa Fe, California; Walter Stender, Gamering, Germany; Jean Christophe Carbonel, Nanis, France; Arthur L. Schoeni, Dallas, Texas; J. Richard Smith, Brimscombe, Stroud, England, Martin G. Winter, Columbus, North Carolina; Dr. Berhard Goethert, Manchester, Tennessee; Stephen Coats, Eastleigh, Hants, England; Frau Sigrid Tank, München, Germany; Dr. Professor Heinrich Hertel, Berlin, Germany; Frau M. Zindel, Bad Hamburg, Germany; Frau Liselotte Henschel, Kusnacht, Switzerland; Frau Clara Voigt, Annapolis, Maryland; Colonel Hugh W. Cowin, Hatfield, Herts, England; Otto Dahlke, Corvallis, Oregon; Dr. Gotthold Matheis, Redondo Beach, California; Walter Horten, Baden Baden, Germany; Dr. Reimar Horten, Córdoba, Argentina; Dr. Patrick Nolan, Wright State University, Dayton, Ohio; Werner P. Henschel, Erlangen, Germany; Mrs. Julie Gustafson, US Defense Audio Visual Agency, Washington, District of Columbia; Sgt. Robert Burten, US Air Force Systems Command, Washington, District of Columbia; Joe Pretsch, Mesa, Arizona; Heinz Nowarra, Babenhausen, Germany; and Dr. Volker Koos, Rostock, Germany.

My gratitude goes to Valerie Northcott-Stephanski, Naples, FL, for her interest and skill in correcting the manuscript from first drafts through final copy. My appreciation goes to Elka Alikhan, Reston, Virginia, for her tireless and eager translation of German language source material.

Finally, my deepest heartfelt thanks must go to four other people: Dan Johnson, Hampton, Virginia, Loretta Dovell, Stephens City, Virginia, and Mario Merino, Dallas, Texas, and Andreas Ott. Dan Johnson is a skilled builder of fine scale model aircraft. Photographs of his scale models appear throughout this book. In addition, Dan encouraged his friends from around the world to provide photographs of their scale models. This book has become a global undertaking. Loetta Dovell is an artist and her outstanding original aviation artwork appears in this book. Mario Mercino's 3D digital images are state-of-the-art and through his skill we can see German project aircraft in a new dimension never seen before. My thanks to each and everyone of you, and especially to Betsy Hertel, the love of my life, thank you dear for all your support.

All drawings in the color gallery are courtesy Mario Merino and Andreas Otte.

Book Design by Ian Robertson.

Copyright © 1998 by David Myhra.
Library of Congress Catalog Number: 97-81279

All rights reserved. No part of this work may be reproduced or used in any forms or by any means – graphic, electronic or mechanical, including photocopying or information storage and retrieval systems – without written permission from the copyright holder.

Printed in China.
ISBN: 0-7643-0564-6

We are interested in hearing from authors with book ideas on related topics.

Published by Schiffer Publishing Ltd.
4880 Lower Valley Road
Atglen, PA 19310
Phone: (610) 593-1777
FAX: (610) 593-2002
E-mail: Schifferbk@aol.com
Please write for a free catalog.
This book may be purchased from the publisher.
Please include $3.95 postage.
Try your bookstore first.

CONTENTS

Preface 006
Introduction 007

Chapter 1	Arado	028
Chapter 2	Bachem	048
Chapter 3	Blohm and Voss	058
Chapter 4	BMW	082
Chapter 5	DFS	092
Chapter 6	Dornier	102
Chapter 7	Focke-Wulf	110
Chapter 8	Gotha	144
Chapter 9	Heinkel	152
Chapter 10	Henschel	180
Chapter 11	Horten	224
Chapter 12	Junkers	243
Chapter 13	Lippisch	268
Chapter 14	Messerschmitt	290
Chapter 15	Sänger	320
Chapter 16	Sombold	326
Chapter 17	Zeppelin	332

Epilogue 336
Glossary 340
Index 346

WALTER HORTEN

A young American aviation historian is at the door.
Shall I let him in?
If I let him in, I'll have less leisure. I'll talk of airplanes. Horten all-wing airplanes
If I let him in, I'll tell him the struggles three brothers from Bonn experienced in their efforts to perfect the all-wing planform.
If I let him in, I'll talk at length of Udet, of Göring, of critics, of friends and supporters.
If I let him in, he'll still see the tears in my eyes as I talk of the agony felt then as now for fellow Lufwaffe boys who with their Messerschmitt mounts fell in the Battle of Britain.
If I let him in, I'll change my daily routine for we'll go for walks along wooded ways with chocolate ice cream cones grasped firmly in hand.
If I let him in, I'll talk hour after hour and when he's gone write long detailed letters.
If I let him in, lives may change forever becoming better persons for having known one another.
If I let him in, I'll grant him my time, strength, and skill responding to his thousand questions.
If I let him in, I'll introduce him around to people I know and love saying this young American is my friend so come and meet him.
If I let him in, I'll begin to think of him as a son.
If I let him in, there may be no limit to the things I'll do for him. Yet I am content within this house, with my life, my memories and old friends.
A young American aviation historian is at the door wishing to know about the Horten brothers.
Thanks and best wishes Walter. Good friend - I wonder often how you're flying.

Preface

This book grew out of many lunch-time conversations in 1950 with a former high-ranking German Air Ministry engineer. As a boy of ten I liked to accompany my father, Olaf Myhra, and his work crew during the summer months while they built grain elevators and livestock barns for the farmers in rural southeastern North Dakota. One day I happened to meet a friendly old man at one of the farms who loved to reminisce about airplanes of the Third Reich. His English was "nicht so gut," but I understood well enough while the rich, black North Dakota soil surrounding one particular new barn twenty to thirty miles outside the town of Wahpeton where we lived, was a complaisant go-between, seeming to welcome an unending number of aircraft designs scratched into it by that expert's stick. Unfortunately, neither my father nor I remember the man's name. The farm on which he was a guest has changed hands several times since the 1950s, so I had little hope of ever learning who he was and in what capacity he carried out his work. My own recollection is that he was a pilot with the Luftwaffe in World War I. He continued his flying as an instructor in sailplane activities throughout the 1920s, and was connected with the Reichsluftfahrtministerium or German Air Ministry from 1933 through 1945.

Whatever this man's background, he knew intimately about the proposed tailless aircraft designs of Alexander Lippisch, the all-wing designs of the Horten brothers and those swept-forward wing bomber designs of Hans Wocke of Junkers. He knew about the twin-fuselage Heinkel He P.1078B and the other exotic designs from Ernst Heinkel AG. He told me about the Hans Multhopp designed Fw Ta 183 turbine-powered fighter and its descendant, the MiG-15, which was just beginning to appear in Korea. He talked about the flights to the edge of space planned by a man and wife design team named Sänger. From there people would travel to the stars in rocket ships pioneered by another German designer named von Braun. Pretty heady stuff for a boy of ten, whose only aircraft experience had come out of a box, the contents of which, when assembled, turned out to be rather conventional P-47s, B-17s, P-51s, and other piston-powered airplanes.

Eventually the barn was completed - or perhaps it was because fall came and school started again. At any rate, all of us went our separate ways. But I didn't forget the turbine-powered and rocket-driven aircraft he told me about, then sketched in that North Dakota dirt. From time to time I would recall those noon-time conversations, but it would be thirty years before I would have the time and resources to investigate these exotic aircraft for myself. To this ancient German Air Ministry engineer whom I had the good fortune to meet in 1950 on a farm in North Dakota, and whose name I no longer remember, thanks for creating within a lifelong interest in the actual as well as proposed turbine and bi-fuel rocket-powered aircraft of the Third Reich.

Introduction

There was a time in the relatively short history of turbine-powered and liquid rocket-driven aircraft when the design and engine technology found in Germany was the world's best and most creative. This was in the 1940s, a Golden Age when designers such as Ernst Heinkel, Willy Messerschmitt, Hans Multhopp, the Günter twins, Alexander Lippisch, the Horten brothers, Felix Kracht, and Woldemar Voigt, and manufacturers such as Heinkel AG, Arado, Messerschmitt AG, and Bachem were producing exotic aircraft that routinely approached the speed of sound (some beyond) and climbed to heights 8 to 9 miles (13 to 15 km) above the earth in a matter of minutes. When it all ended on 7 May 1945 with Germany's surrender, few Allied countries took advantage of either the German designs or their designers. With the German air industry completely closed down (it would remain so for the next ten years before aircraft manufacturing was allowed again in 1955), the very people who had fathered some of the most advanced aerodynamic designs the world had known virtually walked away and were never heard from again.

There were a few exceptions. The Soviets busily dismantled entire German aircraft factories lying within their zone of occupation and hastily moved them to the USSR. Plant and equipment did not work out to be the booty they had hoped for. When train loads of equipment arrived in the USSR no one knew what pieces went where due to improper or no labeling at all of the pieces dismantled. However, the two to three hun-

Hans Wocke Junkers Flugzeugbau-Dessau designer pursuing the idea of the swept-forward wing.

Hans Wocke's six-turbine-powered Junkers Ju 287 V3 (Ju EF 131). All components for its assembly had been completed at the time of Germany's May 1945 surrender. When the Soviets occupied eastern Germany in July 1945 they immediately invited all Junkers people back to the factory in Dessau and the V3 was assembled and test flown. Later, in the summer of 1946 the 287 along with its engineers and families were transported to a new community near Moscow.

Secret Aircraft Designs of the Third Reich

Hans Multhopp, design genius from Focke-Wulf and one of the brightest aviation minds of his generation.

Multhopp's proposed radical fighter design of 1944/45 the Focke-Wulf FwTa 183.

A radical design for its time, too, the Horten Flugzeugbau's all-wing turbojet-powered fighter, the Horten Ho 229 V3.

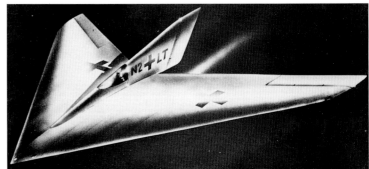

The Horton Ho 13B proposed single seat fighter with a huge vertical fin with an attached hinged rudder of 1944-45. Airbrush by Gert W. Heumann.

The proposed Horton Ho P.10B delta wing fighter design of 1944-45.

LEFT: Multhopp's other proposed radical fighter design, the Lorin ramjet-powered fighter interceptor the Focke-Wulf FwTa 283.

Introduction

A typical 1940's facial expression of Alexander Lippisch.

Dr. Heinrich Hertel and a very capable aircraft designer. Long-time designer with Heinkel and then near war's end with Junkers. Personally interested in the delta wing planform and at Junkers proposed a wide selection of delta winged aircraft for the RLM.

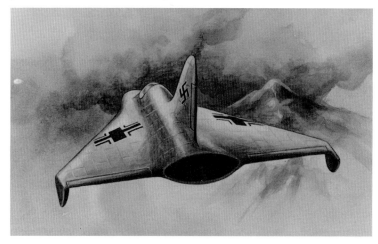

A pair of Lippisch delta wing designs: the Lippisch Li DM-1 and the proposed Li P.13A. Scale models by Steve Malikoff.

The proposed Lippisch Li P.12 delta wing fighter and the mother of all modern delta winged aircraft including the space shuttle.

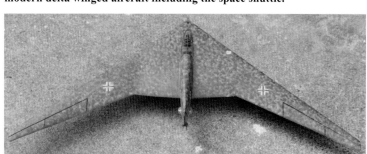

The proposed Junkers "Amerika Bomber" project of 1945, their modified Horten Ho 18A and known to Junkers personnel as the Junkers Ju EF 140.

The Junkers Ju 248 a larger and improved version of the standard Messerschmitt Me 163.

dred specialists of German aviation they collected in their zone of occupation (eastern Germany) after 1945 was another matter. Unlike physical plant and equipment, Heinkel AG, BMW, and Junkers specialists under their control could go on designing no matter where they might be replanted. This windfall of technical personnel with defeated Germany's latest and freshest aviation designs later showed up in a few early turbojet-powered fighters designed by Aleksandr Yakvlev (Yak-15) and the design team of Artem Mikoyan and Mikhail Gurevich (MiG). Sweden's SAAB 29-A is based on German airframe research as was the Douglas F-4D Skyray and Bell Aircraft's X-5 in America, the latter a copy of Messerschmitt's P.1101. After consulting for a time with Lippisch, Convair came out with some delta-wing projects that led to the F-102 Delta Dagger, the F-106 Delta Dart, and the B-58 Hustler bomber, all of which were based on Lippisch's DM-1 delta glider design. Most of Kurt Tank's aircraft design group from Focke-Wulf followed their leader to Argentina where a grateful Juan Péron was presented with a version of Multhopp's Ta 183 prototype jet fighter, known in South America as the Pulque Dos.

For the most part, the world ignored the aeronautical achievements of the Germans. Even today only a handful of pictures of this almost unbelievable collection of turbine-powered and rocket-driven aircraft have found their way into literature on period German aircraft. There are photographs, indeed, several of these very aircraft are with us today in museums - Messerschmitt AG's Me 163 and 262, Arado's Ar 234B reconnaissance bomber, Bachem's Ba 349B, Heinkel AG's He 162 "People's Fighter, and the Horten brothers stealth-like all-wing Ho 229 V3 high-altitude fighter/fast bomber. Several secret designs were captured and after finishing touches applied were flown. The Soviets fell heir to the six turbojet heavy bomber Junkers Ju 287 V3, the liquid rocket-driven Ju 248 interceptor, and the ramjet-powered Ju P.127. The Soviet's may have beaten the United States to the speed of sound in 1947 with their captured DFS 346 liquid rocket-driven high speed research aircraft. Larry Bell of Bell Aircraft, Buffalo, NY took the incomplete Me P.1101 and built an exact copy he called the Bell X-5. But there were others. Literally dozens of high altitude fighter and fast bomber designs were drafted on paper and abandoned. Many other project designs were in various stages of design development when the hostilities ended in May 1945.

Woldemar Voight, the very articulate designer of the Messerschmitt Me 262 in the late 1930s and the Messerschmitt Me P.1101 at war's end. Said post-war that Willy Messerschmitt personally fired him a couple dozen times during his career. Messerschmitt, after cooling off, rehired Voight usually the next day.

The Woldemar Voight design which began the turbojet revolution in 1938: the revolutionary Messerschmitt Me P.1065 and later known to the world as the Me 262.

One of the several proposed streamlined versions of the Messerschmitt Me 262. This one with its turbines tucked into the wing-root was known as the Me 262 HG-2. The "HG" stands for *Hochgeschwindikeit* or high speed.

Introduction

Woldemar Voight's second effort and an entirely new design, the Messerschmitt Me P.1101 swing wing fighter. Post war Voight said that the P.1101's configuration simply wouldn't jell into a workable design despite all the talent of its master craftsman. Scale model by Dan Johnson.

It is not surprising that the United States did not readily embrace the German aerodynamic accomplishments. Americans have seldom been ones to copy designs when they find themselves in second place. Furthermore, after World War II the United States was preoccupied with rocket technology, and to a lesser extent with turbine propulsion. While America was seeking out noted German rocket experts such as Wernher von Braun and Hellmuth Walter, gifted aircraft designers such as Siegfried Günter and Reimar Horten were unable to find work. Political problems were a large factor, but were ignored if the skills of the individual were in demand, as they were in the cases of von Braun and Walter. Airframe designers simply were not needed in the United States and Günter ended up in the Soviet Union while Kurt Tank and Reimar Horten went to Argentina. The loss to the West of not drawing on Günter's expertise, in particular, was substantial. According to his former employer, Ernst Heinkel, one of Germany's largest aircraft manufacturers, Günter was "one of the most important experts on airplane structures and aerodynamics in Europe in the 1940s." His work on the P.1078 and P.1079 turbine series, said Heinkel embraced everything his firm was planning for the 1950s in the way of fresh turbine-propelled airframe development.

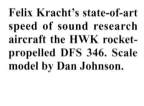

Felix Kracht's state-of-art speed of sound research aircraft the HWK rocket-propelled DFS 346. Scale model by Dan Johnson.

Woldemar Voight's proposed fast bomber project of 1944-45 with numerous components borrowed from his Me 262. Scale model by Jamie Davies.

Dr. Ing. Felix Kracht. Chief designer at Deutsche Forschungsanstalt für Segelflug (DFS).

Dr.Ing Rudolf Göthert.

Dr.Ing. Richard Vogt of Blohm & Voss.

Göthert's proposed replacement for the Horten Ho 229V3, all wing fighter, the Gotha Go P.60B. Scale model by Reinhard Roeser.

Vogt's proposed bomber project, the unusual Blohm & Voss Bv P.188. Illustration by Hugh W. Cowin.

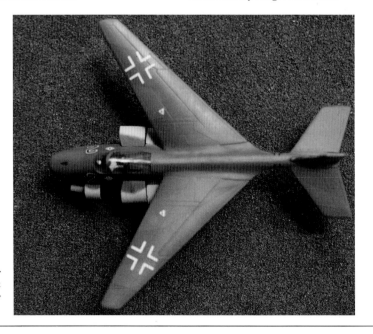

The only known proposed turbojet-powered fighter design of Blohm & Voss with a swept-forward wing, the Bv P.209.02. Scale model by Dan Johnson.

Introduction

Siegfried Günter soon after his return from the USSR in 1954 after eight years of forced labor as an aircraft designer.

The proposed Heinkel He P.1078C. Heinkel liked this one as well as the He P.1079A.

The very modern-looking proposed Heinkel He P.1079A fighter. Scale model by Dan Johnson.

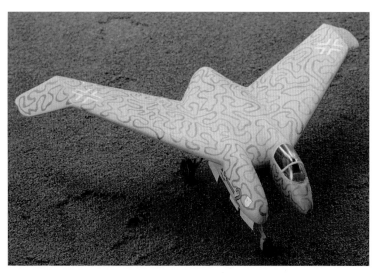

The proposed unusual-looking Heinkel He P.1078B fighter. After the war Ernst Heinkel stated that the He P.1078B was the most stupid aircraft design he'd ever seen. Scale model by Dan Johnson.

The proposed Heinkel He P.343 A-1. Scale model by Dan Johnson.

Walter Blume of Arado: talented designer, good manager, and a fighter ace of WWI with 28 confirmed "kills."

After failing repeatedly post-war to interest the Allies in his abilities, Günter, then living in England, used the last of his money to fly back to Berlin in 1945 to see his father-in-law about work. Neither the Americans nor the British sent for him as he had hoped they would. Lacking money as well as job prospects in the west, he went to work for the Soviet Union's experimental aircraft design unit, OKB-4, in East Berlin, then was transferred to OKB-2 at Podberesje, USSR in October 1946. At OKB-2 Günter was welcomed by Professor Bock, former head of DVL, Hans Wocke, designer of the Ju 287, Wilhelm Benz, co-designer of the He 176 rocket plane and designer of the He P.1078 "Julia." Günter remained at OKB-2 until being returned home with several hundred of his colleagues in 1954. He never spoke about his eight years of aircraft design work in the USSR. Once, a journalists asked him if he had designed the MiG 15 for the Soviets? Günter quietly replied: "I did not design the MiG 15." Well, somebody did a lot for the Soviets post-war because they didn't even have a turbojet engine at war's end in May 1945. It later appeared that a good many captured German aviation and turbojet engine designers helped the Soviet Union leap-frog into the world of supersonic flight but that is another story. Ernst Heinkel once said in 1950, "I am convinced that Günter has worked on those Soviet designs that today (1950) have become a problem for the Western World." After Günter returned to West Germany in 1954 he rejoined his old company Heinkel AG when the industry reopened in 1955 after the World War II ban on manufacturing aircraft in Germany was lifted.

The Arado Ar 234.B2. Scale model by Dan Johnson.

The Arado Ar 234C. Scale model by Günter Sengfelder.

Introduction

The proposed Arado Ar E.581.4 fighter. Scale model by Dan Johnson.

The highly successful MiG design team; Antrem Mikoyan (middle), and Mikhail Gurevich (left).

The Hans Multhopp fighter design sensation; the Focke-Wulf Fw Ta 183.

SAAB of Sweden's shameless copy of Multhopp's Ta 183, the SAAB 29-A.

The MiG design team's shameless copy of Multhopp's Ta 183, known as the MiG 15.

Voight's Messerschmitt Me P.1101 prototype. Shown here in the hangar post-war at Bell Aircraft, Buffalo, New York, being used as a full-scale mock-up for the Bell X-5.

The Bell Aircraft's X-5. America's first fighter aircraft capable of changing the degree of wing sweep-back in flight.

The beautiful delta winged Alexander Lippisch-influenced B-58 "Hustler."

Professor Dr.Ing. Kurt W. Tank former general manager of Focke-Wulf, shown here in retirement.

Introduction

Hans Multhopp's beautiful Focke-Wulf Fw Ta 183. Scale model by Günter Sengfelder.

The Argentine version of Multhopp's Ta 183 by Kurt Tank, known as the Pulqui Dos, (Arrow 2).

Why was Germany so advanced in airframe design and turbojet- and rocket-aeroengine technology in the 1940s? Much had to do with the Treaty of Versailles in 1918 which brought an end to World War I. The details of the treaty meant that manufacture and use of powered aircraft was limited for several years. To continue their flying, all of Germany turned to sailplanes and designers such as Messerschmitt, Lippisch, the Horten brothers, the Günter twins, and others began their aeronautical-work with sailplanes. They and others perfected the flying qualities of their sailplanes through study of aerodynamics - building and flying, rebuilding and flying again, and so on. Messerschmitt frequently boasted throughout his life that one could not build high performance aircraft unless he knew how to build gliders...and that he, Messerschmitt, could build gliders while blindfolded.

Germany was also advanced in airframe design and engine technology because of men like Messerschmitt and Heinkel, men who valued straight line performance above all else. As early as the mid 1930s, the forward thinkers recognized the limitations of piston-driven propellers. Believing that a propeller-powered aircraft probably would never fly faster than 500 mph (805 km/h), Heinkel in particular, and Messerschmitt to a lesser extent, were looking for propulsion units and radically designed air-frames. Not everyone in Germany believed 500 mph was the limiting speed for a propeller. Kurt Tank and Claudius Dornier believed all through the 1940s that supersonic speeds could be obtained by a propeller-driven aircraft. They were wrong. In addition, some men were seeking to disprove the theory that neither man or machine would survive as speeds approached that of sound. Alexander Lippisch, one of Germany's great practicing aerodynamicists, for example, believed that there were no limits on man's ability to achieve forward motion. Of course, testing the limits depended on finding suitable propulsion units and efficient airframe designs.

It took until 1955 to confirm Ernst Heinkel's belief that there were limits on the speed that propeller-driven aircraft could achieve. That was when the US Air Force ended its efforts to develop a propeller-driven strike fighter by substituting an Allision XT40 turboprop in a Republic XF-84H jet airframe. Although the XF-84's mission was not to test the limits of propeller-driven aircraft, it is known that the aircraft suffered tremendous difficulties (propeller reversing), but the most serious problem was caused by the supersonic propeller. When running, it produced extreme nausea in most personnel standing near the aircraft.

Ernst Heinkel, 1960.

Dr. Ing. Claudius Dornier.

The Republic XF-84 H experimental turboprop fighter, designed to test the feasibility of the supersonic propeller.

A Lippisch design idea the Messerschmitt Me 163, taking off.

No performance data on the XF-84H was ever released and it is now displayed on a plinth at Kern County Airport, Bakersfield, California.

In addition to the German designers' quest for speed and performance, an equally compelling reason for their accomplishments in the 1940s was Germany's declining fortunes in World War II. From 1943 onward German designers were called on increasingly to produce defensive weapons, which to the aircraft industry meant missiles, fighters, and interceptors. Every design resulted in an increase in power in turbine and liquid rocket drive units, and the resultant great strides in speed, led to every kind of new aerodynamic problem. German designers such as Hans Multhopp who was the most gifted aerodynamicist to come out of Germany, tackled these problems vigorously.

Just what were the Germans producing? Some of their designs were truly advanced, far too advanced for the technology available to translate them into worthwhile or even flyable machines. These included the rocket-driven aircraft of Lippisch, the orbital bomber of Sänger, the tailless fighters of Heinkel and of Blohm and Voss, and the all-wings of the Horten brothers (even though the Horten's Ho 229 was well advanced for production). Other were old-fashioned, conservative designs for airframes originally housing piston engines of the 1930s and 1940s and later fitted with turbines for propulsion. Especially prominent during the last 2 years of the war were designs for turbine and rocket-driven high altitude interceptor aircraft, designs that under peacetime condi-

The most famous of all German gas turbines, the Jumo 004B. Over 5,000 units had been produced yet in May 1945 production was only at 1,500 units a month.

A German chemical technician puts a few drops of C-Stoff into a dish containing T-Stoff. As the two chemicals combine a violent chemical reaction occurs.

The remains of a Messerschmitt Me 163 after it exploded on take-off. Frequently the bi-fuel rocket was more dangerous to itself and pilot than the heavily armed B-17.

Rudolph Hess and Hermann Göring flank Adolf Hitler at Templehofer Airport. Göring's girth earned him the popular title in the Luftwaffe as "The Fat One."

General Walther Wever.

A Kurt Tank designed Focke-Wulf Fw 200 "Condor." At the beginning of the war Tank said that America could never mass produce B-17's like they mass produced autos. Udet said to him "Tank, you are completely wrong and we will see B-17's driving all over our skies."

Generalleutnant Albert Kesselring.

Ernst Udet doing one of the things he liked best, shown here on a rescue flight over the "White Hells of Pitz Palo," in the Swiss Alps.

A happy looking Erhard Milch on a field trip late in the war with senior staff.

Ernst Udet participating in a children's model airplane competition. This is where he first learned of the Horten Brothers and their all-wing model sailplanes.

Albert Speer listening very intently as Adolf Hitler appears to be asking how he might stop the destruction of the Third Reich by Allied bombers.

Introduction

What Hitler was referring to was how to stop the likes of this B-17 which boldly called itself "Never Satisfied" and showed proudly the tonnage of bombs dropped and Nazi fighters shot down.

In America more "Never Satisfieds" waiting for crew members to be trained in order to fly them over the skies of Germany. Production like this is exactly what Udet knew would happen once America entered the war. Tank's comments at this time have not been recorded.

tions would not have been feasible for all-purpose fighter aircraft. There were also designs for heavy bombers, but only a few prototypes were constructed because heavy bombers were not considered an important part of Germany's war strategy early on.

The organization that carried out the planning function in Germany in the 1930s and 1940s was the Reichsluftfahrtministerium, the RLM or German Air Ministry. The roots of the Air Ministry began with Adolf Hitler. In contrast with his fascination with high-performance automobiles, Hitler generally was unimpressed by high performance aircraft. Nevertheless, shortly after he seized power on 30 January 1933, he announced a bold plan for the creation of a new Luftwaffe with 4,000 airplanes.

History had shown the way for Hitler. By the end of World War I in 1918 it was pretty clear that airplanes had added a new dimension to warfare and had become an important tool for ground troops. In an offensive role airplanes could seek out enemy aircraft in the skies and bomb their bases, thus eliminating the air threat to army ground troops. In a tactical role airplanes could attack enemy ground formations and harass the movement of troops and supplies to the Front. Finally, in a defensive role, airplanes could protect one's own ground troops from enemy attacks.

As Hitler was coming to power, there was no Air Ministry as such, instead he appointed Hermann Göring as the Reich or empire's com-

missioner for Civil Aviation and Erhard Milch as his deputy. The Treaty of Versailles had imposed near-paralyzing air restrictions on Germany, as postwar Europe entered the 1920s. Its terms stipulated that the defeated Germany be left with only 140 aircraft, all of which were to be used for commercial purposes. This situation resulted in the organization of hundreds of sailplane/soaring clubs throughout Germany by ex-flyers and adventurous youth. When Hitler came to power in 1933, officials of the German War Office, or Wehrmacht, under the control of the Army, assumed that they would be in control of the fledgling secret Luftwaffe, just as they had been in World War I. However, Göring had no such intentions, and on 15 May 1933, he got his way by having the War Office, under Wernher von Blomberg, transfer all military air activities to him and his Commission for Civil Aviation. This transfer created the Reichsluftfahrtministerium, and in effect gave birth to an independent Luftwaffe. The birth of the Luftwaffe was pretty much kept secret form the rest of the world (due to the fact that the Treaty of Versailles was still in effect) until 1 March 1935, when it was publically brought out into the open as the official German Air Force. It would be the job of the Air Ministry to develop and supply all the necessary aircraft for the Luftwaffe. This meant working in close collaboration with the German aircraft industry, academic institutions, and the like, to conduct scientific research on aircraft design and efficient management in the production of airplanes.

A collection of British aerial bombs ranging in tonnage from 40 lbs to 22,000 lbs.

A group of B-17s on their way to Germany from bases in England.

Several days after a bombing raid on a German city when all the fires had died out.

Langenhagen airfield near Hannover which had been used by Focke-Wulf Flugzeugbau for aircraft testing. Shown April 1945. The dispersals and flight-test buildings have been destroyed by Allied heavy-bomber attacks leaving this facility virtually useless.

To get the near-dormant German industry active again, and the infant Luftwaffe up to a level of strength equal or superior to that of other nations of the world, Göring needed help. He himself was neither qualified or particularly interested in the details. He had no engineering or technical training, and was totally without corporate managerial experience. Nor was he a student of economics or history, and he lacked a whole-hearted dedication to aviation and the function of air power in a tactical or strategical role. He was, however, a retired captain of the phenomenally successful Richthofen Squadron and the recipient of the Order Pour le Mérite. Official records failed to confirm that Göring had in fact scored the necessary twenty-five kills required to be eligible for Prussia's highest military decoration, he is thought to have obtained only twenty-two kills.

The first Chief of Staff for the new Air Ministry was Generalleutnant Walther Wever, a brilliant officer by anyone's standards. He died on 3 June 1936 in a flying accident caused by his own impatience. Eager to get his Heinkel He 70 "Blitz" into the air at Dresden and headed back to Berlin late one evening, Wever failed to remove the aileron steering

A soldier takes in the aftermath of Allied systematic strategic bombing.

Messerschmitt AG-Augsburg. A huge bomb crater and in the background a Messerschmitt Me 323 "Gigant," also destroyed.

Messerschmitt AG-Augsburg. A bomb crater in the middle of this hangar rendered everything useless junk.

Messerschmitt AG-Augsburg. More bomb craters and more destroyed aircraft in former hangar facilities.

Messerschmitt AG-Augsburg. The high heat generated by bomb explosions has warped and twisted the steel support beams in this former hangar.

system block and the aircraft crashed moments after takeoff. Had Wever lived to mold the Luftwaffe into the fighting unit he believed necessary (he was a proponent of the strategic bomber concept), it is certain that Germany would have had large, four-engine heavy bomers for their attacks on England, as well as the ability to destroy Russian armament factories beyond the Ural mountains. However, the large long-range bomber program was dropped after Wever was succeeded by Generalleutnant Albert Kesselring in 1936, Generaloberst Hans Stumff in 1937, and Generaloberst Hans Jeschonnek in February 1939.

Another important figure in the early days of the Air Ministry was Ernst Udet. Like Göring, Udet was a born fighter pilot and ace from World War I (with 62 confirmed enemy kills Udet was the fourth highest scoring ace of the Great War 1914-1918). Udet was at that time a movie star, a world renowned stunt pilot. Widely popular in the United States, Udet was the sort of man many young Germans wished to imitate and young Germans joined the Luftwaffe by the tens of thousands. Although not a Nazi, Udet was appointed by Göring in 1936 to reorganize the Air Ministry's technical department and to attract new recruits as airmen to the Luftwaffe. The Technical Office, or Technisches Amt, was the main activity in the Air Ministry. In addition, there was the Nachschub Amt, or Supply Office, and the Wirtschafts Amt, or Contracts Office. Within these three directorates all aircraft research and planning, as well as supply and procurement of various airplanes, was carried out. Generaloberst Ernst Udet commtted suicide on 17 November 1941 reportedly over his alleged mishandling of all matters of aircraft production. Later in 1944, Albert Speer was placed in charge of Air Ministry activities as part of his Ministry for Armament and Munitions. All departments and divisions of the Air Minstry connected with the production were transferred to Speer's Ministry and came under Karl Saur, Chief of Technical Air Equipment.

Looking back, it is easy to see the design flaws that would have led to dead ends, the bi-fuel liquid rocket driven Messerschmitt Me 163 fighter interceptor for example, with it short flight duration and its tendency to explode destroying both pilot and aircraft. German design was on the cutting edge (with proposed aircraft such as Hans Multhopp's Fw Ta 183 and Felix Kracht's DFS 346), and anything that appeared possible was tried. Heinkel AG's He 162 "People's Fighter" a small, lighweight interceptor, had a fatal design flaw resulting from poor placement of its wing and air intake for its dorsal-mounted turbine. The product of intensive collaboration by airframe designers, the plane was designed and built in 74 days. The first He 162 was delivered in December 1944 and crashed 2 days later. But persistence was the byword, and

Several new-looking Boeing B-29s.

everything, including experimental designs, was being tried in an effort to turn back the British Lancaster's and the United States Army Air Force's (USAAF) B-24 and B-17 heavy bombers. However, high altitude interceptors, target defense interceptors, fighter aircraft, and so on are not developed quickly and haphazardly.

A fighter aircraft, for example, is basically a design compromise. Ideally, it should be maneuverable and controllable, quick to climb and able to sustain repeated dives on targets, carrying wing-mounted weapons, and be easily serviced under field conditions - but rarely do the military and the aircraft's designers get their wish for all these characteristics in one aircraft. To get superior performance in one area, they may have to give a little in another. For example, an aircraft may not climb as quickly or go as fast as desired if it must carry a lot of externally mounted weapons and still be simple enough to be serviced in remote field operations. The general weakness of nearly all the German turbine and bi-fuel liquid rocket-driven aircraft developed during World War II was a lack of good design compromises. In many instances things that make for a balanced fighter were overlooked; in other instances, they were deliberately overlooked (such as the case with Lippisch's Me 163 and he was the first to admit it). Lippisch's bi-fuel liquid rocket-

A collection of German anti-aircraft missiles which range from ground-to-air and air-to-air, which were to be hurled at passing Allied bombers near war's end but which came much, much too late to be effective. Airbrush by Gert W. Heumann.

An American turbojet-powered fighter prototype from Bell Aircraft; the Bell P-59

A German issued postage stamp of the 1940s, showing ground to air missiles, the caption states "The Great German Empire."

The ground-to-air anti-aircraft missile "Enzian" and its launch ramp.

driven interceptors, for instance, are considered to have had about as much value as a "slingshot," although they were astonishingly quick to climb, attaining perhaps 32,800 ft (10 km) in seconds, they had to rely on their glide momentum to attack and were good for only a few minutes of combat. Unable to stay and fight, they had little value. Consequently, these designs could seldom be regarded as adequate for fighter/interceptor aircraft.

Few of the German turbine and bi-fuel liquid rocket-propelled aircraft demonstrated fighter plane characteristics. Instead, a host of new techniques were designed to increase straight-line performance or interception capabilities. We see all-wing, tailless, delta, variable-angle wings, winglets, swept-forward, swept-backward wings, and other features engineers hoped would increase performance. Individual features, such as the swept-wing and variable angle wing, fitted on Messerschmitt's P.1101, were truly advanced concepts. The Horten's 229 all-wing fighter, full of technical problems (directional stability) that only now are being understood, represent a concept that eventually will be adopted more and more to fighter aircraft such as the Stealth Fighter. The German

The Horten brothers, Reimar left and Walter.

Reichmarshall Hermann Göring. "Hermann" is an ancient German name at least 2,000 years old meaning "Warrior." This man was the Horten's strongest supporter at war's end and was pushing the Ho 229 and the Ho 18B "Amerika Bomber."

Secret Aircraft Designs of the Third Reich

Oberst Siegfried Knemeyer of the RLM.

346. On the following pages, a sample of known German turbojet and bi-fuel liquid rocket driven aircraft designs are pictured. In a few cases the paintings and scale models shown are based on a great deal of information about the aircrafts fuselage shape, wing type, wing location, tailplane, engine, and general performance. But in most cases all there is to go on are design sketches, some wind tunnel scale models, and/or the proposed aircraft's general statistics. In addition, several turbojet and bi-fuel liquid rocket-driven aircraft survived World War II and are now located in museums such as the Messerschmitt Me 163 and Me 262, Arado's Ar 234B, Bachem's Ba 349B, Heinkel's He 162, and Horten's Ho 229.

As a general practice German aircraft were not given names. Instead, as each design was begun the aircraft was assigned a design number selected by the Air Ministry. Turbojet and bi-fuel liquid rocket-driven aircraft started out with a "Project" (usually shortened to "P") number. Most of the "P" series designs were discarded even before a prototype was ordered. Some of these early numbers were dropped after a prototype had been built, while others, having lost in the Air Ministry sponsored design competitions, were eliminated from the firm's portfolio. Several manufacturers used "EF" number designations, meaning "Eprobungsflugzeug," or experimental aircraft. The Junkers company, for example, assigned an EF number to turbine-propelled aircraft. Within the Arado the prefix "E" meaning Entwung or project, was used for their proposed aircraft designs.

designs were on the cutting edge, the right track, but they had no idea what they were getting into in terms of translating their designs into production aircraft and meeting fighter performance criteria.

Nevertheless, many of the German designs were truly awe-inspiring such as the bi-fuel liquid rocket-driven supersonic research DFS

Aircraft that survived to operational status were predesignated, for example, as Me or He and followed by a number. The letters are simply an abbreviation of the manufactures's name, for example "Me" for Messerschmitt or "He" for Heinkel. The major and minor 1933-1945 German aircraft manufacturing companies and the city of their headquarters are as follows:

The proposed Horten Ho 18A, "Amerika Bomber."

Introduction

AGO	Aktien Gesellschaft Otto Flugzeugbau GmbH - Oschersleben/Bode
Ar	Arado Flugzeugwerke GmbH - Berlin/Babelsberg
Ba	Bachem-Werke GmbH - Waldsee/Württemberg
Bf	Bayerische Flugzeugwerke (later Messerschmitt AG) - Augsburg
BMW	Bayerische Motoren Werke GmbH - München
Bü	Bücker Flugzeugbau GmbH - Berlin/Rangsdorf
Bv	Blohm und Voss Schiffswerke-Abteilung Flugzeugbau - Hamburg
DB	Daimler-Benz AG - Stuttgart
DFS	Deutsches Forchunginstitut für Segelflug - Darmstadt/Griesheim
Do	Dornier Werke GmbH - Friedrichshafen
Fa	Focke, Achgelis GmbH - Delmenhorst
Fh	Flugzeugbau Halle GmbH - Halle
Fi	Gerhard Fieseler Werke GmbH - Kassel
Fk	Flugzeugbau Kiel GmbH - Kiel
Fl	Anton Flettner GmbH - Berlin/Johannisthal
Fw	Focke-Wulf Flugzeugbau GmbH - Bremen
Go	Gothaer Waggonfabrik AG - Gotha
Ha	Hamburger Flugzeugbau (later Blohm und Voss) - Hamburg
He	Ernst Heinkel AG - Rostock/Marienhe
Ho	Horten Flugzeugbau GmbH - Bonn
Hs	Henschel Flugzeugbau AG - Berlin/Schönefeld
Hü	Hütter Segelflugzeug-Konstruckteurs
HWK	Hellmuth Walter Kiel - Kiel
Ju	Junkers Flugzeugbau AG - Dessau
Jumo	Junkers Motoren Werke - Dessau
Kl	Hans Klemm Flugzeugbau und Leichtflugzeugbau GmbH - Böblingen
Li	Alexander Lippisch Luftfahrtforschungsanstalt - Vienna
Me	Messerschmitt AG - Augsburg
Nr	Nagler-Rolz Flugzeugbau - Vienna
Pe	Peschke Flugzeugbauwerke - Minden
Sa	Eugen Sänger - Trauen
Si	Siebel Flugzeugbauwerke KG - Halle
Sk	Skoda-Kauba Flugzeugbau - Prag/Cakowitz
So	Heinz G. Sombold - Naumburg
Ta	Kurt Tank (Focke-Wulf) - Bremmen
Wn	Wiener-Neustädter-Flugzeugbau GmbH - Wiene/Neustädt
Ze	Luftschiffbau Zeppelin und Abteilung Flugzeugbau GmbH - Friedrichshafen

It is estimated that there were more than three-hundred proposed turbojet-powered and bi-fuel liquid rocket-driven aircraft. The large number may give the reader the impression that the German aviation industry was somehow superior to any other country. Although it was certainly advanced in many areas, one cannot judge an industry from numbers of "projects" alone. As Hugo Junkers put it so well: "Aircraft ideas are as cheap as blackberries." To an idea, he said, must be added practicality, money, and time - and the last commodity was very, very scarce in Germany during the desperate period in which most of the German projects discussed in this book were conceived. One must also bear in mind that, whereas the details of a great portion of the German work eventually were released to the public, details of the work of the victorious powers are not so easily found for comparison, because in the spirit of competition and national security, aircraft companies generally keep close guard on details of advanced aviation development.

A brief discussion accompanies Loretta Dovell's beautiful rendition of each German aircraft. It is not my intention to criticize the designs, their intended purposes, use or airworthiness of any particular aircraft design. I am amazed by most of them and see each, with a few exceptions, as beautifully designed, well-proportioned pieces of art.

David Olaf Myhra
Naples, Florida, USA
1998

The proposed Horten Ho 18B "Amerika Bomber" and the version which Göring instructed the Horten brothers to construct as soon as possible.

1

Arado Flugzeugwerke GmbH - Berlin, Germany

During World War I he flew with such Luftwaffe greats as Manfred von Richthofen, Ernst Udet, and Hermann Göring. By war's end in October 1918, he had recorded twenty-eight confirmed enemy "kills," and like von Richthofen, Udet, and Göring, and a handful of other men, had received Germany's highest award for valor in the skies - the Order Pour le Mérite (the blue Maltese Cross, or Blue Max). Twenty-three years later (in 1941) he would bring Germany into the turbojet age years ahead of any other country with his long-distance reconnaissance airplane, the Arado Ar 234, which is considered one of the cleanest aerodynamic designs to come out of World War II. The design work for the Ar 234 earned him the coveted title of "Professor," but he seldom used it. Titles were shallow things, he believed. The reward of a thing well done was to have done it, and that was reward enough. A person didn't need a piece of metal with a pretty ribbon around his neck to tell his story, and one should not advertise one's accomplishments so boldly.

This acclaimed but unassuming man was Walter Blume, born on 10 January 1896, in Hirschber/Riesengebirge. While still a student, Blume witnessed Orville Wright's flying demonstrations in Germany during the autumn of 1909. Thereafter he followed the advances of Germany's own aviation pioneers with great enthusiasm, and soon he decided that he, too, would one day become an airplane builder and pilot. But his dreams were interrupted by the start of World War I - or so he first thought.

Blume volunteered for army infantry service on 4 August 1914, at age 18, just a few days after the start of the hostilities. After suffering a gunshot wound in his leg, he was transferred in 1915 to the pilot replacement division, where he was assigned to a pilot training class. By June 1916 he had become a full-fledged pilot in the Luftwaffe. He served for a while with an air reconnaissance group at Schlettstadt in Alsac. Then, becoming bored with aerial reconnaissance, he managed to get transferred to a fighter squadron at Muhlausen, also in Alsac. On 10 May 1917, he recorded his first confirmed air victory by downing a British Bristol bi-wing fighter. Although injured several times in aerial combat, Blume had scored a total of twenty-eight confirmed kills by the end of the war.

With the war over Blume entered the Technical University at Hannover to study engineering. To help pay for his education he worked at a variety of jobs, including flight director for the Reich's mobile defense camp, volunteer pilot of a messenger squadron, and aerobatic pilot in Holland. With the founding of the Akaflieg (flying club) at Hannover, Blume and two of his college friends built the glider Vampyr. The Vampyr went on to become one of the most successful glider designs available in the mid 1920s, the standard by which new models came to be judged in terms of flight performance.

After graduation from the Technical University in 1924, Blume's first employer was the Albatros Flugzeugwerke in Berlin. He remained with Albatros until 1931, when the world wide economic depression forced the firm to close its doors. During the seven years Blume spent working for Albatros, he helped on designs that later were developed into sport and pilot training aircraft. Aircraft built according to several of his team's designs participated in international round-the-world flights in 1930.

Left to right: Ernst Udet, Robert Lucht, (a senior manager at RLM and good friend of Udet), Walter Blume, and Willy Messerschmitt. Berlin, 1938.

Arado Flugzeugwerke GmbH - Berlin, Germany

The Wright brothers, with a great deal of help from German citizens, pull the great Wright bi-wing to its launch site for a demonstration flight. Somewhere in Germany 1909. Walter Blume saw one of these demonstrations by the Wright brothers and decided that he would make aviation his life's work.

Blume left Albatros and in 1932 joined the Arado Flugzeugwerke in Warnemünde. When he became manager of Arado's technical staff a year later, many long-term employees were suspicious because the promotion coincided with Hitler's rise to power and the appointment of Blume's fellow pilot during World War I, flying ace Hermann Göring, as the Reich's Minister of Aviation. Perhaps there was some connection - but no one could complain that it was bad for Arado's business, for the firm immediately received several contracts for a number of new pilot-training aircraft. In fact, Arado was only one of many small companies with aircraft design and manufacturing experience to benefit from Hitler's call for a 4,000-plane Luftwaffe. Most of these firms, as well as many inexperienced ones, were eager to get involved in the lucrative aircraft manufacturing business but were badly capitalized and had only limited funds with which to finance plant expansion. To make certain they were geared up for the massive aircraft production initiated by the Air Ministry, the Reich provided expansion funds for many small businesses. On 4 March 1933, Arado became the Arado Flugzeugwerke GmbH. The firm was under the control of the Air Ministry in terms of the direction it would take, whether it could design its own planes or merely manufacture aircraft designs by larger, more established firms.

Although the Air Ministry did order a large number of Arado's single-seat fighter-trainers, the company was viewed primarily as a licensee for the production of aircraft developed by other companies. From 1935 onward Arado assembly lines were occupied largely by aircraft developed by Heinkel, Messerschmitt, and Focke-Wulf.

In the late 1930s the Air Ministry came to realize that, apart from seaplanes manufactured by Blohm and Voss, Heinkel, and Dornier, Germany still lacked a good, fast naval-reconnaissance aircraft. Although the Dornier Wa1 (whale) flying boat was considered a naval reconnaissance plane and among the best in its class, the Air Ministry viewed it as only a temporary solution to the problem. Discussions about a more suitable naval-reconnaissance aircraft took place among several aircraft manufacturers, and the Air Ministry decided in the autumn of 1940 that any new reconnaissance airplane would have to be fast and able to cruise easily at high altitudes. This ruled out aircraft in the seaplane category. To obtain the high speed the Air Ministry wanted the proposed aircraft to be powered by the new turbine engines then being bench-tested at Bayerische Motoren Werke (BMW) and at Junkers Jumo. The Air Ministry was willing to take a chance with the untried turbines because the staff believed that revolutionary power plants, if they worked as anticipated, would permit high operational speeds and altitudes sufficient to make the proposed reconnaissance plane immune from enemy fighter interception.

Arado was selected to design and build the proposed aircraft, designated the Ar 234, in large part because of Walter Blume's design creativity, but perhaps even more because of his organizational skills and respected efficiency in managing large groups of engineers, designers, and technicians. It was these traits that allowed him to rise rapidly through various management posts within Arado until, by 1945, he had become the firm's general manager. It was also these very traits that may have contributed to the firm's mediocrity in the area of design after the successful Ar 234B-2. Arado's turbine-powered designs grew increasingly conservative, becoming somewhat old fashioned as Blume himself became preoccupied with management tasks and the duties of running a growing aircraft company.

By war's end Arado had a relatively large portfolio of turbine-powered design projects. At least ten different turbine designs existed, about half high-speed reconnaissance bombers and half bad-weather fighter aircraft. But, other than the Ar 234B-2, Blume entered only two of the firm's turbine designs in Air Ministry design competition. The proposed aircraft, both entered in competition in 1945, were the P.NJ-1, a twin-engine tailless night fighter, and the P.NJ-2, a night fighter with a standard tail and similar in appearance to Dornier's Do P.256. Neither of Arado's designs were selected. The Air Ministry cited their relatively heavy weight, poor performance, and conservative designs as reasons for rejection. Blume blamed himself for Arado's poor performance in turbine-powered aircraft designs after his trend setting Ar 234B-2, but in his defense it can be said that it is pretty difficult for one man to lead a large advance aircraft design group while being increasingly given general management responsibilities and eventually the general managership of the company. Time ran out before Walter Blume could rebuild Arado's design team with fresh, creative talent while he concentrated on the day-to-day operations of his Flugzeugbau.

Walter Blume, age 24, in his popular "Vampyr" sailplane, Hannover, 1920.

Walter Blume in the early 1960s. He died in Germany 7 March, 1964, aged 68.

Arado ceased operations at the end of the war and did not reopen in 1955, when the ban prohibiting all aircraft design and manufacturing in Germany was lifted. Blume was interned for a brief time by the Allies in 1945 and was freed. He chose not to leave Germany for career opportunities elsewhere. "I have participated fully in our Luftwaffe's thirty-year lifetime," he said after the war, "I know no other life, I'm staying to help restore what has been undone."

In 1948 Blume became manager of an aircraft metal society in Düsseldorf, helping the metals industry rebuild by providing scientific research and support. In 1953 he opened his own engineering firm for light aircraft construction and flight techniques, Duisburg-Ruhrort GmbH. With others in his firm he designed a four-seat, all metal, sport-and-personal travel aircraft called the Bi 500 which was one of the first civilian aircraft built in Germany after the war; it was manufactured under license at Focke-Wulf's newly opened facilities in 1955. After 1955, when military aircraft manufacturing again was possible, more companies re-entered the aviation business, and Blume enthusiastically assisted them by providing design assistance.

In addition to Blume's aircraft design consulting work, he was elected to the board of directors of the new Deutsche Versuchsanstalt für Luftfahrt (DVL) in 1955 when it was being re-established. He became a co-founder of the Scientific Society for Aviation and was also the editor of the Magazine for Flight Sciences.

Walter Blume died on 27 March 1964 at the age of 68. Although he had earned the Order Pour le Mérite during World War I and was awarded the title of Professor for his aircraft research and development on the Ar 234, he instructed his co-workers never to call him "Professor" nor to mention his many air victories between 1917 and 1918. He viewed the Ar 234B-2 as the most satisfying achievement of his long service to the German nation. Perhaps if he had not risen so fast in management he and Arado would have followed up the Ar 234B-2's success with more creative designs. Still, the aircraft embodied his lifelong philosophy that one's work should speak for itself. Blume wanted to be remembered after his death as an exceedingly loyal servant to his fatherland and an aircraft designer of passing ability. Out of the 210 Ar 234B's built before production was halted in February 1945 in favor of the 234C, only

Arado Ar 234 B-2, photographed at NASM-Silver Spring, Maryland, 1 February 1989. At its first public demonstration since it was fully restored. This Arado Ar 234 is the only known surviving example in the world out of over 200 constructed. However, several were buried at the end of NAS Patuxent River's runway in the 1960s.

NASM's Arado Ar 234 B-2, while it was still under going flight testing in the mid to late 1940s.

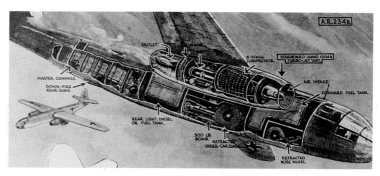

A see-through view by U.S. Army Intelligence, of the Arado Ar 234 showing placement of its various components.

one survives. It has been beautifully and faithfully restored and is on display at the Smithsonian Institution's National Air and Space Museum, Washington, DC.

Blume reflected before his death that all of the honors he'd received, even the title of Professor, would mean little to future generations. However, the fact that a world-class aviation museum had the only existing example of his best work, the Ar 234B, and that millions of people from all over the world eventually would be able to see and touch his accomplishment, that would be his reward for work well done.

PROJECT 234B-2

Although it was only a reconnaissance-bomber aircraft, flying routine rather than glamorous missions, the Ar 234 was a history-maker, for it spearheaded the Third Reich's introduction of turbine engine powered aircraft into service. Introduced in 1944, it was a single-seat, shapely unswept, shoulder-mounted wing, an all-round-vision cockpit, single vertical fin with an attached rudder, and twin Jumo 004B turbines each having 1,980 lb (900 kg) of thrust. Each 004B was housed in its own nacelle tucked under the wing. Conceived by Arado's then chief of design, Walter Blume, his assistant Hans Rebeski, and aerodynamist Rüdinger Kosin in autumn 1940 in response to a design specification issued by the Air Ministry for a high-speed, turbojet-powered reconnaissance aircraft with 1,250 mi (2200 km) radius of operation. It was to be powered by two of the revolutionary new turbine engines being developed by Junkers Jumo and the Bayerische Motoren Werke (BMW). The design was completed in early 1941, and Arado's winning entry, known as the E.370, later designated as the Ar 8-234 when Generalfeldmarshall Milch gave Arado an order to build six test aircraft. This was in April 1942. Milch's initial order was increased to twenty in December 1942 with full production of the 234B started in late 1943.

The Air Ministry's specification for the reconnaissance aircraft required that it have a range of at least 1,340 miles (2,156 km). Arado was able to deliver the range, but because of the aircraft's high, shoulder-level wings, narrow fuselage, and internal fuel tanks, there was no space to store the undercarriage. So the first prototype had no wheels; when the plane flew for the first time on 30 July 1943, it took off from a detachable three-wheel trolley and landed on a built-in retractable skid.

NASM's Arado Ar 234 B-2 retired and in storage, prior to its restoration.

A typical Arado Ar 234 B-2. Scale model by Dan Johnson.

A rear overhead view of an Arado Ar 234 B-2. Scale model by Dan Johnson.

Rear ground level view of an Arado Ar 234 B-2. Scale model by Dan Johnson.

Ground level side view of an Arado Ar 234 B-2. Scale model by Dan Johnson.

Ground level front view of an Arado Ar 234 B-2. Scale model by Dan Johnson.

Front overhead view of an Arado Ar 234 B-2. Scale model by Dan Johnson.

It was not until the ninth prototype, the Ar 234 V9, that a conventional undercarriage was designed into the aircraft. To make the gear fit, the fuselage was widened to accommodate the two main wheels midway along its length, while the nose wheel retracted into a space below the pilot's seat. The track of the undercarriage was narrow, like that of the Messerschmitt Bf 109, and this did create some instability during taxiing; however, it generally was satisfactory. Those Ar 234's with the landing skid arrangement were known as the 234A. When the retractable undercarriage was made standard, this version was known as the 234B.

The first 234 with a retractable undercarriage, the V-9, first flew on 12 March 1944 and was the prototype for the B series. Since the prototypes had so few design and engineering faults, production of twenty Ar 234Bs was completed by 8 June 1944, at assembly lines at Alt-Lönnewitz, which now is on the Czechoslovak-German border. Each of the twenty initial production models included a retractable undercarriage, a pilot-ejection seat, an autopilot, a pressurized cockpit, and auxiliary fuel drop tanks under each of the turbine engines.

At an early stage in the development of the aircraft, the design team found that the basic airframe could readily accommodate more power than the original pair of Jumo 004B turbines, and by the end of the war, the Ar 234's basic frame was being adapted to accommodate a large number of experimental variations. Aircraft in the C series, which production had began in February 1945, had four BMW 003 turbines and a modified cockpit section (a raised cabin roof and less extensive glazing than that of the B series with only two turbines). To be used for several purposes, C series aircraft, with their four turbines, were to have two forward-firing and two rearward-firing 20 mm MG 151 cannon. The C-4, a reconnaissance model, projected to have a maximum speed of 547 mph (880 km/h) at 16,405 ft (4 km), was to be the fastest of the proposed production versions of the basic airframe design was proposed that the Ar 234 C-2 also be employed to tow a Henschel Hs 294 or a Fieseler Fi 103 piloted V-1 flying bomb. There was also a project underway to place an Fi 103 in a fixed cradle on the upper surface of the C2's fuselage. The Fi 103 would be launched by a series of hydraulically operated arms that lifted it clear of its cradle and the Ar 234 C-2's vertical fin/rudder. In another project, Arado engineers were attempting to fit an Ar E.381 rocket-propelled target interceptor be carried to the war's front beneath an Ar 234 C-5 reconnaissance aircraft. Plans called for release of the miniature interceptor high above an Allied bomber formation after the E.381's single HWK 509 bi-fuel rocket drive had been started.

Numerous other experiments were conducted on the basic Ar 234 airframe. One was the use of laminar-flow wings of wooden construc-

Arado Flugzeugwerke GmbH - Berlin, Germany

An Arado Ar 234 B-2 moments after take-off assisted by two HWK solid fuel booster rockets.

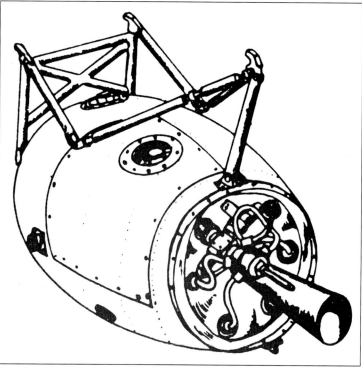

A pen and ink drawing of an HWK solid fuel booster rocket showing its aft-end exhaust thrust nozzle.

Top photo: an HWK solid fuel booster rocket. The white bag on front is where its recovery parachute was stored. The booster rockets were recycled.
Bottom photo: HWK booster rocket without its recovery parachute.

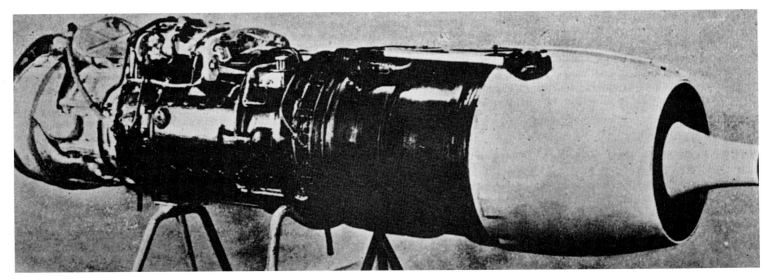

The BMW 003A turbojet engine. Development started later than that of the Jumo 004 and did not proceed as quickly with the result that Junkers produced many more engines than BMW. This engine was capable of producing 1,760 lbs of thrust verses 1,890 lbs for the Jumo 004B.

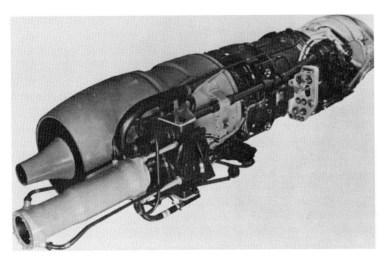

The BMW 003R - bi-fuel rocket combined with a gas turbine. This partaicular model with its bi-fuel BMW 718 liquid rocket engine was intended for use on the Arado Ar 234 V-16 providing extra thrust during take-off and rapid climb during night fighting.

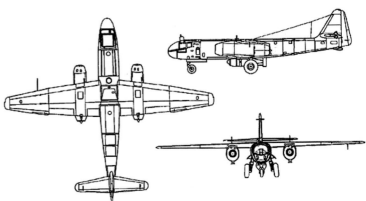

The Heinkel-Hirth HeS 011 turbojet capable of producing 2,866 lb (1,300 kg) of thrust.

Arado Flugzeugwerke GmbH - Berlin, Germany

A production Arado Ar 234C showing the closeness of its twin BMW 003 turbines housed in a single pod per wing.

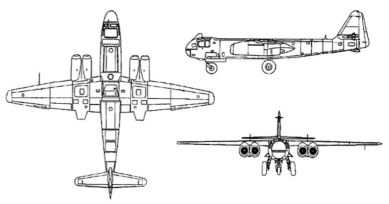

A engine mechanic sitting atop a Jumo 004B belonging to an Arado Ar 234 B-2.

A production Arado Ar 234C with its four turbines placed two within a single pod per wing.

Front side view of a production Arado Ar 234C.

tion such as those that appeared on the Ar 234 V6 and V30 prototypes. Although they were virtually completed for test flying, both aircraft were burned in April 1945 in order to prevent their falling into the hands of Soviet Army troops. An equally interesting prototype, the Ar 234 V16, was intended to test the revolutionary crescent-wing planform designed by Arado's chief aerodynamicist Rüdiger Kosin. With the crescent wing the Ar 234 V16 would have had a leading edge sweep of 37° inboard, decreasing in two steps to 25° outboard. Thrust would have been supplied by twin BMW 003 R composite turbines. "R" meaning rocket drive. A single HWK 509 bi-fuel liquid rocket drive had been attached to the 003 to help inproviding additional thrust during takeoff and/or rapid climb during night fighting.

In April 1945 the experimental crescent-wing was ready to be attached to the fuselage of the V16, but before that could be accomplished the Arado assembly facility was overrun by British troops and the new wing and the aircraft itself was destroyed during the fighting. However, wind tunnel test data and design plans were salvaged and in 1951 the Handley Page Company of England resurrected the crescent-wing idea for use on its Victor bomber. Other 234C prototypes were lost by being

An Arado Ar 234 post-war look-alike, the Consolidated-Vultee XB-46.

An Arado Ar 234 V6 known as the "A" series prototype. This particular aircraft is reported to have flown several reconnaissance missions even though only a prototype.

An Arado Ar 234 post-war look-alike, the Martin XB-48 which at the time in the late 1940s was the largest conventional bomber in the USAAF jet bomber fleet.

An Arado Ar 234 post-war look-alike, the Curtiss XP-87.

Handley-Page "Victor" 8 heavy bomber with its "Crescent" wing of the type Rüdinger Kosin designed for use on his proposed Arado Ar 234 V16 design.

Rüdinger Kosin, well-respected chief aerodynamicist at Arado.

in the middle of cross-fire or through bombing raids such as the 234C V20 which was destroyed 4 April 1944 the aftermath of an air raid on Arado's Wesendorf facilities.

Final sub-types of the Ar 234 under development during the war's closing days were the D- and P-series (models), which were to be used as reconnaissance-bombers and night fighters. All of these were to have been powered by twin HeS 011 turbines. The P-series planes were projected to be night fighters. The nose of the aircraft was to be lengthened and modified to accommodate radar, and each aircraft would have had a two-man crew, the second man being seated aft in the fuselage.

With a typical maximum speed of 461 mph (761 km/h), but frequently reaching more than 500 mph (805 km/h) if the aircraft were lightly loaded and air conditions were suitable, the Ar 234B-2 was able to elude all Allied fighters. When it was put into service during late 1944 and 1945, the aircraft was used to bomb and reconnoiter territory occupied by the Allies. On several occasions it crossed the British coastline in photographic missions. Both British and American intelligence were extremely anxious to obtain a copy of an Ar 234B-2, but it was not until 24 February 1945, that P-47 Thunderbolts of the USAAF were able to force a lone Ar 234B-2 to make a belly landing near Segelsdorf, a village just behind the German lines (one of the Ar 234B's turbines was out at the time). The following day Segelsdorf was captured by the troops of the US 9th Army. Although under fire from German infantry who were trying to prevent the Ar 234B from falling into Allied hands,

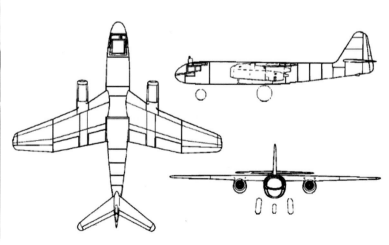

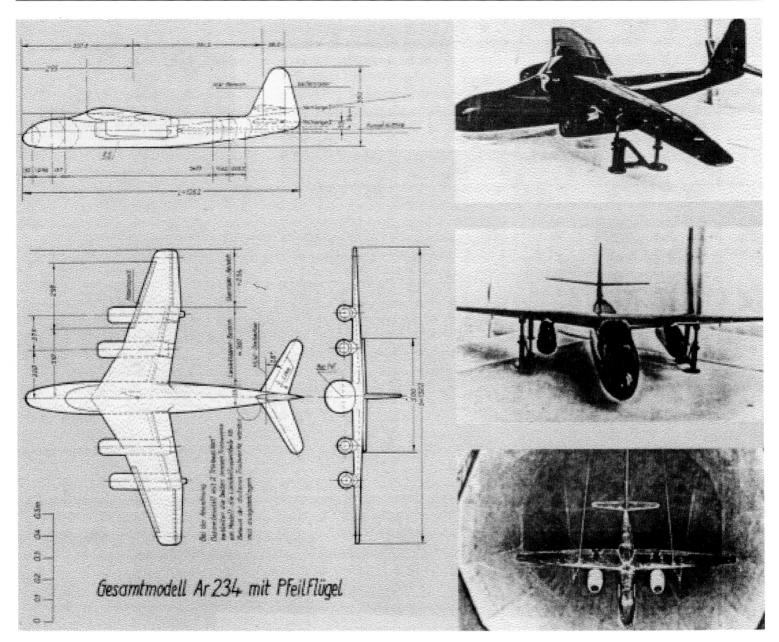

Wind tunnel models of the Arado 234 V16 and to its left are three pen and ink drawings of the same aircraft. *Courtesy of Monogram Monarch Series #1, Arado 234 Blitz.*

9th Army technicians were able to dismantle the prized aircraft, and it was immediately transported back to England for thorough examination.

Overall, 210 Ar 234B-2's were produced between June 1944 and February 1945. In February 1945 Arado's production facilities were overrun by Soviet Red Army troops. However, only a small proportion of the Ar 234 reconnaissance and bomber aircraft completed actually reached operational status, many of the production machines having been destroyed by Allied air attacks on Luftwaffe bases and in accidents at training units. The Arado 234B's basic design style has been copied more often in postwar aircraft designs than has any other German World War II aircraft. Several American aircraft manufacturers built multi-turbine-powered medium range bombers looking surprisingly similar to Walter Blume's Ar 234B, for example North American's B-45C, Convair's XB-46, and Martin's XB-48. It is believed that only one example of the Ar 234 survives. The National Air and Space Museum of the Smithsonian Institution, Washington, DC has an Ar 234B-2, serial number 140312.

Chance Vought XF7U-1 "Cutlass" fighter. This design was first laid down in August 1945. Navy's Bureau of Aeronautics asked for design proposals for a new jet-powered fighter on February 1946. Navy authorized three prototypes April 1946. Its maiden fight occurred on September 29, 1948 at Patuxent River (Maryland) Naval Air Station. Max speed 1,071 km/h (666 mph). A total of 307 were built between March 1950 and July 1954.

Arado Flugzeugwerke GmbH - Berlin, Germany

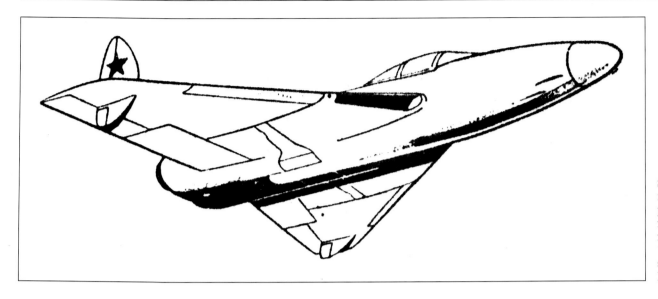

Siegfried Günter formerly of Heinkel proposed design concept for a delta winged fighter for the Soviets 1947-1954, based on suggested Arado's Ar P.NJ-1.

Specifications: Project 234B2

Engine	2xJunkers Jumo 004B turbojets each having 1,980 lb (900 kg) of thrust
Wingspan	46.6 ft (14.2 m)
Wing Area	284.2 sq ft (26.4 sq m)
Length	41.7 ft (12.7 m)
Height	14.1 ft (4.3 m)
Weight, Empty	11.464 lb (5,200 kg)
Weight, Takeoff	18,541 lb (8,410 kg)
Crew	1 (later models such as the C-series had a 2-man crew)
Speed, Cruise	308 mph at 65% thrust (495 km/h)
Speed, Landing	NA
Speed, Top	500 mph (805 km/h)
Rate of Climb	1,870 ft/min (570 m/min)
Radius of Operation	1,013 mi (1,630 km)
Service Ceiling	32,810 ft (10 km)
Armament	2x20 mm MK 151 cannon fixed in rear of fuselage firing to the rear and sighted by periscope
Bomb Load	Various combinations, up to 3,307 lb (1,500 kg)
Flight Duration	NA

Siegfried Günter, former Heinkel design genius with shirt and tie, seated next to his wife boating on a lake with friends. USSR early 1950s.

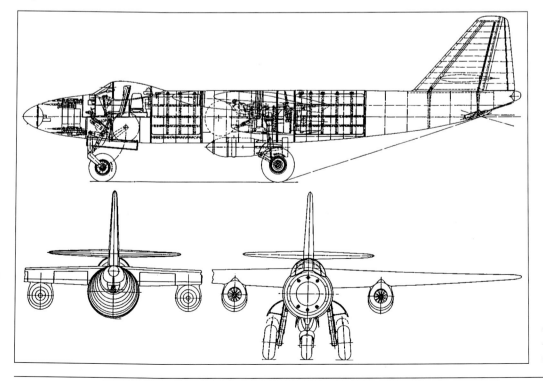

The proposed Arado Ar P.NJ-2, night and foul weather fighter. A three view drawing by Günter Sengfelder.

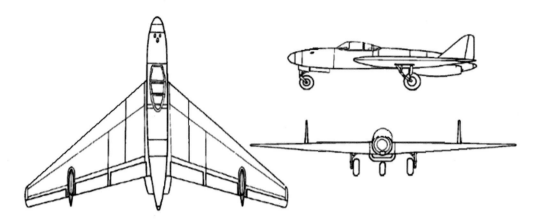

PROJECT NJ-1

The Project NJ-1 (meaning Nacht Jäger, or night hunter) was Arado's entry in the 1945 design competition held by the Luftwaffe's Chef der Technischen Luftrustung (CTL), or Chief of Technical Air Equipment for a twin-engine night fighter. Overall, the Luftwaffe was looking for an aircraft having a flying weight of between 11 and 13 tons, good maneuverability, and maximum speed. Arado's competition included Dornier's P.256, Blohm and Voss's P.215, Focke-Wulf's P.2 and P.3, and Gothaer Waggonfabrik's P.60C. None of the designs submitted was accepted initially because none met the requirements for weight, maneuverability, and performance, and the competition was canceled. But just three weeks prior to Germany's surrender on 7 May 1945, Blohm and Voss was issued a contract by the Luftwaffe to construct several prototypes of its P.215.

Arado's P.NJ-1 was to have been a tailless, sweptback, low-wing monoplane powered by two HeS 011 turbines of 2,866 lb (1,300 kg) thrust, half buried in the stern of the aircraft. The Luftwaffe disliked Arado's placement of the turbines at the extreme rear and felt that the very long air intake for the turbines would cause a loss of thrust equal to about 4 percent (115 lb or 52 kg) of the HeS 011's total output. Overall, the Luftwaffe felt that the P.NJ-1 was physically larger than necessary and that its size would cut down on its maneuverability in pursuit of RAF and USAAF bombers during bad weather and night flying.

As an offensive night fighter, the P.NJ-1 was to have two fixed-forward 30 mm MK 108 cannon with two-hundred rounds of ammunition each, two oblique 30 mm MK 108 cannon with two-hundred rounds each, and two 30 mm MK 108 defense cannon with two-hundred rounds each. Two 1,102 lb (500 kg) bombs also would be accommodated under the aircraft's center section.

A three-man crew sitting in tandem in a fully glazed cockpit, was to operate the P.NJ-1. Bullet-proofing would protect the crew, front and rear, against enemy 20 mm cannon fire. Standard equipment would include radar, direction finding radios, ejection seats for the crew, and fire extinguishers.

Control on the tailless design was to be achieved through the use of ailerons as elevators on the upper side of the wing. Split flaps on the inner section of the wing were to provide additional control. To achieve stability, two vertical control surfaces were to be placed near the wing tips; however, Luftwaffe officials felt that Arado's P.NJ-1's massive size would make these two small rudders inadequate for the constant ma-

The proposed Arado Ar E.381 HWK bi-fuel rocket powered target interceptor, side view. Scale model by Günter Sengfelder.

Front view of the proposed Arado Ar E.381 HWK bi-fuel rocket powered target interceptor. Scale model by Günter Sengfelder.

Arado Flugzeugwerke GmbH - Berlin, Germany

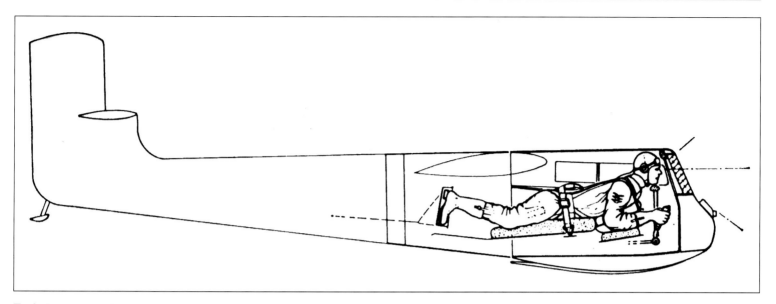

Typical prone position as illustrated in this sketch of a side view of a Blohm & Voss Bv 40. The pilot's chin rests on a padded stick.

A test pilot assuming the prone position for fit and feel in a twin engined Berlin B-9.

neuvering the night fighter would need to do when positioning itself against enemy bombers.

No prototype of the P.NJ-1 was constructed, and all design drawings and plans fell into Soviet hands in March 1945 when their troops captured Arado's facilities near Berlin.

From time to time it is suggested that the P.NJ-1 was a major influence in the design of Chance-Vought's F7U Cutlass shipboard fighter, but the facts indicate that this was not the case. Although the two aircraft are similar, the F7U was developed independently of Arado's P.NJ-1. The F7U's design was completed in early 1945, and the prototype flew in 1947. No German designers were employed at Chance-Vought, and no Chance-Vought designers had access to any Arado design data

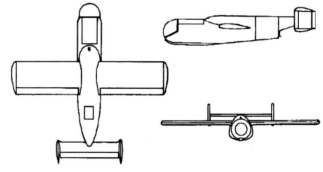

Helmuth Walter-Kiel and his HWK 509 A-2 bi-fuel liquid rocket drive of 3,740 lb thrust which would have propelled the proposed Arado Ar E.381 target interceptor.

Left, Wilhelm Benz designer the Heinkel jet powered He 178 under the leadership of Walter Günter. Benz also designed the proposed Heinkel He P.1077, "Julia." Right, Rüdinger Kosin Arado's chief aerodynamicist who designed the so-called "crescent wing" which was to be a new feature on the Arado Ar 234 V16. Munich, 1988.

The Wilhelm Benz designed proposed Heinkel He P.1077 "Julia". Scale model by Jamie Davies.

in late 1944 through early 1945. The F7U and the P.NJ-1 were coincidental designs in a common attempt to develop an airframe that could fly through the sound barrier with the least difficulty. Designers on both sides of the Atlantic felt that making the aircraft tailless was the answer. As did the designers of the tailless F7U, the P.NJ-1's designers learned early on that its lack of a standard tail section created more problems than it solved. The F7U suffered from poor rearward vision due to the placement of the cockpit and the large wing surface. Landing the aircraft presented another problem. As with most other tailless and delta-winged aircraft, landing was achieved with the nose up, making it extremely difficult for the pilot to get a proper view of the runway. The F7U is considered a design failure; although Arado's P.NJ-1 is similar in design, one can only speculate that it, too, may have suffered from the same shortcomings as the F7U.

Specifications: Project NJ-1

Engine	2xHeinkel-Hirth HeS 011 turbojets each having 2,866 lb (1,300 kg) of thrust
Wingspan	60.4 ft (18.4 m)
Wing Area	807 sq ft (75 sq m)
Length	42.7 ft (13 m)
Height	NA
Weight, Empty	22,487 lb (10,200 kg)
Weight, Takeoff	29,101 lb (13,200 kg)
Crew	3
Speed, Cruise	328 mph (527 km/h)
Speed, Landing	104 mph (168 km/h)
Speed, Top	503 mph (810 km/h)
Rate of Climb	38 ft/min (11.6 m/min)
Radius of Operation	852 mi (1,370 km)
Service Ceiling	44,622 ft (13.6 km)
Armament	6x30 mm MK 108 cannon
Bomb Load	2,204 lb (2x500 kg)
Flight Duration	2.6 hours at 65% thrust at 19,686 ft (6 km)

PROJECT E.381

This proposed HWK 509A bi-fuel rocket-driven target interceptor was entered in the Air Ministry's design competition in August 1944 for a small, lightweight, high-speed interceptor to be used against USAAF heavy bomber formations. Ultimately, Erich Bachem's Ba 349A "Natter" won the competition, after heavy lobbying by Reichsführer SS Heinrich Himmler. Initial winner of the competition was Heinkel's He P.1077 "Julia," which was to use a steep ramp for takeoff and a built-in skid for landing. Junkers EF 127 "Dolly" was runner up.

Arado's proposed target interceptor, like the other entries in the competition, was designed to use the Walter HWK 509 bi-fuel liquid rocket drive with its 3,740 lb (1,700 kg) of thrust. However, the Ar E.381 was different. While the proposed designs from Bachem, Heinkel, Junkers, and Messerschmitt required some sort of ground-launching device, the E.381 was to be carried to the war's front beneath an Ar 234C turbine reconnaissance aircraft. Release of the small interceptor was to be made high above the enemy bomber formations once its HWK 509 bi-fuel rocket drive had been started. After completing its combat mission, the interceptor was to return to its base and land on a built-in skid. A

A proposed Arado Ar E.381 target interceptor parked out in front of an Arado Ar 234C. Scale models by Günter Sengfelder.

An Arado Ar 234C with a proposed Ar E.381 hung underneath. Scale model by Günter Sengfelder.

breaking parachute would be released upon touch down to help slow the aircraft, it would then be hauled back to the flight line, refueled and rearmed, and coupled to another Ar 234C. Because of the Ar 234C's rapid rate of speed coming from its four turbojet engines, it was anticipated that it could quickly overtake the same bomber formation if the first interception point were sufficiently removed from the bomber's target so that each E.381 could carry out several sorties against the enemy bombers. It was the quick recycling, Arado engineers thought, that made their entry more effective than the competition. However, the Air Ministry felt that using an Ar 234C made Arado's design less desirable in that a considerable amount of turbine fuel would be required if fleet of E.381's were to be established. Moreover, maintaining a fleet of Ar 234C's would require a sizable ground crew and an airstrip, not to mention a large number of pilots. Despite the apparent disadvantages of carrying the E.381's aloft by the turbine-powered Ar 234Cs, the Air Ministry gave Arado a contract to develop a prototype interceptor. However, the war ended before a test operation could be tried.

In order to present the smallest possible cross-section diameter to the enemy, the E.381's pilot was to fly the interceptor in a prone position. For defense against a 12-cannon B-17 heavy bomber, the E.381's nose section was to be made of 5 mm armor plate. In addition, behind the one-piece canopy was to be a 5.5 inch (140 mm) armor glass screen for protection. The HWK 509 bi-fuel liquid rocket drive also was to be protected from enemy canon fire through the use of 20 mm armor plating. Up to six 20 mm MK 108 cannon were to be installed in the leading edge of the E.381's rectangular wings, which were to be positioned about half way along the fuselage. Each cannon would have forty-five rounds of ammunition and would be fired over the top of the cockpit canopy.

The Arado Ar 234C carrying aloft a proposed Ar E.381 HWK bi-fuel rocket powered interceptor. Scale model by Günter Sengfelder.

Secret Aircraft Designs of the Third Reich

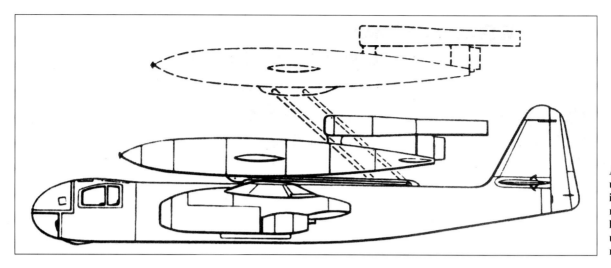

An Arado Ar 234C carrying an unmanned Fieseler Fi 103 flying bomb. The carrying apparatus also included an hydraulic mechanism to lift the Fi 103 up and above the rudder and released.

The tail section was to consist of a one-piece elevator and twin rudders. The rear section of the fuselage would house the HWK rocket drive with its estimated 3,740 lb (1,700 kg) thrust exiting beneath the short tail section down and below the elevator and rudder assembly. Sheet steel on the tail section would help direct the HWK's exhaust away from the tail control surfaces. Estimated top speed of the E.381 was 559 mph (900 km/h).

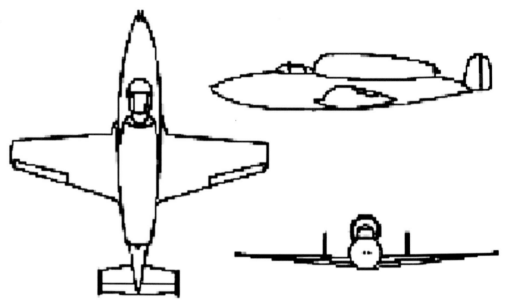

Rear view of an Arado Ar E.581.4. Scale model by Dan Johnson.

Arado Flugzeugwerke GmbH - Berlin, Germany

The proposed Arado Ar E.581.4. Scale model by Dan Johnson.

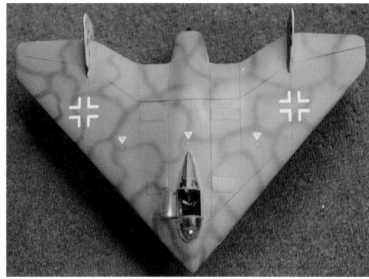

Overhead view of an Arado Ar E.581.4 giving a good view of its delta wing planform. Scale model by Dan Johnson.

Specifications: Project E.381

Engine	1xWalter HWK 509A-2 bi-fuel liquid rocket drive with 3,740 lb (1,700 kg) of thrust
Wingspan	16.5 ft (5 m)
Wing Area	59 sq ft (5.5 sq m)
Length	18.7 ft (5.7 m)
Height	NA
Weight, Empty	1,963 lb (890 kg)
Weight, Takeoff	2,690 lb (1,220 kg)
Crew	1
Speed, Cruise	NA
Speed, Landing	NA
Speed, Top	559 mph (900 km/h) at 26,240 ft (8.0 km)
Rate of Climb	NA
Radius of Operation	NA
Service Ceiling	NA
Armament	6x20 mm MK 108 cannon with 45 rounds of ammunition each
Bomb Load	None
Flight Duration	NA

LEFT: Direct front view of the proposed Arado Ar E.581.4 showing its bifurated turbine air-intake duct. Scale model by Dan Johnson.

45

Front side view of a proposed Arado Ar E.581.4. Scale model by Dan Johnson.

PROJECT E.580

On 8 September 1944, the Air Ministry issued a basic requirement for a lightweight fighter that would use a minimum of strategic materials. The fighter had to be suitable for rapid mass production, yet still have better performance than contemporary piston-powered fighters such as the P-51 Mustang or the P-47 Thunderbolt. Air Ministry engineers also specified a number of design features they desired: a single turbine engine mounted on top for the fuselage directly behind the cockpit, unswept wings, and a tail unit consisting of twin rudders with a full span elevator. (Choice of the twin vertical fin/rudders was dictated by the fact that a single rudder would be right in the hot exhaust of the turbine. With twin rudders the hot gases could pass between without creating a fire hazard.) Six aircraft companies were invited to submit bids: Arado, Blohm and Voss, Focke-Wulf, Junkers, Heinkel, and Messerschmitt. In contrast to the typical 1 to 2 years development time normally allowed aircraft manufacturers on a new design, the Air Ministry wanted a prototype ready in 4 months after the date of award. Arado was unprepared for any such rapid turnaround. Although its designers submitted a proposal, both they and the Air Ministry knew Arado could not possibly get a prototype off the ground in 4 months. Heinkel won the competition, mainly because the firm already had constructed a mockup of its entry, and Heinkel officials were instructed on 15 September 1944, to proceed with a prototype of what eventually came to be called the He 162 "Salamander."

Arado's E.580 design followed the specifications spelled out in the Air Ministry's basic requirements, but it had a softer appearance than Heinkel's He 162, with its more angular shapes on the wings and tail structure. Compared with the He 162, for instance, the E.580's maximum speed was estimated to be 462 mph (774 km/h) or 38 mph (61 km/h) less than the He 162 but its range of 316 miles (508 km) was some 290 miles (467 km) less. Arado's inability to meet the rapid turnaround required by the Air Ministry on the E.580 as well as other proposed projects was a problem that plagued the firm after Walter Blume's rise to upper management following construction of his successful Ar 234-B reconnaissance bomber. Rapid turnaround was a problem he recognized but was unable to correct before the war ended.

Specifications: Project E.580

Engine	1xBMW 003A-1 turbojet having 1,764 lb (800 kg) of thrust
Wingspan	23.8 ft (7.2 m)
Wing Area	NA
Length	26.9 ft (7.9 m)
Height	NA
Weight, Empty	NA
Weight, Takeoff	5,494 lb (2,492 kg)
Crew	1
Speed, Cruise	NA
Speed, Landing	95 mph (153 km/h)
Speed, Top	462 mph (774 km/h) at 19,358 ft (5.9 km)
Rate of Climb	3,347 ft/min (1,020 m/min)
Radius of Operation	316 miles (508 km)
Service Ceiling	NA
Armament	2x20 mm MK 108 cannon
Bomb Load	None
Flight Duration	54 minutes

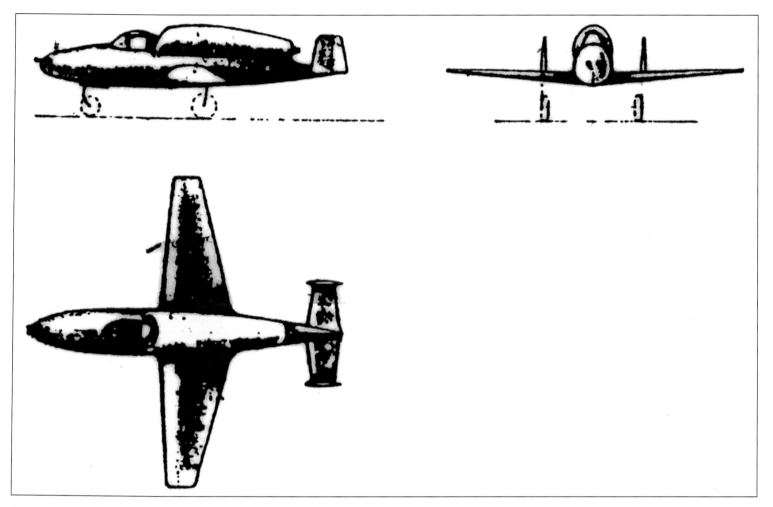

Arado's proposal for the RLM "Volksjäger" competitions. The RLM wished each competitor to submit a design similar to the production Heinkel He 162.

PROJECT E.581.4

This proposed high wing tailless delta single-seat fighter was to have been powered by a single HeS 011 turbine mounted entirely within the fuselage. The E.581.4 was designed with a tricycle landing gear. The main wheels were to retract forward and up into the delta-shaped wing. Upon the wing's upper surfaceat the wing tips were a pair of fin/rudders. Very little is known about this tailless delta design specifications.

Specifications: Project E.581.4

Engine	1xHeinkel-Hirth HeS 011 turbojet having 2,866 lb (1,300 kg) of thrust
Wingspan	29.3 ft (9.7 m)
Wing Area	NA
Length	18.4 ft (6 m)
Weight, Empty	NA
Weight, Takeoff	NA
Crew	1
Speed, Cruise	NA
Speed, Landing	NA
Speed, Top	NA
Rate of Climb	NA
Radius of Operation	NA
Service Ceiling	NA
Armament	NA
Bomb Load	NA
Flight Duration	NA

2

Bachemwerke GmbH - Waldsee, Germany

Secret Project and Intended Purpose:
Project 349B - Manned HWK 509C-1 bi-fuel liquid rocket driven target defense interceptor.

History:
By the spring of 1944 the skies over Germany were seldom free from daily visits by B-24's and the B-17 "Flying Fortress" heavy bombers of the U.S. 8th and 15th Army Air Force. With the battle for air supremacy over Germany long lost by the Luftwaffe, only inclement weather now kept these bombers away - and then not for very long.

A number of ingenious schemes for counteracting the Allied bomber attacks were offered to the Air Ministry by aircraft manufacturing firms, individuals, and Air Ministry engineers themselves. Few of these proposals survived cursory examination of their practicality. One of the

Erich Bachem and his family.

This is the way it was supposed to work. A Bachem Ba 349 "Natter" after a ground launch smashes into its target, a Boeing B-17 bomber.

If the "Natter's" 24 Henschel Hs 217 Fohn air-to-air rockets would not bring down a B-17, the pilot could always attempt to ram the bomber bailing out moments before it hit.

An artist's illustration depicting a "Natter" unleashing its 24 Henschel Hs 217s. The "Natter's" potential fire power was awesome.

The nose of the "Natter" with its plastic cone removed showing its 24 Henschel Hs 217 Fohn air-to-air rockets. During lift off these rockets would be covered by a streamlined, plexiglass nose cone. Prior to attacking a formation of B-17s, or an individual aircraft, the nose cone would be discarded.

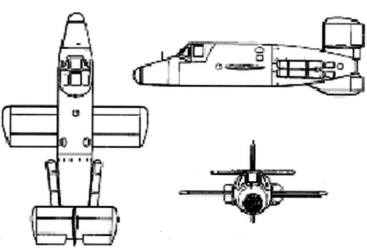

Secret Aircraft Designs of the Third Reich

most radical proposals destined to move from drawing board to prototype was the Bachem Ba 349B "Natter" (meaning the Adder snake). The 349 was a simple, ramp-launched, rocket-propelled interceptor employing an airframe built of low-grade wood and other nonessential materials. It was designed to serve for a single operational sortie, after which the pilot would bail out and the rocket drive of the aircraft would descend to the ground by parachute for retrieval and reuse. The originator of this scheme was Erich Bachem, former chief designer for the Gerhard Fieseler Werke. In 1943 Bachem left Fieseler to form his own company to manufacture airplane parts out of wood in the town of Waldsee. Bachem was well known in aviation circles for his design work on a large number of aircraft produced by Fieseler, including the world famous Fieseler Fi 156 Storch (stork) observation plane. This legendary slow-flying monoplane incorporated extensive high-lift devices which allowed it to make remarkably short takeoffs and landings. At his Waldsee shop, Bachem turned out wooden wings for many of the air-to-air and ground-to-air missiles built by the Henschel Flugzuegwerke. Bachem also manufactured some of the control surfaces for Wernher von Braun's V2 (A4) rocket.

Bachem claimed to have conceived the idea of using a piloted rocket to bring down Allied bombers while standing outside his wood-working factory watching formations of heavy bombers flying overhead day

SS Führer Heinrich Himmler, second from left, shown here at Peenemünde came to Erich Bachem's aid after the RLM totally rejected the "Natter" concept. Himmler was interested in all sorts of exotic weapons and it took only a telephone call to the RLM from this most feared man in all Germany to get enthusiastic support.

Not only did Himmler philosophically support the "Natter" but he sent several hundred of his Waffen SS troops to assist Bachem in building and testing the "Natter."

SS officer Heinz Flessner, center, leading his work-crew to the Bachem Ba 349 test-flight area in early 1945.

SS officer Heinz Flessner in Spring 1945.

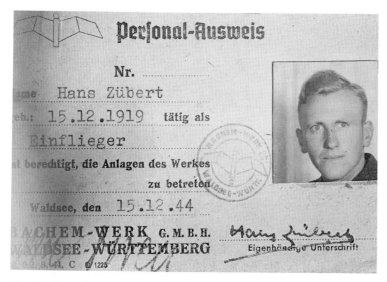

Hans Zübert, formerly of Horten Flugzeugbau, was a highly skilled test pilot for powerless aircraft.

Hans Zübert and Heinz Flessner co-workers on the Bachem Ba 349 test flight program. Germany 1988.

after day with their close-positioning at about 23,000 feet (7 km), with little or no opposition from the Luftwaffe. He had thought about a manually directed missile before, while he was still with Fieseler. As early as 1939 Wernher von Braun had discussed with him the possibilities of a manned rocket that could climb to meet an enemy bomber formation with such great speed that the bombers would have no chance of taking evasion action. This manned rocket would have such a small diameter that airborne gunners would find it nearly impossible to shoot it down. Bachem himself believed that a manually directed missile could do the job. In 1943 he began to outline a project for a cheap, semi-expendable, liquid rocket-driven interceptor that could be built from wood by unskilled labor in small woodworking shops with very little engineering equipment. Low-grade wood could be used for most of the construction, and the few steel fittings required could be crude and simple.

According to Bachem, the normal operating height of the American bomber formations over German targets was generally between 23,000 ft to 25,000 ft (7 km to 7.6 km). Any directed missile would have to rise to at least 30,000 ft (9.2 km), and it would have to do so within sight of the approaching bombers, that is, defend a conical airspace of approximately 12 miles in radius. In addition, the tiny craft would have to reach its operating altitude in less than a minute. Therefore, the start would have to be made vertically. Ideally, the sole duty of its pilot would be to control the tiny aircraft during the last stages of the missile's flight, the attack. As soon as the enemy was sighted, the pilot would take over from the ground-to-air radar launch control and fly to within a few hundred yards of the bomber formation. At this distance he would fire a battery of 24 Föhn RZ-73 mm anti-aircraft rockets (housed in the interceptor's nose) into the oncoming bombers, then break off the engagement. Allowing the aircraft's speed to slow to about 150 mph (241 km/h), the pilot would then jettison the forward part of the craft and parachute to the ground. Bachem called his piloted missile the "Natter." He probably took the name from a line in Shakespeare's Julius Caesar: "It is a bright day that brings forth the Adder (Natter)."

In the spring of 1944 the Air Ministry, too, realized that it had to do something about the Allies' unrelenting bomber offensive. Hoping for some kind of a miracle, the Air Ministry issued a requirement for a small and inexpensive target defense interceptor suitable for mass production. Only Junkers, Heinkel, Messerschmitt, and Bachem responded.

The fuselage of a "Natter" under construction showing its pine wood ribs.

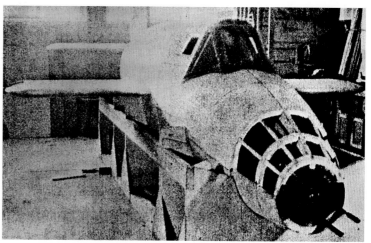

This "Natter" fuselage is almost entirely covered with plywood save for the nose area.

The "Natter's" wooden wing and spar under construction.

The uncovered wing of a "Natter" showing its pine wood ribs.

Lufwaffe Leutnant Lothar Siebert, pilot of the first manned test flight of a "Natter." February 1945.

Bachem's entry, called the Ba P.20 at that time, was rejected out of hand, primarily because the Air Ministry did not envision the sort of craft Bachem proposed - a "throwaway" fighter which was to be launched vertically from an eighty-foot (24.4m) high ramp-structure. Bachem's entry, the Air Ministry officials, joked, was more like a manned anti-aircraft missile than an interceptor. It made no difference to Bachem; manned interceptor or manned anti-aircraft rocket. But it did to the Air Ministry and instead they selected Heinkel's P.1077 "Julia," a bi-fuel rocket driven interceptor designed by Wilhelm Benz that used a steep ramp for takeoff and landed on a built-in skid. Junker's EF.127 "Dolly" was runner-up.

Bachem's proposal might have been given a second look had not his lobbying tactics so annoyed the Air Ministry. Aware that the Air Ministry did not envision a throwaway interceptor, Bachem, nevertheless tried to sway authorities by enlisting Luftwaffe General der Flieger Adolf Galland in his cause. Galland had a reputation for being interested in all manner of novel aircraft but only at this late stage of the war. In the early 1940's Galland had told Walter Horten, technical officer of Jägdgruppe (JG) 26, that his idea for an all-wing fighter was stupid.

Erich Bachem, right, watches workmen readying a "Natter" for its first manned test flight. Lothar Siebert, its pilot, is standing next to Bachem.

Lothar Siebert carefully entering the cockpit of the "Natter."

After listening to Bachem describe his manned missile, Galland fully supported the idea and agreed to go to the Air Ministry to help sell the proposition. Bachem felt he needed all the help he could muster because he had earlier presented a similar idea to the Air Ministry, only to be turned away. Bachem hoped that Galland could influence the Air Ministry to reconsider. It did not work. Despite Galland's tremendous influence and reputation with Air Ministry officials, their opposition remained and, in fact hardened because the Bachem "Natter" proposal had been uninvited and had, in any case, been submitted for consideration through abnormal channels (Galland had walked the proposal through the Air Ministry).

The strongly independent Bachem was not ready to give up so easily. Knowing that the Reichsführer of the Schutzstaffeln (SS), Heinrich Himmler, had a keen interest in exotic military weapons, including aircraft, Bachem sought and received an interview with the most feared man in the Third Reich who at that time wielded more power than Hitler himself. Himmler immediately displayed interest in Bachem's proposal and asked what he needed to put the Natter into operation. Bachem responded that he required skilled woodworkers and other less skilled labor. "How many do you need?" replied Himmler, commander also of the Third Reich's Konzentrationslagers (KZ) or concentration camps, "we'll just open a KZ near you and you'll have access to all the labor you need." Bachem was repelled by the idea of using concentration camp inmates to construct his proposed manned missile and gently suggested to Himmler that disabled skilled workers were what he really

Lothar Siebert fully tucked into the "Natter's" tiny cockpit.

The "Natter" moments after lift-off. With its 4,410 lbs (2,000 kg) of thrust. The "Natter" achieved a very quick vertical lift-off without difficulty.

A "Natter" is delivered to the launch site. The pine tree launch is at far left.

had in mind. Himmler gave Bachem an "as you wish" sort of response and promised his full support in order to get the Natter idea off the ground. The speed at which the situation changed astonished Bachem. Within 24 hours of his meeting with Himmler, he was contacted by Air Ministry engineers who said that they had reconsidered their rejection of the Natter proposal and that it would now receive their highest development priority. Privately, the Air Ministry hated the addition of Bachem's Natter to their target interceptor program - yet no one in the Air Ministry was willing to argue the Natter's merits with Himmler. Besides, the Air Ministry felt, Himmler's interest might be a blessing in disguise because it meant that virtually no one in the Third Reich would attempt to terminate the project now; Himmler's influence was a sort of insurance policy that some type of interceptor would be built, even if it were the less desirable Natter. Within a few days of the Air Ministry's reconsideration, truckloads of disabled skilled workers began arriving at Bachem's Waldsee factory, and the race to build and fly the manned missile before Germany collapsed was on.

By August 1944, only 3 months after the project's initiation, several aircraft were ready for flight testing. The method of launching was also finalized. The Natter was to be attached to a vertical launching ramp 80 ft (24.4 m) high. Four Schmidding solid-fuel booster rockets were to be attached to the rear fuselage, each providing 1,102 lb (500 kg) of thrust during its 6 second burn time. With the booster rockets supplementing the thrust available from the single HWK 509C-1 (auxiliary cruising chamber) bi-fuel rocket drive of 4,410 lb (2,000 kg) thrust, the Natter would achieve sufficient velocity for its vertical takeoff. With some 8,818 lb (4,000 kg) of thrust the 4,920 lb (2,232 kg) Natter was expected to reach an altitude of 35,762 ft (10.9 km) within 60 seconds of lift off.

The Natter's operational plan did not change much from Bachem's original idea: Launch the target defense interceptor as Allied heavy bomber formations were approaching and guide the Ba 329B from the ground to operational altitude by means of a radio-radar link. About 4,922 ft (1.5 km) from the bomber formation, have the pilot take over manual control and jettison the plastic cover over the nose to expose the rockets. After closing to short range, fire the entire load of 24 R4M anti-aircraft rockets in one salvo, clear the bomber formation, jettison the cockpit hood, and, after slowing to about 150 mph (241 km/h), bail out. Simultaneous with bail out, uncouple the rear section housing the HWK 509 bi-fuel rocket drive, lower the whole unit to the ground by parachute, recover it, rebuild it and place the unit into another Natter.

A third version of the 349 with a fixed undercarriage was ready in December 1944. This version was known as the 349-M used the landing gear from a Klemm Kl 35. This allowed the possibility of towed air

A "Natter" being hoisted onto its pine tree launch.

trials to determine the 349's flying characteristics. One of the first Ba 349-M prototypes was towed aloft to 18,000 ft by a Heinkel He 111 to test the flight/gliding characteristics. Test pilot Hans Zübert, who was on loan from the Horten Flugzeugbau, reported that the 349's stability and flight controls functioned within the desired range. Zübert further noted that the 349's sink-rate was extremely high given its 4,000 lb weight at the time plus its very high-wing loading. About this same time

A technician wearing rubber gloves and apron is fueling up a "Natter" with C-Stoff and T-Stoff.

A "Natter" fully secured to its pine tree launcher.

The new and improved "Natter" for 1945 and beyond equipped with optional wheels.

Backem had a small scale model of his 349 in the wind tunnel at DVL. The results of these wind-tunnel testing are not available.

The first manned flight of a Natter took place in late February or early March 1945. It ended in disaster. The HWK 509A-1 bi-fuel rocket drive was started and the Natter began to climb vertically into the morning sky. At about 1,640 ft (500 m) its cockpit canopy (hood) blew off while the Natter continued to climb, but it rolling over on its back. At about 4,920 ft (1,500 m) it performed a half- loop and then plummeted straight down to the ground, exploding upon impact and killing its test pilot Lt. Lothar Siebert. Investigations by a team of SS officers indicated that the cockpit canopy may not have been fastened properly prior to takeoff and that Siebert had suffered a broken neck when his head was forced back into the seat by the air rushing into the open cockpit. All this is speculative given the SS's firm desire to see the Natter in the war-effort. Testing was resumed several days later, and three successful manned flights were carried out. With this the Air Ministry announced the project airworthy and ordered the Ba 349A into operation. The "A" used a single chamber HWK 509A-1 rocket drive while the "B" version obtained more endurance by being powered with the dual chamber HWK 509C-1. By April 1945 thirty-six Natters had been constructed. The first ten were set up at the town of Kirchheim near Stuttgart to await the arrival of the USAAF bomber formations. They never came. Instead,

"Natters" like this one found at Stuttgart, were found in only a few other places in Germany. A mere 36 "Natters" had been constructed by April 1945. None were fired into Allied bomber formations. Only two "Natters" are known to have survived the war and both are in aviation museums: Berlin and Washington.

A "Natter" launch site. One "Natter" is mounted on the launch with another on the truck. Winter 1944-45. Scale model by Jamie Davies.

A factory fresh "Natter" arriving on a flat bed truck while another "Natter" partially assembled, waits on its metal frame launch. Scale model by Jamie Davies.

A "Natter" launch site Winter 1944-45. The launch area is blackened from previous launches. Scale model by Jamie Davies.

A "Natter" on its metal frame launch ready for flight. Scale model by Jamie Davies.

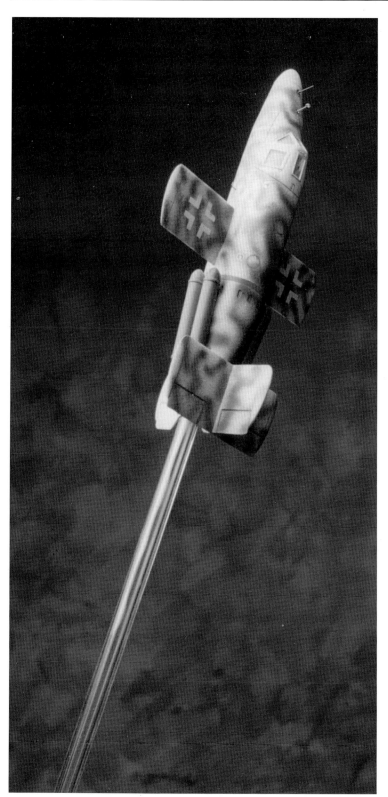

A "Natter" rising vertically moments after launch. Scale model by Jamie Davies.

several days later, American army tanks approached on the edge of the launching site and the German ground crew assigned to launch their Natter's sought to destroyed them on the ground to prevent them from falling into enemy hands. They were not successful but instead were taken prisoner. In addition, it is doubtful that this small group, indeed Bachem himself, in April 1945 had any C-Stoff (methanol) or T-Stoff (hydrogen peroxide) to fuel the HWK bi-fuel rocket drives.

Only two examples of the Natter are known to exist. One is in storage at the Deutsches Museum in München, and USAAF T2-1011 is stored at the Smithsonian Institution's National Air and Space Museum, Silver Hill, Maryland.

Specifications: Project 349B

Engine	1xWalter HWK 509C-1 (auxiliary cruising chamber) bi-fuel liquid rocket drive having a maximum thrust of 4,410 lb (2,000 kg) plus (for take off) 4x1,102 lb (500 kg) thrust Schmidding solid-fuel booster rockets
Wingspan	13.2 ft (4 m)
Wing Area	50.6 sq ft (4.7 sq m)
Length	19.9 ft (6 m)
Height	7.5 ft (2.3 m) fin base to fin tip
Weight, Empty	1,940 lb (880 kg)
Weight, Takeoff	4,920 lb (2,232 kg)
Crew	1
Speed, Cruise	494 mph (797 km/h)
Speed, Landing	Parachute
Speed, Top	620 mph (998 km/h) at 16,400 ft (3 km)
Rate of Climb	37,400 ft/m (11.4 km min)
Radius of Operation	12 mi at 30,000 ft (9 km)
Service Ceiling	39,320 ft (12 km) in about 60 seconds
Armament	24x R4M anti-aircraft rocket missiles or 2x30mm MK 108 cannon
Bomb Load	None
Flight Duration	4.6 minutes at 29,503 ft (9 km)

A "Natter" after lift-off beginning horizontal flight in pursuit of B-17 formations. Scale model by Jamie Davies.

3

Blohm und Voss Abteilung Flugzeugbau - Hamburg, Germany

Secret Projects And Intended Purpose:
Project 607 - An experimental design in the shape of an all-wing aircraft with a 45 degree wing sweep and a lateral placement of the cockpit (to the port, or left, side).
Project 188 - An attractive four turbojet-powered heavy bomber with a "W" wing, that is, a forward sweep followed by a backward sweep.
Project 196 - A twin turbojet-powered close-support aircraft with twin tail booms; the bomb load was to be housed in the forward sections of the tail booms.
Project 197 - A single-seat, twin turbojet-powered high altitude fighter with a projected top speed of 662 mph (1,065 km/h) - the fastest of all Blohm and Voss project designs.
Project 198 - A single-seat, single turbojet-powered high altitude fighter with an operational ceiling up to 50,000 ft (15.5 m) otherwise an ordinary-looking World War II fighter with a rather large unswept wing.
Project 202 - A single-seat, twin turbojet-powered fighter with a unique oblique, or scissors, type of wing, the port side sweeping back and the starboard side sweeping forward.

Project 213 - A miniature, single-seat fighter powered by an Argus-Schmidt-Schubror As 014 pulse-jet engine.
Project 215 - A three-seat, twin-turbojet-powered, night fighter with a 30 degree wing sweep, unique in that its horizontal and vertical control surfaces were mounted at the wing tips but displaced rearward so that the trailing edge extended formed the line of the fin's rear beam; rumored to have won the Air Ministry's last design competition just weeks before all Blohm and Voss operations in Hamburg were stopped by Allied advances.

History:
Like most young men who were getting into aviation between 1910 and 1920, Richard Vogt was eager to make a name for himself. But unlike many of his contemporaries, he was outspoken, blunt, quick to criticize, and prone to disregard orders and instructions from his superiors. As a

Dr.Ing.Richard Vogt soon after he joined Dornier Flugzeugbau in 1923 fresh from the University of Stuttgart with a PhD in aviation engineering.

Blohm und Voss' main administrative office building in Hamburg.

Walther Blohm of Blohm & Voss hoped his firm would capture a large portion of the luxury passenger airline service across the Atlantic by building the largest and most comfortable seaplanes ever designed.

Dr.Ing. Claudius Dornier, mentor to a generation of aircraft designers such as Richard Vogt.

result, he frequently found himself in trouble (he was fired as a World War I aviator for repeatedly disobeying orders and subsequently was sent into the first Battle of the Somme as a foot soldier, where he nearly lost his life from his wounds). But Vogt seemed to grow stronger with each episode. By 1945 he was managing director and chief designer of the airplane division of the famous shipbuilding firm of Blohm and Voss at Hamburg. His floatplane and cargo and hospital flying boats were known worldwide, and his turbojet-powered designs including a host of unusual, asymmetrical, tailless, and oblique-winged aircraft. They were the hallmark of a university-trained, but generally self-taught, aircraft designer.

It was the advice and patience of Dr.Ing. Claudius Dornier that helped establish Vogt's professional life, enabling him to settle down and focus on a single purpose. Soon after Vogt joined the Dornier firm in 1923, fresh from the University of Stuttgart with a Ph.D. in engineering, Dornier took the young, talented, maverick designer aside and offered him some advice, "because I like you and because I want you to have a successful career."

Dornier's advice to Vogt came in two parts. First, "Do not risk more than is needed for a limited step forward; if you overdo it and luck is not

Initially designed for the German airlines "Lufthansa" in the late 1930s, Richard Vogt's, Blohm & Voss' Ha-139 seaplane was converted to Luftwaffe military duties instead. Here workers secure the Ha-139 aboard the German steamer "Schwabenland."

One of two Curtiss F11C-1's bought by Ernst Udet in the USA and shipped by the ocean liner "Europa" to Germany in 1933. Udet successfully demonstrated dive bombing techniques to the new Luftwaffe with his Curtiss F11C-1. The result of these demonstrations was the design and construction of the famous Junkers Ju 87 "Stuka" dive bomber.

Ernst Udet's two Curtiss F11C-1's in containers at a New York City port waiting to be loaded aboard the ocean liner "Europa" in 1933.

Secret Aircraft Designs of the Third Reich

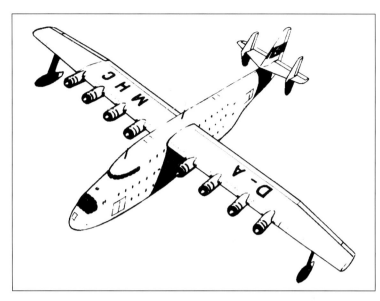

The Richard Vogt designed Blohm & Voss Bv 222 flying boat.

with you, your career can end abruptly." Second, "If you want to become the outstanding designer you can be, then it will be important that you invent or develop a design which has a distinctively personal note, easily recognized as yours." Vogt recalled Dornier's advice throughout his life. His aircraft were different from those of other designers; and, convinced that the designs were worthy, he stuck by them, even when a departure from his concepts might have brought him a short-term business or professional advantage.

Richard Vogt was born on December 19, 1894, in the small town of Schwaebisch Gmuend (near Stuttgart). Known to his brothers and sisters as a "bruttler und noergler" (one who forever criticizes and complains), he was not an easy child to satisfy. When he was 14 he had to leave home in order to attend the Oberrealschule (senior high school) in Stuttgart. About 15 minutes from the school was a large grassy field where a number of early aviators tried out their home-built gliders and powered aircraft. The field, called "Cannstatter Wasen," was used by Hellmuth Hirth, Professor H. Baumann (who later would play a significant role in Vogt's career), and Ernst Heinkel, who was to become the famed head of his own firm (and who happened to be a graduate of the Stuttgart Oberrealschule). When Vogt saw him for the first time, Heinkel was getting ready to test his newly built copy of the French "Farman."

Richard Vogt's Bv 222 flying-boat was the largest of its kind to attain operational status in World War II. Designed for transatlantic luxury passenger service, it could have carried 24 people up to 3,788 mi (6,100 km) nonstop.

The Blohm & Voss Bv 142 shown in the autumn of 1940.

60

Four Blohm & Voss Bv 141s out of the twenty prototypes built.

Vogt also was present when, after a few successful short hops, the aircraft stalled during a turn and crashed. Heinkel was pulled out of the burning bi-wing airplane, alive but badly hurt.

Vogt's own entry into aircraft design began in his typically audacious manner. One Sunday he was witnessing a man attempting to get his home-built copy of the French Bleriot monoplane into the air. The man raced up and down the grassy strip at Cannistatter, getting close to becoming airborne but never quite making it. Afterward, Vogt walked into the hangar and began inspecting the aircraft and its propeller. At the time, Vogt had started work on his own aircraft design. He had no motor yet, but was carving out the propeller on the basis of wind tunnel tests made by the famous Alexandre G. Eiffel, builder of the Eiffel Tower and the first man in France to operate a large-scale wind tunnel. Vogt saw to his surprise that the aviator's propeller had a sharp leading edge. According to Eiffel, such a profile would cause stalling whenever the angle of attack became a bit too large. Vogt concluded that this was what had been happening to the would-be pilot's airplane that day.

Wasting no time, Vogt, a 15-year old schoolboy, began explaining Eiffel's theories to the man; he suggested shaving down the propeller's leading edge by about 4 inches and rounding off the section appropriately. Then, confident of his logic, he took a wood plane and proceeded to shave wood off the valuable propeller. When the job was done a few hours later and the propeller was balanced and polished and put back on, the aircraft lifted off the ground nicely and flew the length of the field.

The pilot crashed the aircraft several days later in a landing accident, ruining the plane. Impressed by Vogt's knowledge of aeronautics, he approached him with a proposition: since Vogt lacked an engine for the aircraft he was attempting to build, he would lend Vogt his 30-horsepower Anzani engine if Vogt, in turn, would make up a set of drawings for a two-passenger airplane. Four months later the two-passenger aircraft was completed and fitted with a large engine and Vogt made the first flight of his life in an aircraft of his own design.

An unusual view from the cockpit of a Blohm & Voss Bv 141 as it banks along with a sister 141 out in front.

A Blohm & Voss Bv P.111. Scale model by Reinhard Roeser.

A Blohm & Voss Bv P.111. Scale model by Reinhard Roeser.

During the following year (1911) Vogt's own aircraft was ready for flight testing but he wrecked it in a ground loop while taxiing. Having no time to repair the aircraft, Vogt gave the engine back to its owner and concentrated on his studies. Graduating in 1913, he took a full-time work-study job as part of the requirements of earning his degree. As he was completing his work-study activities, World War I broke out and, like millions of other young men, Vogt volunteered with the artillery unit in his hometown.

Exposed to a substantial amount of action, Vogt received the Bravery Medal of the King of Würettemberg for leading a group of troops to safety after the commanding officer had been killed. Later Vogt himself was shot and wounded and transferred to a hospital near his hometown. While in hospital he applied for a transfer to the Luftwaffe. Once accepted, he started pilot training in the town of Halberstadt. It was to be an adventure that would later see him fired as a pilot and returned to the Infantry. His file summed up his behavior as a pilot: "He is a good pilot but absolutely unfit for the Luftwaffe because he does not obey the instructions of his superiors, in one case he even offended his officer observer."

Back in the Infantry, Vogt was seriously wounded in the famous Battle of the Somme and hospitalized for a second time. Again a setback led to an unexpected favorable outcome. Since airplanes had become decisive weapons of war by the summer of 1916, the chief commander of the German Defense Forces ordered the release into civilian life of all persons who were trained or otherwise useful, in building airplanes. Vogt applied and was released from the Infantry with an order to report to the Zeppelin Werke at Lake Constance in January 1917.

With the signing of the Armistice in 1919 Zeppelin and Vogt were out of work. With a letter of recommendation from the airplane department of the Zeppelin Works, Vogt started his college education at the Institute of Technology in Stuttgart. He received his master's degree in half the usual time. As an assistant to Professor H. Baumann, the famous designer of the giant Zeppelin airship, he finished his Ph.D. at the end of his third year at the Institute.

Although Professor Baumann was able to guide Vogt through graduate school in record-breaking time, he was unable to help him find work in aeronautics, for the Treaty of Versailles prohibited work on airplanes and in aerodynamics in Germany. Meanwhile, Vogt found employment designing cars and textile machines. In the fall of 1922 Ernst Heinkel started a small aircraft manufacturing company in Warnemünde on the Baltic Sea, and through a chance meeting between Heinkel and Vogt, Vogt was offered a position. Yet he was not sure that a full-time position with Heinkel, who was well known for his flaring temper, would be right for him. Vogt agreed to a week's trial period, during which he learned that Heinkel was planning to build his aircraft of steel tubing, with wood and fabric covering. Convinced that this method had reached the end of the line and that all future planes would be made of metal, as they had been at the Zeppelin Works during the war, Vogt decided not to continue with Heinkel.

Out of work, Vogt returned to his boarding house to find a letter from Dr. Dornier asking him to join him in his renewed aircraft manufacturing activities at Lake Constance. It was a 20-hour train ride to Lake Constance from the Baltic Sea, and Vogt left as soon as he had packed his bags. For Vogt, it seemed as if fate again had intervened, for when he arrived at the Dornier company the firm had just signed a license agreement with the airplane department of the Kawasaki Dockyard in Kobe, Japan. One section of the agreement stipulated that a small number of engineers should go to Japan under the leadership of a man

Dr. Richard Vogt and DFS test pilot Hanna Reitsch share a bit of humor. With them is Mr. Schaudt, Manager of the Fuselage Group at Blohm & Voss - Hamburg.

The proposed Blohm & Voss Bv P.170. Scale model by Reinhard Roeser.

The proposed Blohm & Voss Bv P.192 dive bomber. This was Bv's suggested replacement for the aging Junkers Ju 87. Scale model by Reinhard Roeser.

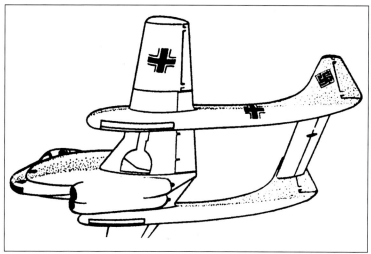

The proposed Blohm & Voss Bv P.196. Illustration by Hugh W. Cowin.

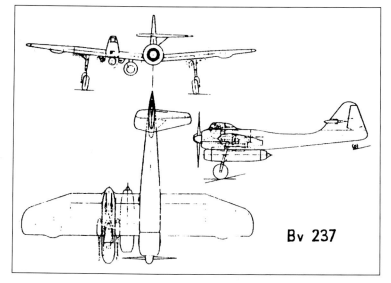

The proposed Blohm and Voss Bv P.237. This was really a Bv 141 with a single BMW 003A or Jumo 004B turbojet engine hung beneath its starboard wing. Never built in this configuration.

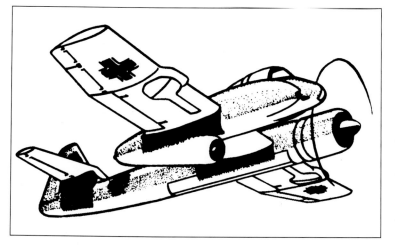

A proposed Blohm & Voss Bv P.194.01. Illustration by Hugh W. Cowin.

with a degree, one who knew the way Dornier's metal airplane were designed and built and had a good background in stress analysis and dynamics. Dornier had Vogt in mind for the Japanese assignment when he hired him. For the next ten years, until 1933, Vogt remained in Japan as Dornier's representative with the Kawasaki company. His principal task was to instruct the Japanese in the design and manufacture of all-metal airplanes, in which Dornier was a pioneer; but Vogt also became responsible for six winning designs for fighters, light bombers, and a private long-range touring airplane which the Kawasaki company had flown in an around-the-world cruise. Vogt frequently boasted that the aircraft he had designed for the Japanese in the 1920s helped them win the Manchurian War in 1931.

In 1933 fate again played a role in Vogt's life. After Hitler took over the government, part of his rearmament program receiving the highest priority was the aircraft industry; he was looking to achieve rapid production as soon as possible. Consequently, aircraft manufacturing suddenly became a substantial business activity with almost unlimited opportunities. Vogt was approached by two companies intent on luring him back to Germany. One was the Klemm Flugzeugbau located at Böblingen near Stuttgart, and the other was the Blohm and Voss shipyard of Hamburg. Blohm and Voss was owned by two brothers, Rudolf and Walther Blohm. They intended to create a new and separate aircraft department in their shipyard, to be called Hamburger Flugzeugbau (HFB). HFB would be owned by three equal partners: Walther Blohm, Elbschiffbau Betelligungs Gesellschaft (a holding company belonging to the Blohm family), and Vogt. Since Klemm was building only small planes and trainers and its ambition did not go much beyond two- and four-seater aircraft, Vogt found that proposal less attractive, and he accepted Blohm and Voss' offer, returning to Germany in the Autumn of 1933.

Enroute to Germany from Japan, Vogt chose a route over the Pacific, across the United States, and over the Atlantic to Hamburg on the German steamship Europa out of New York City. On board the liner was Ernst Udet, the fourth highest scoring ace of World War I with 62 confirmed kills, a stunt pilot, movie maker, and world traveler. He had been to the United States on an "Inspection trip" and was returning to Germany with two bi-wing Curtiss F11C-1 aircraft which had been tied to the deck of the Europa. He was going to use the aircraft to demonstrate the accuracy that could be achieved when bombs were released during a steep dive. Later, Udet's activity would give birth to the "dive bomber," which was so effective during Germany's pre-World War II offensive in Europe.

A proposed Blohm & Voss Bv P.155B. Scale model by Reinhard Roeser.

When Vogt joined Blohm and Voss at the age of 40, he had been designing planes for more than a decade but still had not developed a typical "Vogt design," which Dr. Dornier had told him was a prerequisite for securing an outstanding reputation. His own design, Vogt felt, would come in due time. For the moment, however, there were planes assigned to Blohm and Voss by the Air Ministry to be modified and produced under license. In addition, there were the flying boats the Blohm brothers wanted to develop as an extension of their passenger liner business.

Walther Blohm's primary interest was commercial transportation, which meant commercial because he believed that the transport of passengers by ship was not going to be profitable much longer. Eventually, Blohm believed, commercial aviation would make considerable inroads into the domain of ships. Blohm wanted his firm to develop long-range commercial aircraft. It was up to Vogt to produce the design that would allow Walther Blohm to see his ideas to fruition. Because of Walther Blohm's focus on commercial aviation, generally, Blohm and Voss' contribution to the Luftwaffe in terms of numbers of aircraft produced was small; by war's end its entire aircraft works employed only 5,000 workers, whereas Heinkel employed 50,000, and Junkers, more than 147,000.

From time to time Vogt would reflect on the apparatus people had designed and built in order to lift themselves into the air. Most of the attempts, beginning with the drawings of Leonardo da Vinci in the fifteenth century and continuing right up to the mid 1930s and Vogt's time, resembled the example given by nature - the bird. Nature, of course, never made birds like bi-planes, with tension wires and support rods reaching into the air. Still Vogt observed, designers kept improving their work. Many aircraft designs now were looking rather bird-like, and many aircraft were able to fly long and well. However, the demands of the Air Ministry for weapons of war meant that some airplanes had to do more than simply fly. Reconnaissance aircraft, fighters, bombers, and interceptors had to be able to fight and to shoot with immovable, front-facing guns. Yet the very design of many planes - a single engine with a pilot and observer placed well back toward the center of the fuselage - made defense from attack quite difficult, while at the same time, carrying out the function for which the craft was designed. Moreover, a single-engine aircraft exhibited poor performance and control characteristics. Having been a pilot in World War I, Vogt knew what every pilot of single-engine airplanes knows - that symmetrical airplanes frequently react asymmetrically. For instance, when the pilot of a powerful, high-seed, propeller-driven interceptor tried to go into a steep climb from straight flight, the aircraft turned immediately to the left, and the pilot had to counter this by stepping down hard on the right rudder pedal. Not only did the engine's output of power and torque require heavy use of the rudder, the circling propeller also caused problems. As it turned it continuously threw great volumes of air toward the rear of the aircraft, all of it whirling about in the form of a huge circle. As the spiraling wind moved back along the fuselage, it hit the rudder from the side. This had the same effect as if the rudder had turned, with the result that the aircraft had a tendency to go in circles.

The job for aeronautical engineers was to come up with a design in which these effects were neutralized. Designing an aircraft that looked like a bird would not do. A different configuration was required, one that moved away from placing an engine in the nose and the pilots midway back on the fuselage. Vogt's solution turned out to be a truly unconventional aircraft design - an asymmetrically mounted crew nacelle - first seen on the Bv 141, a short-range reconnaissance plane. The Bv 141 brought considerable notoriety to Vogt, and he was pleased because he believed he had the perfect design for a fighter, reconnaissance aircraft, or interceptor. For one thing the design afforded the pilot an outstanding all-round view; the crew likely would not be surprised by enemy planes because it was housed in an extensively glazed nacelle offset to starboard. Morever, Vogt believed, this arrangement would cancel out the propeller's torque, which had presented problems to every designer of single-engine aircraft.

Ernst Udet, by then in charge of aircraft development for the Air Ministry, first flew the Bv 141 in February 1938 and was so impressed

A Messerschmitt ME Zerstörer Project II. Scale model by Reinhard Roeser.

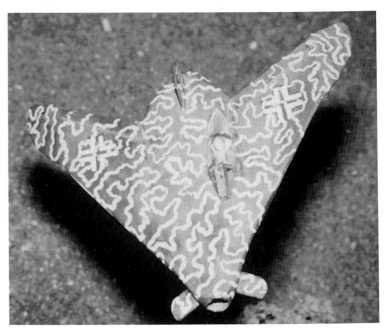

A proposed Blohm & Voss design concept, the Bv Ae 607.

Blohm und Voss Abteilung Flugzeugbau - Hamburg, Germany

The proposed Blohm & Voss Bv P.211 "Peoples' Fighter." Scale model by Frank Henriquez.

The proposed Blohm & Voss Bv P.211 "Peoples' Fighter." Scale model by Frank Henriquez.

The proposed Blohm & Voss Bv P.211 "Peoples' Fighter." Scale model by Frank Henriquez.

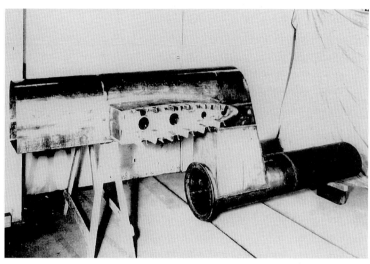

Dr. Richard Vogt's novel metal turbojet air intake duct as the main structural member of a series of proposed turbojet-powered aircraft, such as the Bv P.209, P.211, P.213, and so on.

with its handling that he ordered 500 to be manufactured. Vogt, like most other design engineers, found Udet a sociable, humorous man and an extremely qualified pilot, but hardly an aircraft engineer. Nevertheless, Udet, a daredevil and flying ace of some renown, liked the asymmetrical arrangement and it appeared that Vogt had at long last achieved a "Vogt Design." But only about twenty Bv 141s had been delivered when the Air Ministry in the spring of 1941, decided not to continue using the aircraft in an operational reconnaissance squadron. The aircraft did suffer from several minor teething problems; in addition, the role for which it had been intended was being filled adequately with the reliable twin-engine Focke-Wulf Fw 189. But the main reason for its cancellation probably had little to do with the Bv 141 itself. Some 80 percent of available Blohm and Voss assembly shop space at Hamburg had been taken over for production of the Fw 200 after the Focke-Wulf plant was damaged in a bombing raid, and the Air Ministry wanted to ensure that production would not be delayed by attention to the Bv 141.

Later in the war Vogt put forth several other asymmetrical designs similar in layout to the Bv141, including one for a single-seat dive bomber to be built substantially of wood. The bomber was never built because the BMW 801D radial engines it was to use, were already allocated to the extensively used twin-engine Dornier Do 217 bomber and the Focke-Wulf Fw 190. In 1944 Vogt tried again with an asymmetrical design, proposing a variation of the Bv 141, designated the P.194, in response to an Air Ministry call for a new dive bomber having a welded steel skin covering and extensive armoring. To provide extra bursts of speed for climbing out of a dive, the Bv P.194 was to have, in addition to a single BMW 801D 14-cylinder radial engine, a Jumo 004 turbojet having 1,980 lb (900 kg) of thrust, to be placed under the redesigned cockpit nacelle. Had it been built as planned, the P.194 would have been the world's first asymmetrical turbojet-powered aircraft.

Whether or not Richard Vogt ever felt he had achieved a "Vogt Design" with his Bv 141, he never said. But he was very proud of the fuel-carrying tubular spars he had invented for use in his flying boats. Vogt believed that rather than supporting the wing with two or three spars running from wing-tip to wing-tip, as was the custom then, a single hollow tube could serve as the only supporting wing structure. It would simplify enormously both the design and the construction of the airplane. The volume inside the tube, especially in the case of large airplanes, Vogt found adequate for carrying all the fuel needed for even the longest distances - even for a flight across the Atlantic. Divided into a few sections by bulkheads, the tube replaced the many small tanks and

The proposed Blohm & Voss Bv P.212.03. Scale model by Dan Johnson.

The proposed Blohm & Voss Bv P.212.03. Scale model by Dan Johnson.

the complicated pumping and valve systems then being used. Vogt designed an equally simple internal structural member for his proposed turbojets. For his P.209 he built an intake tube for the jet engine that doubled as the main structural member, with another tube above and running all the way to the tail as a boom support. The upper tube/tail boom support would hold 620 liters (164 gallons) of fuel, and by adding only 176 lb (80 kg) to the overall weight of the aircraft, it would provide a measure of bullet-proofing to the fuselage fuel compartment, the wing tank fuel being consumed first during fight. The same arrangement would serve to protect the pilot from the rear.

The wartime efforts of the Blohm and Voss organization and those of Richard Vogt probably are best remembered in connection with the development of floatplane and flying boats that culminated in the six-motor Bv 238 of 1944. Only one of the company's designs, the twin-boomed Bv 138 reconnaissance flying boat, was built in quantity, a total of 279 aircraft of this type being produced between 1938 and 1942.

The Bv 138 achieved international recognition - which really amused Vogt. The British periodical "*The Aeroplane*" from time to time caricatured new German airplanes under the heading "Oddentifications." when the British became aware of the Bv 138, the editors presented it to readers with a drawing and this poem:

Richard Vogt, that original man,
turns out aeroplanes uglier than
most any other designer can.
Here is shown on Baltic Sea
a typical Vogt monstrosity.
The One-Three-Eight by B. and V.

Vogt viewed the caricature as striking proof of the originality of his designs. The world's other asymmetrical aircraft the Bv 141 was described in "*The Aeroplane*" in the following way:

Your nose is here, your engine's there,
Your tail's just stuck on anywhere.
You symbolize the grotesque Hun,
Begone, distorted One-Four-One!

One of Vogt's most successful airplanes was the six-engine, 50-ton flying boat, the Bv 222. Started in 1938 as a purely commercial vehicle, the Bv 222 was designed to be able to cross the Atlantic between Hamburg and New York City with about twenty passengers, all of them having sleeping accommodations similar to those of railway sleeping cars. It was supposed to be able to reach its destination or to return to the starting point should one engine fail in the middle of the Atlantic. Only fifty of these flying boats were built before 1945, and none was used commercially. Some were used for long-range reconnaissance missions over the Atlantic and for transporting wounded German solders from Norway when the Germans were fighting the Russians, but most of them transported gasoline, ammunition, general supplies, and troops across the Mediterranean to Rommel's army in northern Africa.

"*The Aeroplane*" also caricatured Vogt's big flying boat:
Not distant far with heavy pace the foe
Approaching gross and huge, in hollow cube
Which to our eyes discovered, new and strange
two triple-mounted rows of engines laid
On wings (for like to engines most they seemed,
Embowelling with outrageous noise in the air),
The flying boat and transport of Richard Vogt
The Hun "Short" Blohm and Voss Two-Double-Two.

The proposed Blohm & Voss Bv P.212.03. Scale model by Ed Bailey.

The proposed Blohm & Voss Bv P.212.03. Scale model by Dan Johnson.

The proposed Blohm & Voss Bv P.213. Scale model by Steve Malikoff.

The proposed Blohm & Voss Bv P.212. Scale model by Steve Malikoff.

The proposed Blohm & Voss Bv P.212. Scale model by Steve Malikoff.

The proposed Blohm & Voss Bv P.215. Scale model by Reinhard Roeser.

The proposed Blohm & Voss Bv P.208.03. Scale model by Reinhard Roeser.

The Bv 222 flying boat was the last Blohm and Voss aircraft design to attain production status, and only one new design, the Bv 238, reached the prototype stage. When the prototype of the Bv 238, anticipated to serve as a long-range transport, maritime patrol, and flying boat, was tested in 1944, it weighed nearly 100 tons and was the world's heaviest aircraft.

From mid 1942 onward the engineering and production capacity of the Blohm and Voss company increasingly was absorbed by sub-assembly work and detail design development for Junkers and Messerschmitt aircraft - notably, for the high-altitude Bf 109H and its successor, the Me 155 (later redesignated the Bv 155 when its production became Blohm and Voss's responsibility. Although sub-assembly work did make full use of the plant's production facilities, it left Vogt and his design staff with a great deal of time to pursue design studies of their own. For this reason alone Blohm and Voss had the largest portfolio of turbojet-powered aircraft designs of any German aircraft manufacturer. However, although Vogt's team may have been prolific, none of its proposals progressed beyond the sketch stage, nor did any of Blohm and Voss's proposals win any of the design competitions in which they were entered (although Vogt claimed to have been awarded a contract to produce his P.215 by the Air Ministry three weeks before the war ended). The main reason for this was the virtual lack of interest on the part of the Blohm brothers; they simply were not interested in producing aircraft other than commercial transports. When the market for transports, flying boats, and other large aircraft disappeared in 1942, the brothers kept their factories and employees occupied with sub-assembly work. Vogt and his designers, frustrated and unable to move or transfer to a Messerschmitt, Heinkel, or Focke-Wulf operation, had to content themselves with theoretical doodling. No complete turbojet-powered prototypes were built by Blohm and Voss up through 1945, and Vogt and the entire organization were criticized that their tailless aircraft designs were based not on fact (except for results of wind-tunnel tests on small-scale models), but on hunch rather than on theory or evidence.

In February 1944 the Vogt team responded to an Air Ministry call for a ground-support bomber to replace the aging Junkers 87 Stuka dive-bomber. Blohm and Voss's entry comprised four widely differing designs. The P.192, with its engine mounted near the aircraft's center of gravity behind the cockpit, allowed the pilot an unobstructed view forward and downward during close support activities. The P.193 was a "pusher" type of aircraft having an inverted fin and rudder arrangement to protect the propeller should the pilot attempt a landing with a nose-high attitude. Again, the pilot would have a good view of the ground from the cockpit. Vogt's third design, the P.194, represented a return to the asymmetry he had pioneered with his Bv 141. In this design Vogt for

Dr. Richard Vogt's proposed swept forward wing fighter, the Blohm & Voss Bv P.209.02. Scale model by Dan Johnson.

The proposed swept forward wing fighter, the Blohm & Voss Bv P.209.02. Scale model by Dan Johnson.

the first time used a turbojet in conjunction with a fuselage-mounted BMW 801D radial engine having 2,200 horsepower. In addition, a huge BMW 018 axial-flow turbojet, to have been one of Germany's most powerful, with a rating of 7,500 lb (3,402 kg) of thrust, would have been fitted beneath the pilot's gondola. Vogt believed that mounting the turbojet below the pilot would help cancel out the torque resulting from the powerful BMW radial engine. The fourth project in the series, the P.196, would have been powered by two BMW turbojets. Unusual because of its twin booms, the P.196 was also different in that its undercarriage would have had a four-wheel arrangement instead of the more common tricycle arrangement. None of the four proposed aircraft was successful in the design competition. Instead, the Air Ministry chose a design by Vogt's arch rival, Messerschmitt's Me 262.

In addition to his inability to convince the Blohm brothers to get into turbojet-propelled fighters in a serious way (they repeatedly denied him funds with which he might build prototypes), Vogt was unable to obtain any commitment for turbojet engines from BMW, Junkers, or Heinkel. Blohm and Voss was on the waiting list for the massive 7,500 lb (3,402 kg) thrust BMW 018 engine, but the entire engine project was canceled in 1944 due to design difficulties. Consequently, many of Vogt's turbojet-powered designs were suddenly without engines. Since the promised performance of the BMW 018 was so great, Blohm and Voss's designs would be substantially underpowered were Heinkel or Junkers units to be substituted. It appeared that Vogt's famed good fortune had run out.

Vogt's last entry into an Air Ministry competition, the P.211 came in December 1944 with the Air Ministry's call for a Volksjäger (People's fighter) aircraft. Interested companies had only one week in which to respond with an aircraft that was to be easy for inexperienced pilots to fly, inexpensive to manufacture, and easy to service in the field. Considerable political influence entered into the awarding of the Volksjäger contract, which went to Heinkel and his He 162. It appears that National Socialist (Nazi) Party officials had more influence in selecting Heinkel than did the Air Ministry itself. Several weeks after Heinkel had been awarded the contract and was well under way with the prototype, Vogt and his design team were approached by SS officers. According to Vogt the SS team felt Blohm and Voss's aircraft was far superior to Heinkel's and they wanted to place an order, through Vogt, with Blohm and Voss

The proposed swept forward wing fighter, the Blohm & Voss Bv P.209.02. Scale model by Dan Johnson.

Another Richard Vogt concept rediscovered. North American Aircraft Division of Rockwell International proposed in the late 1970s that the fighters of the year 2000 and beyond be designed with swept forward wings due to their high potential for better lift, better distribution of internal volume in the fuselage, and lower supersonic drag.

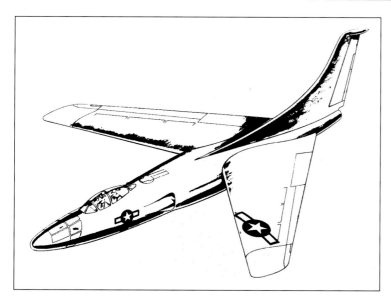

Convair's proposed XB-53 swept-forward winged bomber. In 1948 Convair engineers, using captured German tests data of aircraft with swept-forward wings, was given a USAF contract to develop two prototype bombers with a 30 degree forward wing sweep powered by three General Electric J35 turbojets. The project was canceled in 1948 and only a wooden mock up of its fuselage had been completed.

to produce their P.211 as an SS Volksjäger. When the Blohm brothers heard about the SS plan, they emphatically declined any such arrangement and at substantial personal risk to themselves and Vogt, told the SS to go through the Air Ministry if they really wanted to see the P.211 fly. The SS did not pursue the matter, and the P.211 was shelved.

Despite Vogt's team's poor performance in design competition for turbojet-powered aircraft, Blohm & Voss' owners would not allow him to build any prototypes, to leave the company for employment elsewhere, or to stop their design activities (idle designers might be sent to the Russian Front if they really had nothing to do). So Vogt and his team continued to pursue the concept of turbojet-powered aircraft even though they knew their work would not appear in the skies.

The Vogt family lived in Hamburg throughout the war, including the summer of 1943 when British and American bombers were attempting the complete destruction of the city. Although an estimated 40,000 people died and another 100,000 fled the city during the especially intense bombing on four successive nights in the summer of 1943 when about 75 percent of greater metropolitan Hamburg was destroyed, Vogt, his wife and two sons were unharmed. They remained in Hamburg until the summer of 1946 when Richard Vogt was sent to Wimbledon, England, for interrogation. It was at Wimbledon that fifty to one hundred former German scientists, engineers, aircraft designers, manufacturers, and offices of all branches of the military were brought before similar groups of people from England and America.

Shortly after Vogt's return to Hamburg, after a stay of five weeks at Wimbledon, he received an offer of employment from the British government. The job had nothing to do with aircraft design, but, instead, would allow him to develop the flat, fish-shaped torpedo he had been working on at Blohm and Voss a year earlier. Vogt declined the offer. At about the same time he was approached by an American officer asking him if he would like to become a U.S. civil service employee. In December Vogt left for America and a job that would pay a salary the equivalent of 30,000 German marks, about one third the amount he had earned as director of Blohm and Voss. His family joined him the following November.

Like many of his German colleagues, Vogt came to America under "Operation Paperclip." He arrived in New York City in January 1947 and spent several weeks of interrogation at nearby Mitchell Field before being sent on to Wright-Patterson Air Force Base in Dayton, Ohio. For a number of years Vogt occupied his time working on his "Floating Wing," an arrangement whereby F-84 fighters could attach themselves while in flight to the wingtips of a B-29 bomber (later a B-36 jet-powered bomber was substituted for the B-29). Such an arrangement would save a great deal of fuel during long bombing runs. Should the enemy appear, the fighters could detach themselves and ward off or destroy the enemy. After the danger had passed, the fighters could reattach themselves and reduce their engine RPMs to idle. During a test flight in the early 1950s, a prototype B-29 with its two F-84 fighter aircraft attached went into a dive from which it could not recover. All three aircraft were lost along with their crews. The accident ended any further testing of Vogt's "Floating Wing." Tests of a B-36 jet-propelled bomber continued, but eventually the Air force canceled the B-36 bomber in favor of the B-52, and the entire "Floating Wing" concept was abandoned.

In 1954 Vogt joined the Air Force's nuclear-propelled bomber pro-

A novel ideal by Dr. Richard Vogt in the early 1940s, was a coupling experiment whereby two aircraft (escorts) attached themselves in flight at the wing tips to conserve fuel over long-range flights. The results were inconclusive and the experiments were discontinued.

When Richard Vogt arrived in the USA with the "Operation Paperclip," he convinced the US Air Force to experiment further with fighter aircraft (escorts) attached to a bomber's wing tips. The idea being that the fighters would be there to protect the bombers over long range flights. They could detach themselves from the wing tips and re-attach when the danger was gone. These experiments ended in disaster with the loss of a B-29 and its two escorts. Experiments were canceled.

A wind tunnel scale model of the proposed Blohm & Voss Bv P.188 bomber undergoing tests. 1943.

gram. The atomic bomber, due to its projected range (great enough to attack targets deep inside Russia) was being pursued actively by the Pentagon. The plan was to have the bomber cruise at subsonic speed on power supplied by its atomic reactor, then, as it approached its target, to speed up, through the use of ramjets, to supersonic speed to avoid Russian fighters and rockets. When the nuclear-propelled bomber project was canceled in 1955 because of difficulties with the nuclear reactor, Vogt took a position with the Aerophysics Development Corporation in Santa Barbara, California. While he was at Aerophysics he worked on rocket development, winning a design competition from the U.S. Army for a vertical takeoff and landing "Aerial Jeep." Aerophysics experienced difficulties with its projects, and by 1960 the corporation was dissolved for lack of work.

At this time, Vogt joined the Boeing Company under a consultant agreement whereby he was able to carry out his work from his home in Santa Barbara. For the next 7 years, 1960 through 1966, Vogt worked on such Boeing projects as hydrofoils, commercial transport aircraft, and missiles. For a time he was involved in Boeing's supersonic transport aircraft, concentrating on problems associated with its complex swing-wing. In 1966, at the age of 72, Vogt retired. While in retirement he was awarded honors by the Japanese government for his work in Japan between 1923 and 1933. The Society of German Engineers awarded him an honors medallion, and the Lilienthal Society awarded him its gold medal for aircraft design achievement. He died on 23 January 1979 at the age of 85.

Eventually all of the Third Reich's secret aircraft design projects were published between 1945-1950 and circulated throughout the United States. Although it seemed reasonable that someone would be interested in his work, Vogt frequently wondered while he worked at Wright Patterson Air Force Base why no representatives of American aircraft manufacturing companies ever contacted him to discuss his World War II-era designs. His conclusion was that few, if any, manufacturers or their chief designers read the many reports. For a decade after he arrived in the United States in 1947, Vogt was ready to describe how his oblique (swiveling) wing concept could be applied to aircraft in order to minimize the negative effects of a pure swept wing. With the swiveling wing, a pilot could position the wing into whatever sweep angle was appropriate. One side of the wing would then have a forward sweep and the other a backward sweep. In 1970, about 35 years after he had published his views, the National Aeronautics and Space Administration

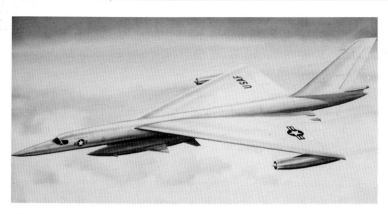 

Richard Vogt worked on these nuclear-powered bomber projects in the United States in the mid 1950s. These projects were abandoned when problems with the nuclear reactors could not be resolved.

(NASA) rediscovered Vogt's idea. Although a prototype eventually was built and test flown with satisfactory results, Vogt did not participate in NASA's oblique-wing activity.

Throughout his fifty years of aircraft design and manufacturing, Vogt was granted thirty-three patents. One of them, the jet flap, received worldwide attention for its use on short-takeoff-and-landing aircraft. However, he did not achieve the recognition as an outstanding designer for which he had labored in Japan, Germany, and later in the United States. If Claudius Dornier was right, that to be considered great a designer had to evolve a distinctly personal model, then Richard Vogt may have fallen short. During his most creative years (in Germany), the Blohm brothers placed severe limits on his design activities. As a result, Blohm and Voss, like Focke-Wulf, was among the few medium-to-large aircraft manufacturers hat did not produce at least one prototype of a turbojet-powered aircraft. Although they missed out on all opportunities to build jets, and none of their designs ever won an Air Ministry competition, Vogt and his design team did dream big dreams about turbojet designs and turned out a considerable number of ideas for others to consider.

It was the Blohm brothers who drew the line, and with no enthusiastic backing, Vogt saw other more aggressive, more politically astute firms such as Heinkel and Messerschmitt come away with the lion's share of the new design business. As a result, Vogt was unable to test any of this tailless and oblique-wing designs in any place other than is creative mind, or as scale models in his company's wind tunnel at Hamburg.

When the ban on manufacturing and selling military aircraft in Germany was lifted in 1955, Walther Blohm wrote to Vogt, then living in California, requesting that he come back to Hamburg to head up Blohm and Voss's aircraft activities once again. Vogt turned down the offer. At the time he was 60 years old and an "American." He also knew from experience that the American aircraft industry was far ahead of any in Europe and was convinced that Germany, if it could get going again at all, could start only on a small scale and only with specialized aircraft. He explained to Blohm that he just could not get interested in returning to Germany and its lifestyle. Later Walther Blohm merged his aircraft manufacturing activities (called the Hamburg Flugzeugbau) with Messerschmitt and Bölkow, forming the firm now known as

Between 1960 and 1966 Richard Vogt worked as a consultant to Boeing Aircraft Company on hydrofoils and commercial transport aircraft. Here he is pictured with George Schainer, Boeing Vice President for Research and Development.

Left: Dip.Ing. Hans Pohlmann, designer of the Junkers Ju 87 "Stuka." Center, Richard Vogt and wife.

A beautiful overhead view of NASA's AD-1 pivoting wing in flight. Early 1980s.

Messerschmitt-Bölkow-Blohm (MBB) GmbH. About the same time (1955) Vogt was asked to become head of the department of aeronautics at the University of Stuttgart, where he had earned his college degrees, including his Ph.D. He refused the offer for many of the same reasons he had turned down Walther Blohm.

Although Richard Vogt found California more to his liking than any other place on earth in his later years, he is known as a German, perhaps one of the better known German airplane designers in this century. His fame is due to his imagination and skill at conceptualizing and making inventions that worked - albeit in war, a war that overwhelmed the country of his birth. Vogt looked on life as a great adventure, a battle he enjoyed - whether facing hardships or good times. From time to time he liked to speculate about the turns his life might have taken had he accepted an offer of employment in 1930 from the Boeing Company after it had lost its chief design engineer. "One thing is certain," he would say, "the town of my birth would certainly not have named a street 'Richard Vogt Way' after me like they did!" Richard Vogt, no stranger to personal battles, believed that an individual's greatest feats came out of conflict, that a person should be committed to achieving quality in his own life - not quality as defined by a bureaucracy, but quality in relation

Richard Vogt's pivoting wing concept had been rediscovered by America's NASA. The NASA prototype AD-1 was capable of having its wing rotate in flight up to 60 degrees and in doing so the aircraft experienced less drag and increased fuel efficiency during is initial testing in 1980 and 1981.

Blohm und Voss Abteilung Flugzeugbau - Hamburg, Germany

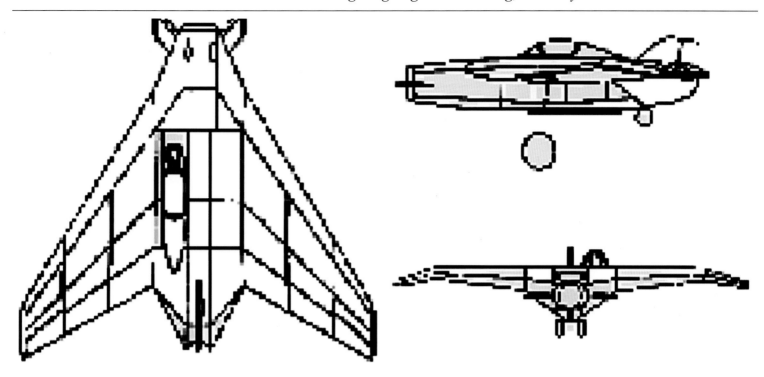

to his own capacity. He believed that it did not matter how fast or how far a person ran against others, only how he ran against his own capacity to run. Consequently, Vogt could be a thorn in the side of co-workers, managers, and government bureaucrats whom he felt were not working up to their potential. His attitude toward his own tumultuous life was simple and puritanical. This was how he once summed up the entire affair: "The whole joy of life is in its continuous encounters - not in winning. Challenge is the joy of life." And what was he most proud of? "My family. And in my working life commanding the services of a most devoted and able team."

PROJECT 607

This delta-wing, turbojet-powered aircraft was not classified as a "design project," but as an unofficial experimental design study by Richard Vogt of Blohm and Voss. Appearing as a tailless aircraft with only a rudder for a tail, it had a sweep to its wings and two unique features making it look unlike any other turbojet-powered aircraft in Germany. First, it had nose-mounted canards to serve as elevators (it was one of the few advanced designs that featured canards. The Horten brothers had incorporated a pre-wing on their Ho 3-B sailplane for the 1938 Rhön Wasserkuppe.). Second, its single-seat cockpit was off-center, far to the port (left) side of the aircraft. Blohm and Voss engineers had decided that, with the large diameters of the internally mounted twin HeS 011 turbojets, lateral placement of the cockpit would lead to more aerodynamic efficiencies.

The undercarriage was to consist of four main landing wheels. The two front wheels were to be steerable and retract to the outside under surfaces of the fuselage wings. The rear wheels, which were to bear the larger percentage of the aircraft's total weight, were to retract into the rear trailing edges of the fuselage wings. Armament was to consist of three 30 mm MK 108 cannon. Very little performance or specification information about the Ae 607 is available.

Specifications: Project 607

Engine	2xHeinkel-Hirth HeS 011 turbojets each having 2,866 lbs (1,300 kg) of thrust
Wingspan	NA
Wing Area	NA
Length	NA
Height	NA
Weight, Empty	NA
Weight, Takeoff	NA
Crew	1
Speed, Cruise	NA
Speed, Landing	NA
Speed, Top	NA
Rate of Climb	NA
Radius of Operation	NA
Service Ceiling	NA
Armament	3x30 mm MK 108 cannon
Bomb Load	NA
Flight Duration	NA

PROJECT 188

By early 1943 Air Ministry officials had pretty much given up hope that the Heinkel He 177 heavy bomber would ever become a workable design, the aircraft had become one of the true 'white elephants' of the German aircraft industry during World War II. From the time it first flew in 1939, the He 177 suffered all kinds of troubles. Its main problem arose from the way its two Daimler-Benz DB 601 engines were joined by a common crankcase. So complex was this arrangement that the engine heads had to be removed in order to replace the spark plugs. As he considered replacing the He 177, Siegfried Kneymeyer, Chief of Technical Air Armament for the Air Ministry, proposed that all conventionally powered bombers be phased out, for the world's first turbojet-pow-

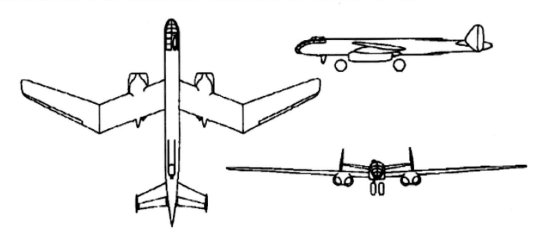

ered bomber, the Arado Ar 234, had already flown on 15 June 1943. Subsequently, the Air Ministry, through Knemeyer's office, issued a design specification calling for a heavy bomber powered by up to four turbojets. Junkers proposed to fill this design requirement which came to be known as the Strahlbomber, or jet bomber, with its swept-forward-wing Ju 287. Blohm and Voss responded with a series of four designs, all designated P.188 and all featuring a W-shaped wing.

The inner section of the P.188's wings swept back about 20 degrees and the outer section swept forward. Blohm and Voss's choice of this so-called "reflex-swept" wing was based on extensive development work on wing sweep, both forward and rearward, by the firm's engineers, who by 1943 had obtained a number of patents on forward-sweeping wings. (The P.188's chief competitor in the Strahlbomber competition, the Ju 287, in fact was based on one of Blohm and Voss's patents.)

Work on the first of the four designs drawn up around a common main frame and bomb load was started late in the summer of 1943. Blohm and Voss believed a prototype P.188 could be started in October 1944 and its first flight attempted by 15 December 1944. The fuselage of all four versions was to have been built around an armored steel fuel tank which was to comprise the largest part of the fuselage center section. The lower part of the fuselage center section was to form the bomb bay, which was designed to hold up to 4,409 lb (2,000 kg) of conventional high-explosive bombs or two controlled-trajectory anti-shipping (Fritz X) bombs. Immediately fore and aft of the bomb bay were to be the wheel bays, into which the two twin-wheel legs of the main undercarriage were to retract vertically, straight up. Small outrigger wheels were to be housed in the out sections of the wings.

The first version of the P.188 was an unarmed aircraft that would rely on its superior speed to elude interception. It was to be fitted with a single fin and rudder, and its four Junkers Jumo 004C turbojets were to hang beneath the wings in individual nacelles. The so-called P.188.04 version was to be a stretched-range aircraft, with much of the fuel carried in wing tanks. For defense the P.188.04 was to have remote-controlled dorsal and ventral twin-gun barbettes, plus four fixed MG 151 cannon, two in the extreme rear of the fuselage and two in the nose.

The P.188.04 represented an extensive redesign of earlier versions. For one thing, the nose section was changed, allowing the fuselage as a whole to be made slimmer (the P.188.04 had the smallest frontal area of any in the series). In addition, the fuselage-mounted air brakes of the earlier versions were replaced by a sturdy (spoiler) system (involving) flaps and lift devices. The P.188.04 design also had a pronounced dihedral on its twin-fin tail unit to raise it out of the turbojet's exhaust. It also had an improved nacelle design, the turbojets being grouped in pairs within one nacelle on each side. This not only reduced drag, but also, since the outboard turbojets were now closer to the fuselage, afforded better asymmetrical engine handling.

None of the four proposed P.188 heavy bomber variations was constructed.

Specifications: Project 188

Engine	4xJunkers Jumo 004C turbojets, each having 2,690 lb (1,220 kg) of thrust.
Wingspan	88.6 ft (27 m)
Wing Area	646 sq ft (60 sq m)
Length	57.1 ft (17.4 m)
Height	NA
Weight, Empty	29,321 lb (13,300 kg)
Weight, Takeoff	52,470 lb (23,800 kg)
Crew	2
Speed, Crusie	NA
Speed, Landing	105 mph (170 km/h)
Speed, Top	542 mph (873 km/h) at 26,248 ft (8 km)
Rate of Climb	2,244 ft/min (684 m/min)
Radius of Operation	1,419 mi (2,285 km)
Service Ceiling	42,653 ft (13 km)
Armament	4x30 mm MG 151 cannon fixed in the front and rear of the fuselage and 6x30 mm MG 131 cannon remotely controlled in the tail of the fuselage and on the aft sides
Bomb Load	8x551 lb (250 kg) bombs, or 4x1,102 lb (500 kg) bombs, or 1x4,409 lb (2,000 kg) bomb, or 2xPC 1400X (Fritz X) controlled-trajectory, anti shipping, armor-piercing, high-explosive bombs
Flight Duration	6 hours

PROJECT 196

Project 196 was a proposed twin-turbojet ground-support bomber featuring a dual boom system and a twin fin-and-rudder assembly. With this design, Richard Vogt put aside for the moment his unappreciated asymmetrical concept and submitted to the Air Ministry a plan for a conventional (though turbojet-powered) bomber. Two interesting features of the P.196 include installation of its two BMW 003 turbojets, each having 2,425 lb (1,100 kg) thrust, in a single nacelle beneath the cockpit and its two 1,101 lb (500 kg) bombs which were to have been carried inside the forward sections of its tail booms.

However, the story of the P.196 began some time before its design was conceived. For a long time the Air Ministry had been looking for a replacement for the badly underpowered Junkers Ju 87 dive-bomber, which had been designed in 1936. At first Vogt sought to interest the Air Ministry in a single engine aircraft that afforded its crew an excellent all-around view because it had an asymmetrical layout, with the engine at the forward end of the tail boom and a crew nacelle mounted sepa-

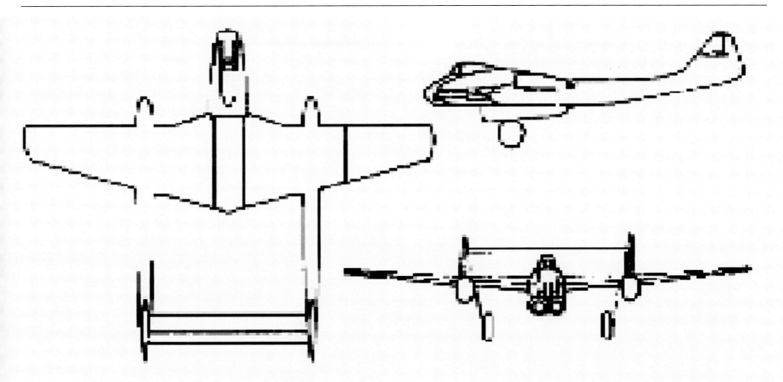

rately on the wing to starboard of the engine. Although twenty prototypes had been manufactured, Air Ministry official refused to finance a production run of the asymmetrical aircraft, which came to be called the Bv 141, despite the fact that they, as well as Ernst Udet, who was in charge of aircraft development for the Air Ministry, were quite interested. In addition to the lack of assembly space at Blohm and Voss (80% of assembly shop space had been turned over to manufacture the Fw 200 after Focke-Wulf assembly facilities had been damaged in a bomb raid) it was the aero-engine which powered the aircraft that also helped doom its future. The Bv 141 prototypes utilized the BMW 801A, 14-cylinder, radial, air-cooled engine - the engine used in the Junkers Ju 388 bomber and the Focke-Wulf Fw 190 single-engine fighter. The Air Ministry felt that the war effort could get along without the Bv 141 ground-support aircraft, but could not do without the Ju 388 or the Fw 190.

Before scrapping his Bv 141 idea, Vogt tried one more time to interest the Air Ministry in his asymmetrical layout by proposing to hang one BMW 003 or Jumo 004 beneath the crew's nacelle. Estimated top speed was 452 mph (727 km/h) at 22,967 ft (7 km), decreasing to 397 mph (638 km/h) at sea level. An operational radius of action of 630 miles (1,041 km) was anticipated with the aircraft carrying a normal complement of bombs and ammunition. However, the Air Ministry remained uninterested and the aircraft was not built.

Rejected for the second time, Vogt completely redesigned the Bv 141 along more conventional lines, and called the results the P.196. In addition to its unconventional twin tail boom arrangement, the P.196 also would have had a rather unusual landing gear. Location of the twin turbojets beneath the cockpit would not allow for a tricycle type of landing gear, so Vogt and his design team devised a four-wheel arrangement to keep the aircraft level during its takeoff run. When the craft was airborne, the two tail wheels would have retracted into the rear of the booms, and the two front wheels into the wings.

Although the P.196 showed promising performance (no details are available) and development potential, it did not gain official Air Ministry approval and no prototypes were built. Close air support for troops remained for the Fw 190 to carry out, pending introduction of another turbojet-powered fighter - the Me 262.

Specifications: Project 196

Engine	several combinations featuring piston and/or turbojets. A favored combination included 2xBMW 003 turbojets each having 2,425 lb (1,100kg) of thrust
Wingspan	NA
Wing Area	NA
Length	NA
Height	NA
Weight, Empty	NA
Weight, Takeoff	NA
Crew	NA
Speed, Cruise	NA
Speed, Landing	NA
Speed, Top	NA
Rate of Climb	NA
Radius of Operation	630 mi (1,041 km)
Service Ceiling	NA
Armament	NA
Bomb Load	2x1,101 lb (500kg)
Flight Duration	NA

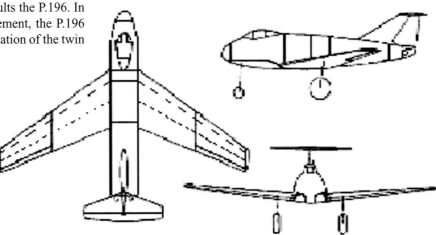

PROJECT 197

During the summer of 1944 the Air Ministry informed the German aircraft industry of a need for a twin-turbojet, single-seat interceptor capable of operating above 40,020 ft (12.2 km). Richard Vogt and his design engineers had a proposed design ready by August 1944. The P.197 represented Vogt's attempt to design the smallest and most efficient aerodynamic shape possible around two Junkers Jumo 004B turbojets, each having 1,984 lb (900 kg) of thrust. Projected to be one of the fastest Blohm and Voss fighters ever built, the P.197 offered the Air Ministry an estimated level flight speed of 662 mph (1,065 km/h) at 26,248 ft (8 km). Vogt's engineers claimed that the aircraft's effective ceiling would be 41,013 ft (12.5 km), exceeding Air Ministry's specifications. Anticipated flight weight of the proposed P.197 is not known, but these expectations for performance appear to be optimistic in view of the thrust provided. Messerschmitt's Me 262 was powered by the same Junkers Jumo 004B turbojets, and its maximum speed was only 514 mph (827 km/h) at 21,327 ft (6.5 km), while its service ceiling was 36,080 ft (11 km).

The P.197's turbojets were to be mounted side by side in the fuselage, and the air intakes were to be flush with the fuselage sides (similar to those adopted for the Lockheed P-80 "Shooting Star"). This engine arrangement would allow the overall frontal area of the P.197 to be kept to a minimum and would provide for more aerodynamic stability, its designers believed, should the aircraft be forced to fly on only one engine.

Vogt gave the P.197's wings a 40 degree sweep on the leading edge and a strongly sweptback vertical rudder. Mounted on the very top of the fin was a Multhopp T-Tail with a moderately sweptback, triangular elevator. One of Vogt's major concerns in designing a fighter was visibility. In the case of the P.197, Vogt placed the cockpit well forward on the fuselage near the nose of the aircraft. This forward location, Vogt believed, and the use of a clear vision canopy would afford the pilot an excellent field of view. Each of the P.197's four 30 mm MK 103 cannon were to be located in the nose. The landing gear was to be a tricycle arrangement, with the nose wheel retracting back into the fuselage below the cockpit and the two main wheels retracting up into the wings.

No other specifications or performance data are known. However, records show that the P.197 was expected to reach its service ceiling of 41,013 ft (12.5 km) in less than 29 minutes from a standing start. This would have meant an initial climbing rate of 4,985 ft/min (1,519 m/min). This design project was not selected by the Air Ministry for series production nor were any prototypes constructed.

Specifications: Project 197

Engine	2xJunkers Jumo 004B turbojets each having 1,984 lb (900 kg) of thrust
Wingspan	NA
Wing Area	NA
Length	NA
Height	NA
Weight, Empty	NA
Weight, Takeoff	NA
Crew	1
Speed, Cruise	NA
Speed, Landing	NA
Speed, Top	662 mph (1,065 km/h) at 26,248 ft (8km)
Rate of Climb	4,985 ft/min (1,519 m/min)
Radius of Operation	4,985 ft/min (1,519 m/min)
Service Ceiling	41,013 ft (12.5 km)
Armament	4x30 mm MK 103 cannon
Bomb Load	NA
Flight Duration	NA

PROJECT 198

In 1944 the Bayerische Motoren Werke (BMW) began work on a twelve-stage, axial-flow turbojet engine projected to have a thrust of 7,500 lb (3,402 kg). Known as the BMW 018, it was to be considerably larger in diameter than the widely used BMW 003, weighing 5,060 lb (2,295 kg) compared with the 003's 1,342 lb (609 kg) and having an overall diameter of 4.1 ft (1.3 m) compared with 2.3 ft (0.7 m). In August 1944 the Air Ministry issued a request for proposals for an airframe that could house the huge new turbojet. What authorities had in mind was a high-altitude interceptor-fighter.

To accommodate the BMW 018, the most powerful of all Germany's turbojet designs during World War II, Richard Vogt proposed a very basic, rather orthodox design with rugged simplicity throughout, the P.198. The design carried an unswept wing with a rather generous surface area and standard tail, although the elevator was placed beyond the leading edge of the vertical fin. Wingspan of the P.198 would have been 49.2 ft (15 m), and overall length would have been 49 ft (14.9 m).

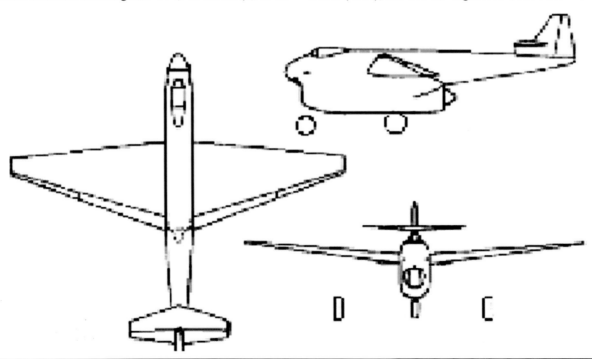

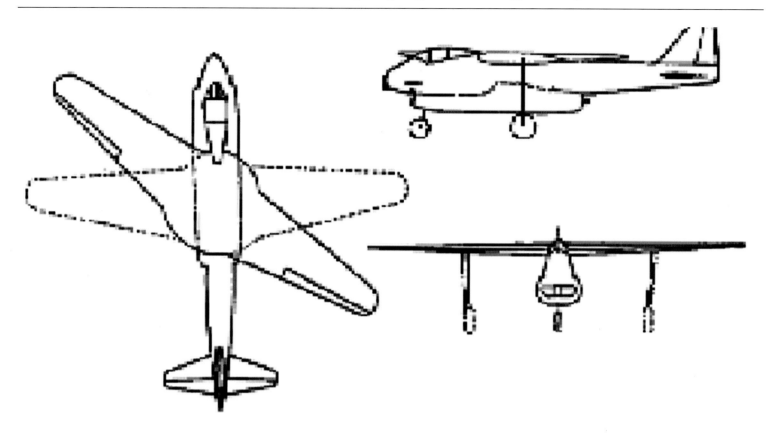

Anticipated speed for the P.198 equipped with the BMW 018 turbojet was 554 mph (891 km/h) at 11,484 ft (3.5 km). Expected to operate up to a service ceiling of 50,850 ft (15.5 km), the P.198's initial rate-of-climb was projected to be 8,901 ft/min (2,713 m/min). Its range was estimated at 900 mi (1,448 km), and its typical flight duration at 110 minutes.

With the P.198 Vogt and his design team made sure the pilot had an excellent field of vision by placing the pressurized cabin well aft on the nose of the fuselage. The tricycle landing gear, characteristic of Vogt's designs, had a wide track, with the main wheels retracting into the wings and the nose wheel retracting into a recess in the air intake duct.

Development of the P.198 was still in the design stage (a swept wing version had been planned) when BMW shelved its 018 turbojet project late in 1944 due to design difficulties and the scarcity of suitable heat-resistant metals.

Specifications: Project 198

Engine	1xBMW 018 turbojet having 7,500 lbs (3,402 kg) of thrust
Wingspan	49.2 ft (15 m)
Wing Area	NA
Length	49 ft (14.9 m)
Height	NA
Weight, Empty	NA
Weight, Takeoff	NA
Crew	1
Speed, Cruise	NA
Speed, Landing	NA
Speed, Top	554 mph (891km/h)
Rate of Climb	8,901 ft/min (2,713m/min)
Radius of Operation	900mi (1,448 km)
Service Ceiling	50,850 ft (15.5 km)
Armament	NA
Bomb Load	NA
Flight Duration	110 min

PROJECT 202

Richard Vogt called the P.202 research aircraft his "Swiveling Wing" project. It was to be a single-seat fighter powered by twin HeS 011 turbojet engines. Its distinctive feature was a wing that was to rotate 35 degrees horizontally on its axis, giving a backsweep to the starboard wing (right side) and a forward sweep to the port wing (left side). The arrangement was intended to delay the occurrence of shock waves during flights at transonic speeds.

In the early 1940s aeronautical engineers found that there were both advantages and disadvantages to using a sweptback wing on high-performance aircraft. The major disadvantage was that such a configuration resulted in instability at low speeds. Using a symmetrical wing (a straight wing having no backsweep) would avoid such problems - but at high speeds it was desirable to have a sweptback wing to help minimize compressibility problems. Vogt had the idea of using a symmetrical wing that would swivel around a vertical axle, or shaft, placed in the middle of the fuselage. During takeoff and landing and at low speeds the wing would be set at right angles to the fuselage. As speed increased, the wing would be rotated up to a maximum of 35 degrees. Thus one part of the wing could maintain a positive contour position, the other a negative position. Vogt believed that rolling and sliding forces could be stabilized through the use of such swiveling wing.

The P.202 was not built before the war ended. However, the idea was "rediscovered" and used by the National Aeronautics and Space Administration (NASA) 30 years later on an aircraft that first flew in 1980. Known as the AD-1, the craft could pivot its wing in flight up to 60 degrees to reduce drag and increase fuel efficiency. Flight tests at NASA's Dryden Flight Research Center at Edwards Air Force Base, California, indicated that a swiveling-wing transport aircraft flying at 1,000 mph (621 km/h) might use half the fuel required by more conventional supersonic transport aircraft. Just as Vogt proposed, the AD-1's wing was set perpendicular to the fuselage for low-speed flight and pivoted to decrease drag during high-speed flight.

Overall, the fuselage of Vogt's P.202 appears similar to that of many other of his turbojet-powered designs, such as the P.198, except that the wing of the P.202 was to be placed on top of the fuselage. A conventional tail configuration was to be used on the P.202, and its twin HeS 011's were to be placed in a single nacelle under the pilot's cockpit, thus eliminating the need for a long intake duct. The twin HeS' exhaust were to exit beneath and down a long, slender tail boom.

Because of the aircraft's high-mounted wing, Vogt chose a wide-track tricycle landing gear. The nose wheel was to retract upward into a space between the two turbojets, and the main gear was to retract inboard into the wing. A single-piece bubble cockpit canopy, a Vogt trademark, was to be used. Although the P.202 was intended to be a research aircraft to test the pivoting wing principle in flight, Vogt envisioned its ultimate use as a fighter-interceptor thus he provided two 30 mm MK 108 cannon in the nose of the aircraft.

Specifications: Project 202

Engine	2xHeinkel-Hirth HeS 011 turbojets, each having 2,866 lb (1,300 kg) of thrust
Wingspan	39.4 ft (12 m)
Wing Area	215.3 sq ft (20 sq m)
Length	32.8 ft (10 m)
Height	10.5 ft (3.2 m)
Weight, Empty	NA
Weight, Takeoff	11,904 lb (5,400 kg)
Crew	1
Speed, Cruise	NA
Speed, Landing	NA
Speed, Top	NA
Rate of Climb	NA
Radius of Operation	NA
Service Ceiling	NA
Armament	2x30 mm MK 108 cannon
Bomb Load	NA
Flight Duration	NA

PROJECT 209

Several designs for turbojet-powered aircraft with forward-sweeping wings came out of Germany during World War II. One was Junkers' Ju 287 V1 heavy bomber, and another was Blohm and Voss' P.209.

In November 1944 the Air Ministry had invited representatives of Messerschmitt, Focke-Wulf, Heinkel, and Blohm and Voss to meet at Messerschmitt's Advanced Design Office at Oberammergau to discuss a Schnellstjäger, or very high speed fighter. The purpose of calling the four manufacturing firms together was to establish a uniform starting point for the design. The firms were encouraged to work together on a common design but were told they could submit individual design proposals if they wished. Richard Vogt chose the latter course, and by 13 November 1944, he had a design ready for the Air Ministry.

The aircraft Vogt proposed, the P.209, was a single-seat fighter powered by a single HeS 011 turbojet having 2,866 lb (1,300 kg) of thrust. What made the design uniquely "Vogt" despite its wings, which were swept forward 35 degrees, was the use of the turbojet air intake duct as the main structural member, or chassis, for the entire aircraft. With a another steel duct or pipe superimposed above it, the duct would extend inside the fuselage all the way to the aircraft's tail. The fuselage portion of the steel structural member would also serve as a fuel tank holding about 164 gal (620 liters) of fuel.

With the exception of its swept-forward wings, the P.209, looked very much like other Blohm and Voss turbojet designs, especially the P.211 and the P.213. The pilot's cockpit was to be placed well forward on the fuselage and enclosed in a one-piece bubble canopy. Tricycle landing gear and tail surfaces were to be conventional except that the elevators were to have a slight negative dihedral (i.e. to bend downward). Armament was to include three 30 mm MK 108 cannon placed around the air intake in the nose.

Richard Vogt had been interested in forward-sweeping wings for a long time (they first appeared on his P.188 heavy bomber), and he believed that the high speed desired in the Schnellst-jäger could be achieved more quickly through such a configuration. He also saw other benefits from sweeping the wings forward. First it would make the aircraft extremely stable and especially safe from uncontrolled dives, and second, it would allow a more "arrowing" (higher degree of sweep) of the wings, which, he believed, would reduce internal stress greatly, thus allowing him to build the entire aircraft, particularly the wings, much lighter than would be normally possible.

Vogt expected that his P.209's speed would be 615 mph (990 km/h) at 29,529 ft (9 km). So confident was he of his design that he guaranteed that the aircraft would perform within 3 percent of predicted speed and range and 4 percent of predicted rate of climb. The Air Ministry was not impressed, however. It is not known whether a design for the Schnellstjäger was ever selected - but it is clear that Richard Vogt had every intention of constructing a prototype of the P.209. The main structural member which was to be the carrying chassis for the turbojet and the tail unit had been constructed by war's end.

Specifications: Project 209

Engine	1xHeinkel-Hirth HeS 011 turbojet having 2,866 lb (1,300 kg) of thrust
Wingspan	26.6 ft (8.1 m)
Wing Area	151 sq ft (14 sq m)
Length	28.9 ft (8.8 m)
Height	34.4 ft (3.2 m)
Weight, Empty	NA
Weight, Takeoff	NA
Crew	1
Speed, Cruise	NA
Speed, Landing	NA
Speed, Top	615 mph (990 km/h) at 29,529 ft (9 km)
Rate of Climb	5,078 ft/min (1,548 m/min)
Radius of Operation	NA
Service Ceiling	39,700 ft (12.1 km)
Armament	3x30 mm MK 108 cannon
Bomb Load	NA
Flight Duration	NA

PROJECT 211

On 8 September 1944, the newly reorganized Air Ministry, now under the day-to-day control of Albert Speer's Ministry for Armament and Ammunition, issued a requirement for a Volksjäger, or "People's Fighter." The new fighter was to weigh no more than 4,410 lb (2,000 kg), use a minimum of strategic materials, be suitable for mass production - and still have a performance superior to contemporary piston-engined fighters. In addition, it was to have a minimum endurance of 30 minutes, take off within 1,640 ft (500 m), and be armed with two 30 mm MK 108 cannon. It was to be powered by a single BMW 003D turbojet having 2,425 lb (1,100 kg) of thrust, an engine originally planned for the Me 262 but by late 1944 reserved chiefly for the proposed Volksjäger. A critical requirement was that the aircraft be easy to construct, for it was planned that the aircraft components would be manufactured throughout the Reich and later assembled by unskilled workers. To get the first aircraft off and flying as soon as possible, all design material was to be submitted by 14 September, and the aircraft itself had to be ready in less than four months, by 1 January 1945.

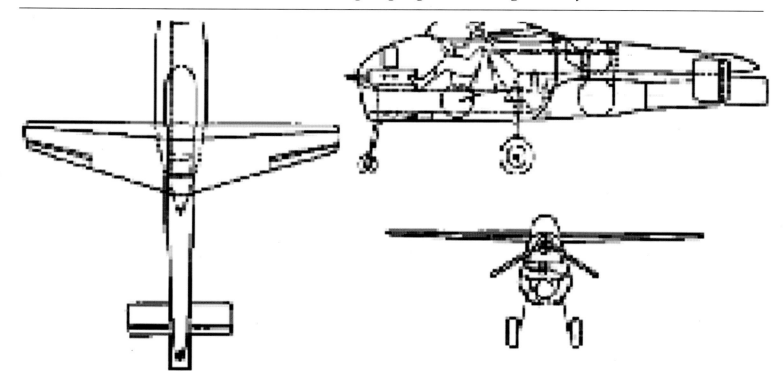

On 15 September the Air Ministry officials met to consider the various proposals. It did not take them long to select a winner. On 19 September the design submitted by Blohm and Voss, the P.211, was judged the best. However, Heinkel had already constructed a mockup of the runner-up, the He P.1073 (later re-designated as the He 162), and when Air Ministry representatives went to Heinkel's Vienna-Schwechat plant on 23 September to see the mockup they were impressed enough to reconsider their initial decision. A week later they re-awarded the project to Heinkel and his P.1073, instead.

Consistent with the Air Ministry's requirement, the P.211's design called for light-weight material such as Duralumin, which was scarce by September 1944; specifically the aircraft was to use 58% steel, 13% Duralumin, and 68% miscellaneous materials. Vogt also kept to Air Ministry requirements by designing the craft with an eye toward rapid assembly of components. The turbojet intake tube (duct) was the main structural member. Wings were constant chord, with all ribs the same profile and with no backsweep, an arrangement that would facilitate construction. Even the landing gear was designed to be uncomplicated. The pilot would crank up the gear by hand after takeoff. Lowering the gear would be even more simple. All the pilot had to do was unlatch the gear and it would lower itself into position by its own weight, and since it had originally retracted forward, it would be locked into place by the pressure of the air stream.

Estimated speed of the P.211 was put at 537 mph (864 km/h) at 26,248 ft (8 km). No other specifications are available.

Specifications: Project 211

Engine	1xBMW 003D turbojet having 2,425 lbs (1,100 kg) of thrust
Wingspan	NA
Wing Area	NA
Length	NA
Height	NA
Weight, Empty	NA
Weight, Takeoff	4,410 lbs (2,000 kg)
Crew	1
Speed, Cruise	NA
Speed, Landing	NA
Speed, Top	537 mph (864 km/h) at 26,248 ft (8 km)
Rate of Climb	NA
Radius of Operation	NA
Service Ceiling	NA
Armament	2x30 mm MK 108 cannon
Bomb Load	NA
Flight Duration	30 minutes minimum

PROJECT 213

In November 1944 the Air Ministry issued a requirement for the simplest possible type of fighter, one that could be produced more rapidly than the Heinkel He 162 Volksjäger. The new aircraft was to be called Miniatur-Jäger, or lightweight fighter, and was to be built around a single Argus As 014 pulse jet engine of 660 lb (300 kg) thrust. The Argus 014, considered the simplest and cheapest aero-engine in the Reich, initially had been built by Fieseler for its Fi 103 (VI) flying bomb (commonly known as the "buzz bomb").

The Miniatur-Jäger was not to be a semi-expendable weapon, in the manner of the Bachem Ba 349 (which was a lightweight target interceptor built of wood and other easy to obtain material). Nor was the P.213 to be a conventional fighter aircraft which could be built simply with an absolute minimum of strategic materials, yet still provide a stable gun platform. Firepower and performance were of secondary importance; the specifications called for only one 30 mm MK 108 cannon with 135 rounds of ammunition, and the fighter was to carry no radar or radio in order to keep its weight down. The P.213 was more of an target (bomber formation) interceptor.

As a target defense interceptor, the P.213 came under the Air Ministry's doctrine of target defense. Under this concept, fighters such as the P.213 would make visual contact with enemy bomber formations, and engagement would take place within a small radius of action. The number of enemy bombers shot down would depend on the number of fighters put into action. The idea was to overwhelm the bombers' defenses with large numbers of P.213's swarming over each bomber. Several other German aircraft manufacturers submitted designs in response to the Miniatur-Jäger requirements, including Junkers with its Ju EF 126 "Elli" and Heinkel with its Argus-powered version of the He 162 Volksjäger.

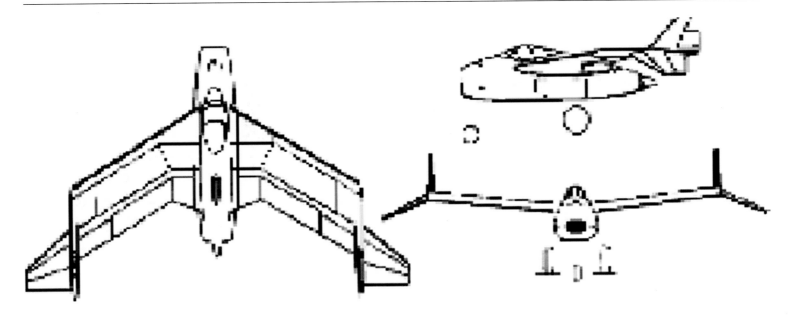

Blohm and Voss' P.213 design called for extensive use of armored steel throughout the fuselage, wood being confined to the wings and tailplanes. Takeoff weight was 3,435 lb (1,560 kg). A boom was to lead from the fuselage to support the inverted butterfly-type rudder-elevator combinations. The boom was also to support the exhaust pipe of the Argus 014 pulse jet. Experience with Messerschmitt's Me 328 had shown that the pulse jet pulsated at a frequency capable of setting up destructive mechanical resonance, and one Me 328 actually had been destroyed in flight due to the resonance. To avoid this problem Vogt planned to support the Argus As 014 inside the fuselage with free-swinging brackets. The air intake was to be separated from the pulse jet engine by a section of flexible tubing.

To cut down on weight-increasing equipment, the P.213's retractable tricycle landing gear was to be operated by compressed air taken from the supply used to operate its MK 108 cannon. Gravity would pull the gear down, and it was to lock automatically. A full-bubble canopy would be provided for the pilot. The P.213's wings were to be unswept with a straight leading edge, while the trailing edge was to sweep forward slightly.

Estimated performance of the P.213 was modest, failing to reach the figures already achieved by piston-engined Me Bf 109Ks and Fw 190Ds. This was to be expected because the Air Ministry wanted a target defender that could be produced inexpensively and then thrown in great numbers against bomber formations. Top speed was estimated at 435 mph (700 km/h) at sea level, with a range of 775 mi (124 km). To propel the P.213 up to its pulse jet operating speed of 149 mph (240 km/h), a solid-fuel assist takeoff rocket would have been required; catapult launching also was considered as a way to boost speed.

Neither the P.213 nor other proposed Miniatur-Jäger target defense fighters were built; nor were any prototypes constructed. The Bachem Ba 349 powered by a single HWK 509 bi-fuel liquid rocket drive; a semi-expendable target defender being developed at the same time, appeared to be more economical and promised better performance after takeoff than any of the proposed Miniatur-Jäger fighters, anyway, the entire Miniatur-Jäger program was abandoned in December 1944.

Specifications: Project 213

Engine	1xArgus As 014 pulse jet having 660 lb (300 kg) of thrust, plus solid-fuel rocket boosters during takeoff
Wingspan	19.8 ft (6 m)
Wing Area	53.8 sq ft (5 sq m)
Length	20.4 ft (6.2 m)
Height	NA
Weight, Empty	NA
Weight, Takeoff	3,435 lb (1,560 kg)
Number in Crew	1
Speed, Cruise	NA
Speed, Landing	NA
Speed, Top	4435 mph (700 km/h) at sea level
Rate of Climb	3,936 ft/min (1,200 m/min)
Radius of Operation	775 mi (124 km)
Service Ceiling	32,810 ft (10,000 m)
Armament	1x30 mm MK 108 cannon with 135 rounds of ammunition
Bomb Load	None
Flight Duration	NA

PROJECT 215

An unorthodox design, the P. 215 was entered in a design competition set out by the Chef der Technischen Luftrüstung (Chief of Technical Air Equipment) of the Air Ministry on 27 February 1945. The Air Ministry was looking for twin-turbojet night and bad-weather fighters with exceptional performance and aerodynamic stability, and Richard Vogt believed that this requirement would benefit substantially from a tailless configuration he called the "Arrow Wing."

Vogt had developed his ideas about a tailless configuration in 1942. He believed a fighter would be more maneuverable if the tail were removed from the rear of the fuselage and placed in the free airstream. Vogt did not plan to do this in the conventional way, that is, by tying the two vertical end plates together with a central elevator, like the tail on the proposed P.196. He had found, through wind tunnel tests, that sweeping the wing aback far enough and placing a small rudder and elevator on its tip could result in enormous benefits: the aircraft would exhibit better overall performance and control and, because of its small tail, would experience less drag - leading naturally to increased speed and improved maneuverability.

The P.215 was not truly a tailless design, but a configuration in which the standard tail had been displaced away from the rear fuselage outboard to the wingtips. Unlike the Horten brothers, Vogt strongly believed that an aircraft required a standard rudder-elevator arrangement for directional control and stability. The only question was how to retain the standard tail but at the same time improve its functions, achieve less drag, and gain greater maneuverability. Placing the rudder and elevator outboard on the wingtips was the answer, he believed. Although he viewed lateral placement of the tail to the wingtips as appropriate for a fighter, Vogt did not advocate a similar arrangement for his larger aircraft. Larger and heavier aircraft required a strong rudder and elevator system, and he believed that the structural strengthening necessary to place this configuration on the wingtips of a larger aircraft such as his P.188 bomber would deprive it of any improvement in performance.

Blohm and Voss was notified by the Air Ministry that the P.215 had been selected for series production just as Hamburg and the Blohm and Voss facilities were falling to advancing American armor. In anticipation of a favorable response from the Air Ministry, Vogt and his staff had started construction of the wings for several P.215's. With a 40 degree sweepback and small booms at the wingtips to support the outrigger rudder and elevator, this configuration was expected to contribute to the total lift of the wing and simultaneously to reduce the possibility of wingtip stall. The fuselage was to be long and extending well aft of the trailing edge roots, and the turbojet exhaust pipe was to be almost in line with the trailing edge of the wingtip rudders. The pressurized cockpit was to have a bubble canopy and would be positioned above the air intake duct just forward of the wing's leading edge. The nose wheel undercarriage was to retract into the fuselage, making a 90 degree turn to lay flat up against the air intake duct. The two main wheels were to retract into the sides of the fuselage.

As a projected all-weather fighter, the P.215 was to have a three-man crew - a pilot, a navigator, and a radar/radio operator. The three-man crew was justified on the grounds that the heavy armament on the P.215, including an obliquely upward and rearward-firing cannon, and its two 1,101 lb (500 kg) bombs might place too great a workload on a standard two-man crew. Vogt proposed a variety of interchangeable armament packs, including forward-firing armament, normally comprising five 30 mm MK 108 cannon with 200 rounds each and two 30 mm MK 108 cannon fitted in the nose and firing upward at an angle of 80 degrees to the line as the P.215 passed beneath enemy bombers.

Although the P.215 was never completed, Vogt remained optimistic bout the validity of his idea for taking the rudder off the fuselage and placing it in the free airstream. He believed it to be a concept that eventually would be proven and applied to aircraft design. After all, it had taken 35 years for his swivel wing (as seen on his oblique-wing P.202) to be proved (by the National Aeronautics and Space Administration) as a means of delaying the occurrence of shock waves during flights at transonic speeds. Vogt felt that his P.215, with its displaced horizontal and vertical control surfaces mounted at the wingtips eventually would be proved a valid concept as well.

Specifications: Project 215

Engine	2xHeinkel-Hirth HeS 011 turbojets, each having 2,866 lb (1,300 kg) of thrust
Wingspan	61.7 ft (18.8 m)
Wing Area	592 sq ft (55 sq m)
Length	38.1 ft (11.6 m)
Height	NA
Weight, Empty	16,314 lb (7,400 kg)
Weight, Takeoff	32,628 lb (14,800 kg)
Crew	3
Speed, Cruise	352 mph (566 km/h)
Speed, Landing	106 mph (170 km/h)
Speed, Top	550 mph (885 km/h) at 29,530 (9 km)
Rate of Climb	4,232 ft/min (1,290 m/min)
Radius of Operation	NA
Service Ceiling	41,669 ft (12.7 km)
Armament	Fixed forward: 5x30 mm MK 108 cannon with 200 rounds each; Oblique: 2x30 mm MK 108 cannon with 100 rounds each; Defense: 1x20 mm FHL 151 cannon.
Bomb Load	1,102 lb (2x500 kg)
Flight Duration	1.7 hours at 100% thrust at 32,810 ft (10 km), or 2.7 hours at 64% thrust (cruise speed) at 19,686 ft (6 km)

4

Bayerische Motoren Werke GmbH - München, Germany

Secret Projects and Intended Purpose:
Project Strahlbomber I - A four-BMW 018 turbojet-powered tailless heavy bomber concept.
Project Schnellbomber II - A twin-BMW 028 turboprop-powered swept-forward winged high speed bomber design concept.

History:
In 1944 the Bayerische Motoren Werke (BMW) at München commenced work on what was to be the most powerful turbojet-engine in Germany, the BMW 018, designed to provide 7,500 lb (3,402 kg) of thrust. Plans called for a prototype by 1945 and serial production in early 1947. When the war ended in May 1945, BMW engineers at Spandau (Berlin) had finished handcrafting all the component parts for their first prototype 018 engine. As American army troops were entering Berlin, engineers at BMW-Spandau tried to destroy the unassembled component parts

but were unsuccessful. Most of the engine's components were salvaged and flown immediately to the United States for examination. The primary purpose of the BMW 018, according to the Air Ministry, was to power a fast, hard-hitting bomber force composed of Arado Ar 234Cs, Junkers Ju 287s, Heinkel He 343s, and Messerschmitt Me 264s. Beyond 1947 these turbines were to power the huge all-wing aircraft of the Horten brothers.

One interesting modification of the BMW 018 turbine was a model having a propeller drive, the BMW 028. Known today as a turboprop, the 028 was designed to produce 4,700 s.hp., enough to drive two contra-rotating propellers, and to produce 4,850 lb (2,200 kg) of thrust. BMW officials felt that their 028 turbine would be a versatile power plant, and they were surprised that no aircraft manufacturer had produced an airframe design that could take advantage of the potential of a turboprop engine. BMW had not built any aircraft since World War I, so in an effort to nudge the German aircraft designers' imaginations, they decided to expend a bit of effort and market their new turbine engines. In 1944 BMW drew up a series of four heavy bomber designs showing how their 018 and 028 turbines could be applied to large aircraft.

As far as it is known, these BMW heavy bomber designs were merely concepts in how their turbines might be applied and were never were entered in any aircraft design competition. It is not known either whether BMW considered building the proposed bombers, but it appears that their intention was only to present the designs to Air Ministry officials and, if they liked what they saw, to let them pass the ideas on to existing aircraft manufacturers for further development. It is believed that no

Dr. Hermann Oestrich, BMW's Director of Turbojet Engine R&D. Picture shows Dr. Oestrich standing in the remains of their bombed-out BMW turbojet engine factory in Spandau-Berlin, 1944.

Bayerische Motoren Werke GmbH - München, Germany

Freshly made BMW 003 turbines litter a bomb crater at a bombed-out aircraft manufacturing factory somewhere in Germany. 1944.

A pair of proposed Heinkel He P.343 bombers. The "A" version was to come with a single vertical rudder and below is the "B" version with double vertical rudders. Scale models by Reinhard Roeser.

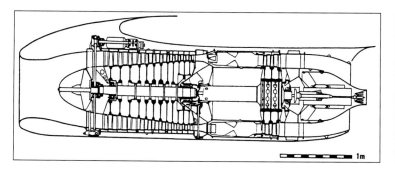

The BMW 018 turbine designed in 1940 along the same general lines as the BMW 003. Thrust for the 018 was to be rated at 7,525 lbs.

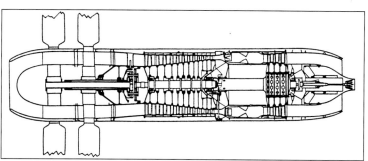

The BMW 028 a modified BMW 018 turbojet engine to drive contra-rotating propellers through planetary reduction gears from the main compressor shaft, 7,000 hp was expected at the propeller.

The Junkers Ju 287 V3 (Ju EF 131). Scale model by Günter Sengfelder.

The Arado Ar 234C. Scale model by Günter Sengfelder.

German aircraft company became interested enough in the construction of any BMW-designed heavy bombers.

Each of the four turbine-powered heavy bombers proposed by BMW had their own unique airframe configuration so that Air Ministry officials and others could have a range of design planforms from which to choose. Two suggested uses for the 018 turbojet included a six-engined tailless bomber and a twin-engined all-wing bomber. Two suggested uses of the 028 turboprop included a huge twin-engine conventional unswept winged aircraft having a 166 foot (50.5 m) wingspan. The fourth proposed aircraft, also using twin 028s, was to be a bit larger than a USAAF B-17 "Flying Fortress" (wingspan of 116.5 feet or 35.5 meters, vs. 103 feet or 31.4 meters for the B-17). BMW called this suggestion their "Schnellbomber II," or fast bomber II. This aircraft was unusual in that its wings were to sweep forward 45 degrees and its twin BMW 028 turboprops were to be placed high above the wing on a pylon. No other BMW turbojet- or turboprop-powered designs are known to exist.

PROJECT Strahlbomber I

The Strahlbomber I project was a design for a tailless heavy bomber. Initially it was suggested to be powered by six BMW 003 turbojet each with 1,764 lb (800 kg) of thrust. It would later be fitted (in 1947) with four BMW 018 turbojets, each having 7,500 lb (3,402 kg) of thrust. BMW engineers originally proposed that one 003 turbine be placed one each side of the Strahlbomber I's forward fuselage, which also served as the cockpit for the two-man crew. Additionally, four 003's were to be housed inside the wings. BMW believed this arrangement would minimize the possibility that all six turbines be lost at same time should their heavy bomber suffer cannon fire from enemy fighters. Due to the P.Strahlbomber I's anticipated high speed (510 mph or 820 km/h), it was thought unlikely that any Allied piston-powered fighters would shoot it down under normal conditions. With the substitution of the 018s for the 003s, maximum speed of the BMW's P.Strahlbomber I would be even greater, making attack from the rear all the more unlikely. Nevertheless, to protect the turbines against Allied fighters, the P.Strahlbomber I was to be fitted with two fixed, rear firing 30 mm MK 108 cannon.

It was anticipated that the P.Strahlbomber I would have a flight range of 1,243 miles (2,000 km) when carrying 4 tons of bombs; with a 2-ton bomb load it was expected to have a range of 1,616 miles (2,600 km) at 31,170 feet (9.5 km) and a flying time of approximately 4 hours.

In keeping with Air Ministry preference, this proposed tailless bomber would have come with a huge vertical stabilizer with an attached hinged rudder. The P.Strahlbomber I would have been somewhat smaller than the USAAF B-17 heavy bomber shorter in wingspan by about 16 feet (4.9 m), and shorter also in length by about 15 feet (4.6 m). Yet because it was powered by turbines, the P.Strahlbomber I would have been able to carry up to 4 tons of bombs, vs. 2 tons for the B-17,

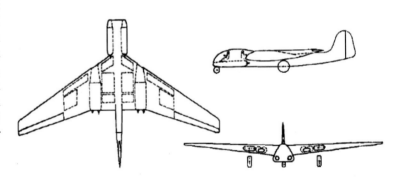

LEFT: The proposed Focke-Wulf Fw "Kamikaze" mother carrier. Initially proposed by Kurt Tank to Daimler-Benz, and to be powered by six Daimler-Benz turboprop engines, however, in 1943, given the technical difficulties DB was experienced in developing the 007, the RLM told Daimler-Benz to give it all up and cease any further work on their entire turbine program. Kurt Tank then sought out BMW for the possible use of their 018 turboprop for use on his proposed "Kamikaze" mother carrier. Scale model by Reinhard Roeser. RIGHT: Kurt Tank's proposed Focke-Wulf Fw "Kamikaze" mother carrier. Scale model by Reinhard Roeser.

Bayerische Motoren Werke GmbH - München, Germany

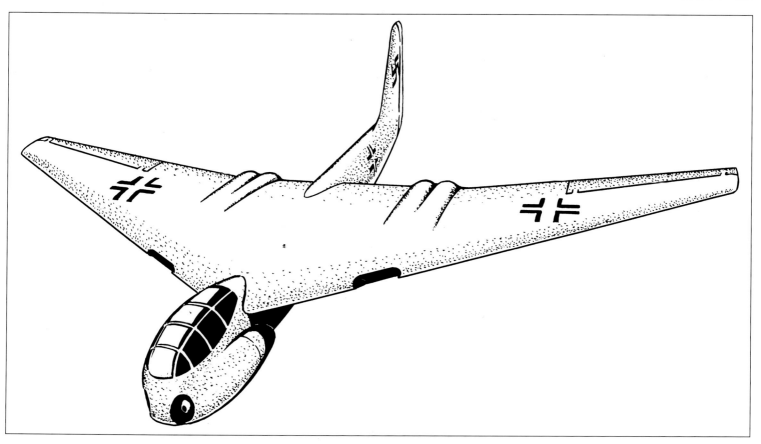

The proposed "Strahlbomber I," a turbojet-powered tailless bomber, was to have been propelled by six BMW 003 turbojets up to 1,616 mi (2,600 km) at 510 mph (820 km/h). None was ever built.

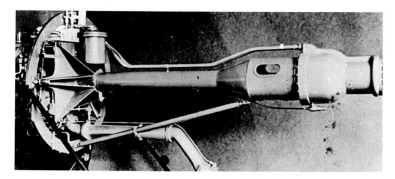

HWK 509 bi-fuel liquid rocket engine.

BMW "parallel entwick lung" BZW 718, an alternative rocket engine developed for the HWK-powered bi-fuel rocket-powered Messerschmitt Me 163 and other proposed rocket powered interceptors.

and to fly at a speed of nearly 200 mph (321 km/h) faster than that of a B-17. The P.Strahlbomber I was to have a two-man crew seated in a pressurized cockpit cabin; they likely would have been rather busy, for, if comparisons can be made, they would have been carrying out duties which on a B-17 occupied a crew of eleven.

Specifications: Project Strahlbomber I

Engine	6xBMW 003 turbojets, each having 1,764 lb (800 kg) of thrust, for a total thrust of 10,584 lb (4,800 kg). Later 4xBMW 018 turbojets, each having 7,500 lb (3,402 kg) of thrust for a total thrust of 30,000 lb (13,608 kg)
Wingspan	86.9 ft (26.5 m)
Wing Area	328 sq ft (100 sq m)
Length	59 ft (18 m)
Height	16.4 ft (5 m)
Weight, Empty	NA
Weight, Takoff	25 tons (22,680 kg)
Crew	2
Speed Cruise	NA
Speed, Landing	NA
Speed, Top	510 mph (820 km/h)
Rate of Climb	31,170 ft (9.5 km) in 24 minutes
Radius of Operation	1,616 mi (2,600 km) with a 2 ton bomb load
Service Ceiling	36,090 ft (11 km)
Armament	2x30 mm remote-operated rearward-firing MK 108 cannon
Bomb Load	variable: 2 to 4 ton
Flight Duration	4 hours

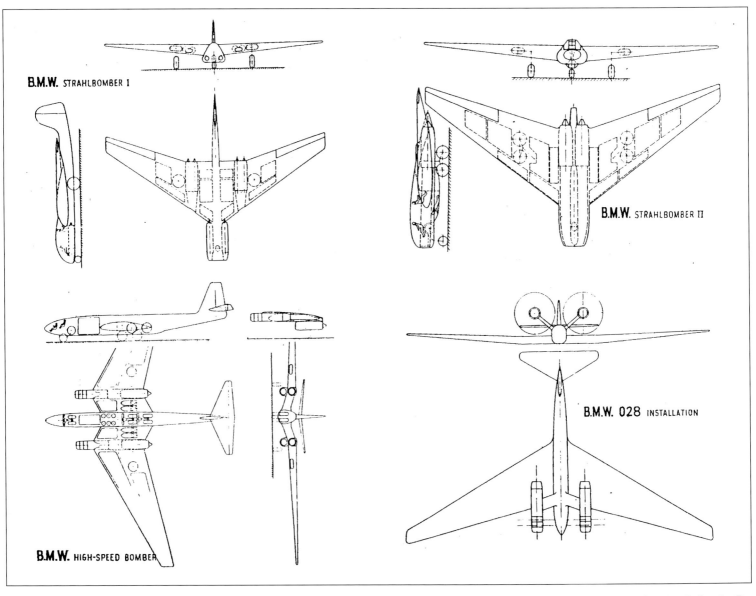

The aircraft design suggestions of Daimler-Benz were handled by EZS, a subsidiary of BMW within the Development Studies section, headed up by Dr. Ing. Huber where proposed bomber and fighter designs were made. These designs emphasized simplicity and ease of construction in order to reduce the strain on the German aircraft industry. The proposed BMW aircraft suggestions included:
Schnellbomber I - heavy bomber with compound swept wings, turbojet and propeller-turbines. Two 3,400 kg. BMW 018 and two 6,570 ehp BMW 028.
Schnellbomber II - Propeller-turbine bomber with swept forward wings. Two 6,570 ehp BMW 028.
Strahlbomber I - Jet bomber. Six 800 kg BMW 003.
Strahlbomber II - Tailless jet bomber. Two 3,400 kg BMW 018.

The proposed Daimler-Benz aircraft suggestions: (only good to 1943 because all aircraft design was canceled in 1943 by order of the RLM).
Project A - Combination or composite aircraft. Upper component: propeller turbine powered carrier aircraft. Four or six 3,300 ehp HeS 021. Lower component: jet bomber with V-tail. Two 3,400 kg BMW 018.
Project B - Carrier aircraft (upper component) with four tractor and two pusher airscrews. To carry one Project C bomber or five Project D flying bombs. Six 1,900 hp DB 603G piston engines.
Project C - Jet bomber. One 1,275 kg DB 007 turbojet engine.
Project D - Ramjet-powered flying bomb. One ramjet engine.
Daimler-Benz also proposed several turbojet and mixed propeller-turbine powered fast bombers, reconnaissance and fighter aircraft, but no project designations have been recorded.

Bayerische Motoren Werke GmbH - München, Germany

The "Julia" would have used the BMW BZW 718 bi-fuel rocket engine which promised more duration and power than the HWK 109-509. Scale model by Jamie Davies.

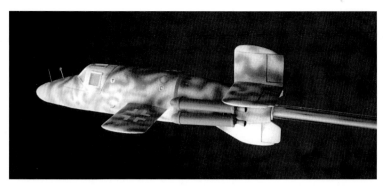

Latter "Natter's" would have been re-powered with the BMW BZW 718 bi-fuel rocket engine. Scale model by Jamie Davies.

PROJECT Schnellbomber II

When BMW proposed their P.Schnellbomber II or fast bomber concept in 1944, it was a highly unusual design. Not only did it call for the wings to be swept forward 45 degrees, but it showed the 028 turboprop engines placed atop the wings instead of beneath them or in a cluster inside the wings. This placement was chosen because BMW engineers feared that the 028's contra-rotating propellers would suck up small stones and other debris. With its turboprops mounted high atop the wings, the bomber would be suitable for dirt-strip operation. Not many aircraft designs have had this engine mounting arrangement, one was the VFW-Fokker 614 commercial turbojet aircraft, introduced in Germany in 1971.

Little is known about the P.Schnellbomber II beyond some basic measurements. Maximum performance was anticipated to be 404 mph (650 km/h) at 26,250 feet (8.0 km). As promising as this design study appeared to be, with its forward 45 degree wing sweep (like the Junkers Ju 287), it appeared to be ignored totally by the Air Ministry. BMW's P.Schnellbomber II is the only turboprop-powered design concept known to have been submitted to the Air Ministry by an firm. In addition, the fact that it would have had swept-forward wings made this design-concept all the more unique.

Professor List developed the Daimler-Benz 007 (ZTL) ducted-fan turbojet engine. It was tested in the Autumn of 1943 but work was terminated by the RLM because of its over-complexity and poor thrust. ZTL stands for "Zweikreis-Turbine Luftstrahl" or a two circuit turbojet. Weight was 2,870 lb vs. 2,090 lbs for the HeS 011. Static thrust for the DB007 was only 1,363 vs. 2,860 for the HeS 011.

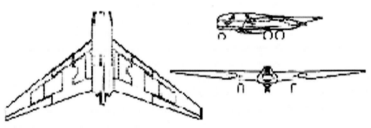

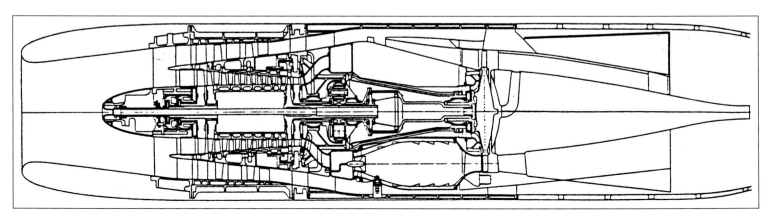

A pen and ink drawing of the Daimler-Benz DB 007 ducted-fan turbojet.

Bayerische Motoren Werke GmbH - München, Germany

A Daimler-Benz suggested Project "A" releases a proposed fast bomber.

Professor Dr. Wagner was one of the leading turbojet pioneers at Junkers Jumo who designed the Jumo 004. After leaving Jumo, Wagner joined Henschel and proceeded to design and build a large number of successful ground-to-air and air-to-air missles. With Dr Wagner, Henschel became the "high tech" leader in German remote controlled missles at war's end.

Specifications: Project Schnellbomber II

Engine	2xBMW 028 turboprops producing 4,700 s.hp each plus 4,850 lb (2,200 kg) of thrust each and turning eight-blade contra-rotating propellers
Wingspan	116.6 ft (35.5 m)
Wing Area	NA
Length	72.2 ft (22 m)
Height	21.3 ft (6.5 m)
Weight, Empty	NA
Weight, Takeoff	NA
Crew	NA
Speed, Cruise	342 mph (550 km/h) at 20,014 ft to 32,154 ft (6.1 km to 9.8 km)
Speed, Landing	NA
Speed, Top	404 mph (650 km/h) at 26,250 ft (8.8 km)
Rate of Climb	NA
Radius of Operation	2,112 mi (3,399 km)
Service Ceiling	36,090 ft (11 km)
Armament	NA
Bomb Load	2 ton
Flight Duration	5 hours

A Jumo 004B-2 turbojet engine.

BMW turbojet engine designer, Horst Schneider, shown here post-war at Kuibbyshyev, USSR. Schneider, along with other BMW engineers, were moved to the Soviet Union in 1946 and forced to continue design development on the BMW and Jumo turbojet engines.

Horst Schneider in retirement in the U.S. Covina, California, 1989.

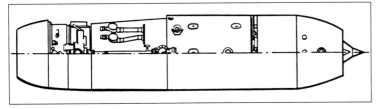

LEFT: A pen and ink drawing of a Jumo 012 turbojet engine with an anticipated static thrust of 6,000 lbs.

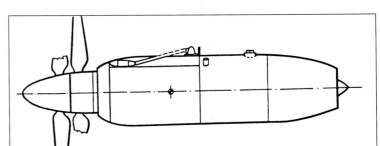

A pen and ink drawing of a Jumo 022 turboprop engine.

Bayerische Motoren Werke GmbH - München, Germany

The Jumo 022-K and known by the Soviets as the NK-12M. This turboprop powered the Russian Tu-20 "Bear" long-range bomber and was the most powerful turboprop ever produced.

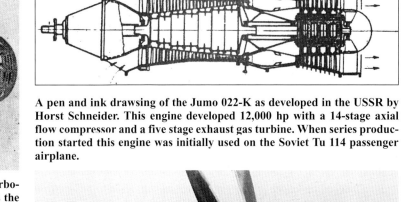

A pen and ink drawsing of the Jumo 022-K as developed in the USSR by Horst Schneider. This engine developed 12,000 hp with a 14-stage axial flow compressor and a five stage exhaust gas turbine. When series production started this engine was initially used on the Soviet Tu 114 passenger airplane.

The Jumo 022 turbojet would have had contra-rotating propellers as shown in this photograph.

A Soviet Tu 20 "Bear" long-range bomber powered by four Ju 022-K engines each producing 12,000 hp or a total of 48,000 hp.

The Russian Tu 20 "Bear."

A Russian flying test bed for the Soviet-redesigned Jumo 022-K turboprop.

5

Deutsches Forschungsinstitut für Segelflug - Darmstadt, Germany

The DFS "Kranich" a well respected DFS-designed sailplane of the 1930s.

Secret Project and Intended Purpose:
Project 346 - 2xWalter HWK 509 bi-fuel liquid rocket-drive research aircraft to test flight control problems at and above the speed of sound.

History:
The Deutsches Forschungsinstitut für Segelflug (DFS), or German Research Institute for Gliding Flight, came to be somewhat of a misnomer in 1940 Germany. Initially the work of DFS centered on sailpane research, construction techniques, and flight-testing of sailplanes. But from an association between the Rhön-Gesellschaft establishment at the town of Wasserkuppe and a group of sailplane enthusiasts begun in 1925, the Institute grew to a highly organized team of aeronautical scientists with

Professor Walter Georgii, Director of DFS. Under his leadership DFS grew from a loose-knit organization involved initially with sailplanes to an aeronautical research institute with an international reputation.

A British officer inspects the wing spar of a DFS 230 A-1 glider downed in North Africa.

Deutsches Forschungsinstitut für Segelflug - Darmstadt, Germany

DFS engineers and designers converted this Fieseler Fi 103 "Buzz Bomb" into a piloted version to make it more accurate-effective. It was flight tested by Hanna Reitsch of DFS.

Perhaps the world's most famous female test pilot ever, Hanna Reitsch was associated with DFS.

an ambitious research program in supersonic flight. In 1933, under the leadership of Professor Dr. Walter Georgii, the organization moved its research group to Darmstadt-Griesheim, and activities expanded even more rapidly.

In the beginning DFS devoted attention exclusively to sport glider and sailplane research, its design group evolving a variety of sailplanes such as the well-known Kranich and Seeadler and also some novel freight-carrying gliders. The latter would later be used for meteorological research and provide inspiration for military assault gliders such as the troop-carrying DFS 230 A-1. As the development of the art of sailplaning advanced within DFS, researchers started venturing beyond the day-to-day needs of the military. Supported by the RLM, DFS quickly changed its purpose from one of developing aircraft for sport and the Luftwaffe, to developing special planes for testing aeronautical innovations and concepts for high-speed flight.

By 1941 DFS had grown into a number of specialty research/design groups, among them bureaus focusing on aerodynamics, special projects, construction statistics, and aircraft testing. Famed female test pilot Hanna Reitsch was employed as a full-time test pilot by DFS's test flight bureau. The construction bureau had a large, efficient, and productive workshop, although the mass production of DFS aircraft was subcontracted to such firms as a Gothaer Waggonfabrik AG and Siebel Flugzeugwerke KG. Only experimental or prototype aircraft were constructed in DFS's central workshops at Ainring.

The Messerschmitt's Me 163 (Komet) by Dr Alexander Lippisch with its single HWK 509 bi-fuel liquid rocket-drive owed its existence to work originally undertaken at DFS. The practicality of coupling aircraft "piggy back style" was first demonstrated by DFS's Dr.Ing. Felix Kracht. DFS was also responsible for evolving a piloted version of the Fieseler Fi 103 "flying bomb." The Me P.328 long-distance combat aircraft powered by twin Argus pulsejets was first flown at DFS by Eric Klöckner, although DFS designers questioned the placement of the pulsejets on the Me P.328 in terms of efficiency and safety. As flying speeds increased, DFS was requested by the RLM in the early 1940s to investigate the mechanical behavior of airplanes flying beyond the speed of sound. The research vehicle they proposed was their DFS P.346, a twin HWK 509 bi-fuel liquid rocket-driven aircraft which they expected to reach Mach 2.6, or 1,615 mph (2,599 km/h) at an altitude of 114,835 ft (35 km). All plans had been worked out by DFS's Kracht, and the aircraft's construction had been contracted out to Siebel Flugzeugbauwerke, at Halle. At the time of Germany's surrender in early May 1945, the P.346 was about 30% flight ready. It was completed un-

Ernst Udet and Hanna Reitsch speaking. Reitsch is seated in a Focke, Achgelis-Delmenhorst Fa 224, shown inside the Deutschlandhalle, Berlin.

DFS test pilot Hanna Reitsch hovering inside the Deutschlandhalle, Berlin, in a Focke, Achgelis-Delmenhorst Fa 224.

A DFS "Mistel" being ferried to its intended target by a Messerschmitt Bf 109. Upon reaching the vicinity the Bf 109 would release the bomb-laden aircraft and allowing it to proceed alone to a selected target.

DFS senior staff greets a visitor by the name of Ernst Udet. Left to right: Meyer, Georgii (white hair), Udet, Jakobs, (the tall one), and Kracht with hat. Darmstadt 1940.

The wooden Messerschmitt Me 328, one of two built by DFS-Ainring for aerodynamic flight testing. The twin Argus pulse jets attached to the rear of the fuselage were not operational at this time.

The DFS designed and built DFS 228 HWK bi-fuel liquid rocket-powered, high altitude research aircraft.

Deutsches Forschungsinstitut für Segelflug - Darmstadt, Germany

The DFS 228. Numerous gliding tests were conducted as well as several rocket-powered flights.

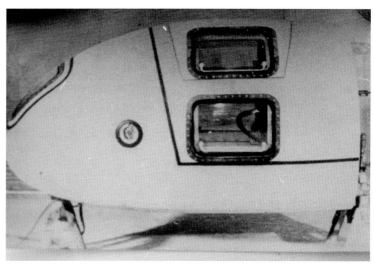

The cabin of the DFS 228. It was modified slightly for use on the DFS 346. The DFS 346's cabin's nose was in the form of a cone built out of plexiglass.

A DFS 228 making a landing approach with its fuselage-mounted landing skid extended.

The pilot cabin of the DFS 228. Same arrangement was to be used for the DFS 346. Four explosive bolts can be seen as well as several quick disconnects for electrical and flight controls.

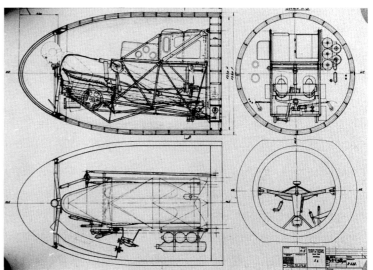

A three view drawing of the cockpit arrangement and layout used in piloting the DFS 228 and 346. This cockpit was designed by DFS's Dr.Ing. Felix Kracht.

Secret Aircraft Designs of the Third Reich

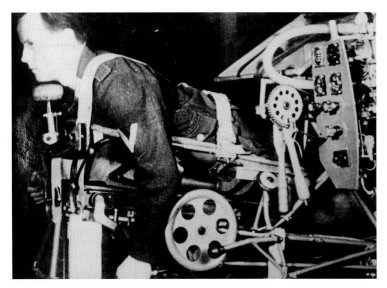

DFS test pilot Rolf Mödel trying out the prone piloting position in a lab at DFS Ainring.

An illustration of the DFS 346 by Hugh W. Cowin.

The DFS 346-P. The 346-P was a sailplane built out of wood. It was without a sealed cabin, liquid rocket motor (LRM), and fuel tanks. Ballast was used to adjust the weigh of the machine. Aerodynamic fences were incorporated into the wings. Scale model by Dan Johnson.

Rear view of the DFS 346-P. Scale model by Dan Johnson.

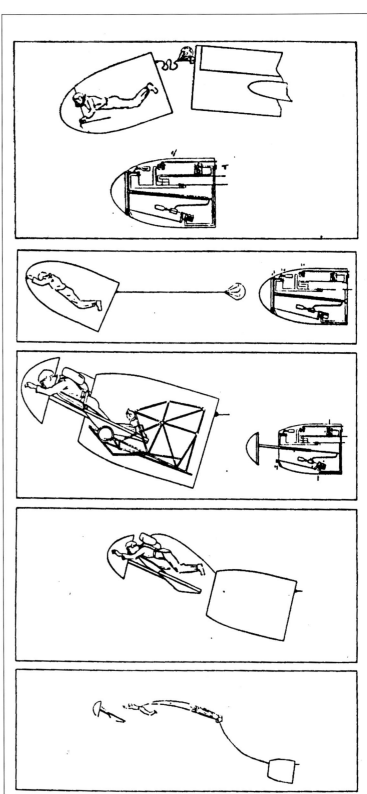

In an emergency the pilot of the DFS 228 or the 346 would fire the explosive bolts separating the cockpit from the fuselage. The separated cockpit would then release a parachute and at a lower altitude the pilot and his prone apparatus would eject from the cockpit cabin and descend further with the aid its own parachute. Finally the pilot would separate himself from the prone mechanism and descend to the ground with the aid of his own parachute.

Deutsches Forschungsinstitut für Segelflug - Darmstadt, Germany

Front view of the DFS 346-P. Scale model by Dan Johnson.

The fuselage outline of the DFS 346 hanging from the starboard wing of an interned USAAF B-29. *Courtesy of Dmitri Sobolev.*

Side view of the DFS 346-P. Scale model by Dan Johnson.

Photograph of a head-on view of the DFS 346 mounted under the starboard wing of an interned USAAF B-29. *Courtesy of Dmitri Sobolev.*

The DFS 346-P sailplane as completed in the USSR.

In this photograph, the DFS 346 can be seen attached to the starboard wing of an interned USAAF B-29 bomber. *Courtesy of Dmitri Sobolev.*

A very poor quality photograph of the DS 346 gliding back to earth after a non-powered flight.

Dr.Ing. Felix Kracht, Chief Designer at DFS.

Several former Junkers Flugzeugbau-Dessau design members shown in the USSR: Left to right: Backhaus, Lehmann, Wocke, Bock of DFS, Baade, and Kunzel at Sawjolovo, USSR, about November 1953.

Wolfgang Zeise on far left. Oranienburg, about 1944/45.

Wolfgang Zeise's Arado Ar 234 B-2 at Oranienburg, 1944/45.

Three aluminum DFS 346s were built in Halle in 1945-46 by the Soviets. One was used for static stress testing and discarded, the remaining two aircraft were taken to Moscow where one was placed in the great wind tunnel at ZAGI. The third went to the air-strip where German designers were working: a small rural community known as Podberesje.

Wolfgang Zeise's casket. He died on 28 August 1953. On the left unknown man, Siegfried Günter, unknown, and fellow test pilot, Barnard Trauter.

Wolfgang Zeise observing a Soviet aircraft demo overhead, early 1950s.

der Soviet occupation and later taken to the USSR in mid 1946 along with several hundred German aircraft designers and their families. The DFS P.346 was divided into three sections: The nose section was a fully pressurized cabin and which could be detached from the remaining fuselage through explosive bolts. The pilot then left the cabin after a suitable altitude had been reached for safety parachuting to ground. The P.346's mid section contained the bi-fuel tanks while the tail section housed the one of a kind, twin mixing chamber HWK 509 bi-fuel liquid rocket engine. An all-metal Multhopp T-tail completed the P.346 with its elevator mounted on top of fin and rudder. Overall, the P.346 was an all metal monocoque design of circular cross-section. Wings were all metal with a symmetrical cross-section and a 45 degree sweep-back at quarter chord. Inboard and outboard ailerons were used with the outboard aileron used additionally for trimming during supersonic flight. The P.346 was carried aloft under the starboard wing of an interned USAAF B-29. At an altitude of 30,000 ft (10 km) the P.346 was released at a forward speed of (520 km/h) after starting the HWK 509 bi-fuel liquid rocket drive. A metal landing skid with outriggers on the underside of each wing served to stabilized the aircraft during landing. Both the landing skid and the outriggers were fully retractable.

The Soviet's are rumored to have built their own P.346's and flight tested them extensively in 1947 without the knowledge of their German prisoners. The Soviets claimed to have broken the speed of sound several months prior to the United States' successful effort with Chuck Yaeger and Bell Aircraft's X-1. Whether or not this is true is not the issue. The issue is that the Soviets reached and surpassed the speed of

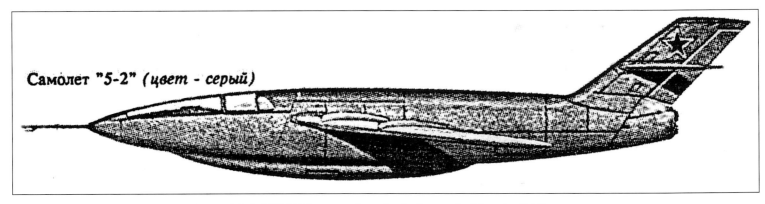

Artist's impression of the Soviet, highly modified, DFS 346, known only to the Soviets as the "Samolot 5-2."

sound with a modified DFS 346. The original DFS P.346 was flown numerous times as an unpowered glider by former Ar 234B-2 reconnaissance pilot Wolfgang Ziese. The Soviet's would not allow a powered flight of the DFS P.346 until September 1950. Soviet aviation experts forbade Ziese to push their P.346 close to Mach 1.0, explaining that the rocket-ship would become uncontrollable. Ziese discounted the advice given by his Soviet masters telling his German colleagues that since this may be the only time he'd be allowed ever again to pilot the P.346 under power that he intended to see just how fast the aircraft would fly. Zeise reached 0.95 Mach but had to leave the aircraft when the P.346 went into uncontrollable flutter and could not be brought out. The Siebel-built P.346 crashed and was totally destroyed. Wolfgang Ziese landed by parachute and with only minor injuries. An investigation determined that at near Mach speed, the air flow over the controlling surfaces would separate leaving these surfaces ineffective. Ziese was blamed for the loss of the P.346. He died mysteriously in the hospital several days after his powered flight in the 346.

Specifications: Project 346

Engine	2x Walter HWK 509 custom-designed bi-fuel liquid rocket drive with an estimated 4,000 kg thrust
Wingspan	(9.0 m)
Wing Area	(19.9 sq m)
Length	(16.6 m)
Height	NA
Weight, Empty	NA
Weight, Takeoff	5,230 kg
Crew	1
Speed, Cruise	NA
Speed, Landing	220km/h
Speed, Design Top	(1,700-2,000 km/h - top recorded speed: 590 mph (950 km/h)
Rate of Climb	100 m/sec
Radius of Operation	NA
Service Ceiling	39,000-43,000 ft (12,000-13,000 m)
Armament	none
Bomb Load	none
Flight Duration	NA, however HWK rocket drive provided thrust for approximately 2.8 min

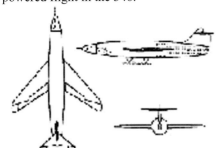

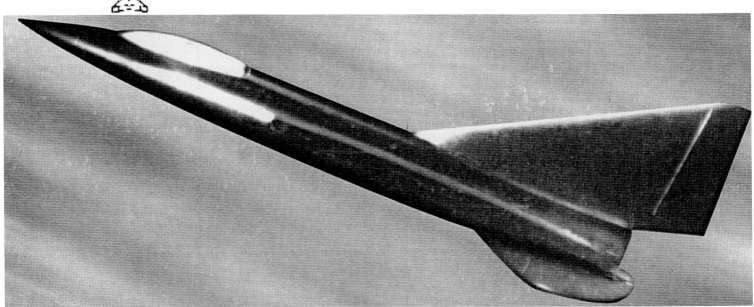

Allegedly this aircraft was known as the DFS 486, with a delta wing planform, which the Soviets claimed had been designed by Siegfried Günter 1946-47. The Soviets also claimed to have broken the speed of sound in 1947 with an aircraft of this type, some five months prior to Chuck Yaeger's successful attempt to break the speed of sound on 14 October 1947.

Shown at Edwards Air Force Base, California in late 1947, are two fresh Bell rocket-powered aircraft which are hoped to break the speed of sound after air launched from the belly of a B-29 bomber.

One of the Bell X-1s with an unknown pilot, late 1947.

Muroc Air Force Base, California, January 1948, Captain Charles K. Yeager, standing beside his Bell X-1 following powered take-off of the supersonic research airplane.

6

Dornier Werke GmbH - Friedrichshaften, Germany

Secret Projects and Intended Purpose:
Project 254 - A variant of the powerful front and rear-mounted twin pistoned engined Do 335 with its rear piston engine being replaced by a HeS 011 turbojet.
Project 256 - A conservatively designed straight-wing night fighter powered by twin HeS 011 turbojets. Each of the HeS 011 units were placed in individual nacelles widely spaced and hung under each wing. It had an unusually large rudder.

History:
Perhaps the longest continually operating aircraft manufacturing company in the world today, the Dornier Werke dates back to the summer of

The Dornier Do 17 known as the 'flying pencil' one of the smoothest designs to come out of pre-war Germany.

Perhaps the fastest piston powered fighter ever built, the Dornier Do 335 "Pfeil" or Arrow, joined the Luftwaffe just days before the end of the war. The Dornier Do 335 was powered by two Daimler-Benz DB 603 engines, one in the nose and the second buried in the rear fuselage that drove a propeller behind the tail via an extension shaft.

Although he seldom piloted an aircraft himself, Claudius Dornier frequently flew as a co-pilot on his new models. He is shown here checking out a new Do 17 with Dornier Werke test pilot Egan Fatli.

Famed North Pole explorer, Roald Amundsen, relied on Dornier seaplanes on his many expeditions.

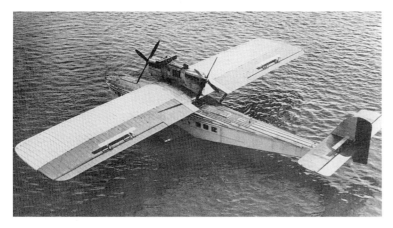

Immediately after the Armistice on 11 November 1918, Claudius Dornier turned his attention to civil aviation, especially seaplanes. The Dornier Do 15 (Wal) first flew in November 1922.

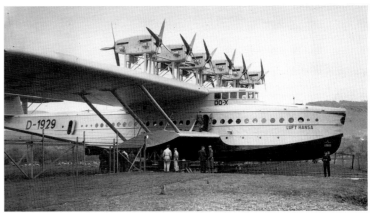

The twelve-engine Dornier X seaplane appeared in 1929 and a year later made a 28,000 mile round trip journey from Berlin to Rio de Janeiro, the West Indies, New York City, and Spain.

Dornier "Superwal," pictured in 1930, had four 600 h.p. Hisspano-Suiza engines.

General Walther Wever.

The Dornier Do 17 and a close up of this aircraft's famed "beetle-eye nose."

The Dornier Do 26, which flew for the first time in May 1938, was used as general reconnaissance aircraft and for flights in and out of the fjords in Norwegian waters during the war.

The Dornier Do 24K three-engine seaplane, designed in 1935, was converted to general reconnaissance duties during the war.

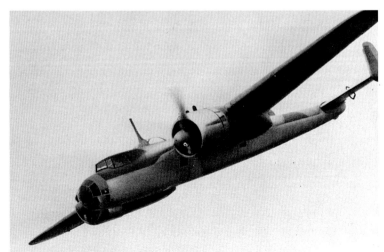

An early Dornier Do 17 showing its characteristic split-pattern paint job: green camouflage upper surface and sky blue lower surface.

Dornier Werke GmbH - Friedrichshaften, Germany

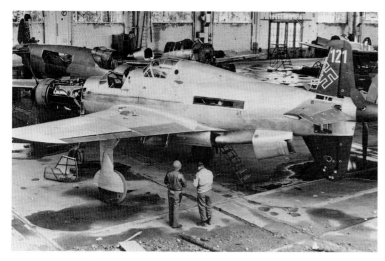

A number of unfinished Dornier Do 335s at war's end. Note that manufacturing still continued in spite of the fact that the aircraft assembly hangar had only a partial roof and no window glass at this time of the war.

Several unfinished Dornier Do 335s at their assembly facility at Oberpfaffenhofen about 15 miles southwest of Munich.

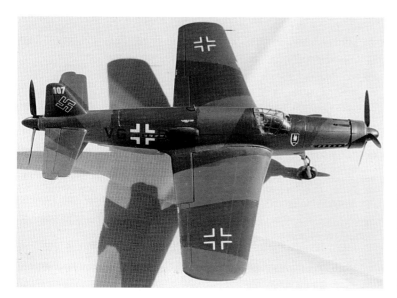

The Dornier Do 335 standard day-time fighter interceptor. Scale model by Reinhard Roeser.

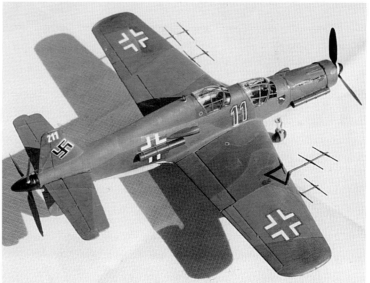

A Dornier Do 335 radar equipped night fighter. Scale model by Reinhard Roeser.

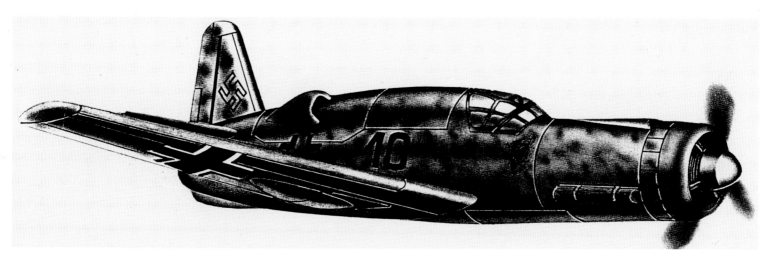

A Dornier Do P.254 mixed power plants: DB-603LA in nose and HeS 011 in tail. *Courtesy Luftwaffe Secret Projects: Fighters 1939-1945.*

105

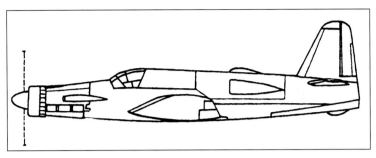

Line drawing of the Dornier Do P.254 with one DB-603LA of 2,300 hp and one HeS 011 rated at 2,865 lbs static thrust. *Courtesy of Luftwaffe Secret Projects: Fighters 1939-1945.*

1914. Known widely for its commercial aircraft and the development of superb flying boats, Dornier also produced (unintentionally) what was to become one of the most famous (and certainly one of the most attractive) bombers of World War II - the twin-engine Dornier Do 17. But Dornier had no interest in the turbine engine and, like Kurt Tank of Focke-Wulf, believed that super-sonic speed could be achieved with piston power plants. His unique front and rear-engined Dornier Do 335 fighter-bomber could have been powered by turbines when it appeared in 1945 but Dornier would not allow any such decisions until Spring 1945 and then it didn't matter anymore.

Claudius Dornier was one of the pioneers of metal aircraft, and, along with Hugo Junkers, he dominated German aircraft designs in the 1920s and 1930s. Overall, the Dornier Werke produced nearly 150 different types of aircraft between 1914 and 1945 - but it was the Dornier flying boats that gave the company its earliest fame by setting standards for others to match. Dornier flying boats were pioneers, crossing the Atlantic and other great stretches of water long before aircraft from other countries. In 1912 Dornier wrote an article (the first of many) on air screws (propellers) and how to calculate the proper pitch depending on the aircraft's intended use. Despite his highly active business life, he found time to be a mentor to generations of aircraft designers. Richard Vogt, Alexander Lippisch, Adolf Rohrbach, Hans Klemm, and numerous others got their start in aviation by working with Dornier.

Claudius Dornier was born on 14 May 1884, in Kempten, in the German Alps. His father was French and his mother was German. In 1910, three years after Dornier's graduation from München's Technical High School, Germany's famed airship designer Count Ferdinand von Zeppelin hired him as a statistical calculator for his airship works at Friedrichshafen by Lake Constance in mid, southern Germany. In 1911 he designed the world's first all-metal plane, and by 1914 he had sufficiently impressed the Zeppelin people with his abilities that Count Zeppelin let him found a separate airplane division, the Dornier Metallbauten GmbH. It was here that construction and stressed metal skin proved to be a success. Throughout World War I, the Dornier factory produced both wooden and metal-skinned fighter aircraft and very large seaplanes. Dornier's first seaplane, known as the RS-1, had a wingspan of 143 ft (43.5 m), making it the largest aircraft in the world. Because existing knowledge and previous experience were limited, Dornier spent a considerable amount of money carrying out scientific research in the area of stress, aerodynamics, hydrodynamics, and metallurgy. There were problems in every one of these areas that had to be resolved before many of Dornier's ideas could be implemented. In the process of resolving the problems, Dornier sought out the best young talent he could find.

With Dornier's ideas and the steady stream of young talent, Dornier Metallbauten turned out to be a sort of postgraduate training ground where everyone benefitted. Owing to the outstanding performance and reliability of Dornier aircraft, in 1925 alone twenty world records were set by the seaplane Dornier Wal (whale). At other times throughout the 1920s, aviation pioneers and explorers took advantage of Dornier aircraft to help them achieve their goals. Walter Mittleholzer used a Dornier Merkur (Mercury) equipped with floats on his famous research flight form Zürich over the Mediterranean and all over Africa down to Cape Town. Spanish Major Francisco Franco for the first time crossed the South Atlantic from east to west in a Dornier Wal, and Wolfgang von Gronau flew the North Atlantic route in a Wal. Roald Amundsen's North Pole expeditions relied extensively on Dornier aircraft.

As plans were being drawn up in the early 1930s for expansion of the German aircraft industry, the Dornier company was seen as part of the nucleus around which the growth was to take place. Dornier cooperated fully, allowing the Reich to finance associate companies under his management and leadership. The Luftwaffe, through its leader at the time, Generalleutnant Walther Wever, looked to Dornier to produce Langstrecken-Grossbombers (long-range large bombers) that could carry a large bomb load to England or could bomb Russian industry over the Ural mountains from German bases. However, the untimely death of General Wever in 1936 in his own crashed airplane and his replacement by Generalleutnant Albert Kesselring resulted in a reappraisal of the "Ural Bomber" program. By 1937 a decision had been made to terminate the "Ural Bomber." That decision has been the subject of controversy ever since. It was Kesselring who believed that small, twin engine bombers would suffice for any wars such as was envisaged in Western Europe. The primary role of the Luftwaffe, Kesselring believed, must remain tactical rather than strategic. A key decision they later would come to regret.

With the cancellation of the "Ural Bomber" program, Dornier's star dimmed considerably. Instead of being asked to produce new types

The proposed Gotha P.60C fighter. This is the radar version with the bulbous nose. Scale model by Reinhard Roeser.

The proposed Blohm & Voss Bv P.215 fighter. Scale model by Reinhard Roeser.

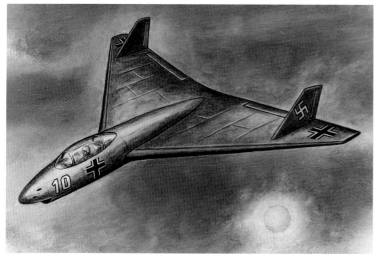

The proposed Arado Ar NJ.I (night hunter).

The proposed Focke-Wulf Fw P.2 fighter. Scale model by Steve Malikoff.

of aircraft, the firm was ordered to produce aircraft designed by Messerschmitt, Junkers, Heinkel, and Focke-Wulf. This included Me 410s, He 111s, Ju 88s, and Fw 190s. One of the few Dornier designs to be ordered into production during the war (in 1945) was the twin piston powered Do 335 night fighter. It may have been the fastest piston-powered fighter of World War II, but its manufacture in 1945 meant it was already too late to make a difference in the war effort. An unusual looking aircraft, it had one engine in the front and another in the tail. Twenty production models were ordered, and they performed better than Dornier had predicted, achieving a maximum speed of 474 mph (763 km/h).

The Dornier Werke produced no turbine-powered aircraft or even prototypes during the 1940s, and only three turbine design projects are known to have existed: the P.232, the P.256, and the P.257. The P.232 was a projected variation of the Do 335 that called for the rear piston engine to be replaced by a HeS 011 turbojet. The P.256, proposed night fighter, was characterized by two HeS 011 turbines placed in nacelles under the wings. Details of the P.257 are very sketchy. The P.257 is thought to have been a fighter with a canard-wing layout and powered by three HeS 011 turbojet engines.

After the war the Dornier Werke became dormant for a time. All its factories located in what was to become East Germany were completely lost, and the plants in the Western Zone had been destroyed by Allied bombing attacks. The company had the opportunity to liquidate, but Dornier, out of personal pride, instead struggled to keep his factories open. His long-range goal was to restore them once the Allied ban on military aircraft manufacturing was lifted. In the meantime Dornier designed aircraft in Spain for a number of Spanish companies and used the profits to start rebuilding his German factories for the temporary production of textile machinery and other non-aviation equipment. The manufacturing of textile machines offered a good opportunity because most of the German textile industry had been concentrated in eastern Germany. Now cut off by the Russian occupation, new production facilities had to be created in the West. Despite Dornier's contribution to the German war effort, he was "exonerated" by an Allied denazification court after the war. He maintained that he had joined the Nazi Party in 1940 only "under pressure" and that his intention had been to construct civilian, not military, aircraft.

When the Allied-imposed ban on military aircraft manufacturing was lifted in 1955, Dornier opened a new factory hear München and started licensed production of Lockheed F-104 Starfighters for the new Bundes Luftwaffe. Dornier's headquarters in Friedrichshafen-Manzell had been taken from him after the war. In 1959 Dornier acquired a large tract of land near Immenstaad on Lake Constance and constructed new buildings to house his firm's administration offices and engineering departments. In 1966 Dornier developed the Skyservant, a short take-off-and-landing executive aircraft that became a sales success throughout the world. The following year, the Dornier Werke introduced the world's first turbine-powered, vertical takeoff-and-landing transport, the Do 31, which attracted considerable attention at international air shows.

Dornier died on 6 December 1969, at his home in Zug, Switzerland. He was 85 years old. Throughout his aircraft design years, Dornier sought to impress on his fellow workers his idea of the proper role of the airplane and its designer:

"During the World War and the years following it, the science of aviation made notable progress. The modern airplane has become faster, safer, and cheaper. Great as all this progress has been, the main problem of aviation has not yet been satisfactorily solved. New speed records can certainly be attained if a specially powerful engine is used, but the

Claudius Dornier prior to his death.

Dornier Werke's modern facilities near Immenstaad on Lake Constance.

sensational speeds of which we read occasionally in the American Press make little impression on me or the European traveler, who is more concerned with reaching his destination with the same certainty when he travels by airplane as when he travels by rail or sea."

PROJECT 254

The Do P.254 was to be a further development of the twin nose and tail mounted DB 603A piston-powered Do 335 "Pfeil" or arrow. The major change being the substitution of the rear-mounted DB 603A 2,300 hp piston engine for an HeS 011 turbojet engine of 2,86 lb thrust. What this change of engines would have accomplished is not known. It is known that the high-speed bomber (its intended purpose changed frequently) based on prototypes which first flew on 26 October 1943, produced only unsatisfactory results. Potentially the Pfeil would be the fastest piston-engined bomber built by anyone during the war - one engine was conventionally mounted in the nose while the second was placed in the rear fuselage, driving a pusher type of propeller allowing the Do 335 to have a maximum speed of 474 mph (763 km/h). However, its unsatisfactory results appeared to be several. First it could not reach its design altitude goal of 25,000 ft (7.8 km). Although it was much more powerful than any Allied fighter aircraft experience in night-fighting combat didn't greatly show this advantage. Finally, the Do 335's fuel consumption was much too high, production and repair costs much too expensive. In addition, the Air Ministry's experience with the Focke-Wulf Ta 152 showed that the 152 was more efficient in terms of combat operations, fuel, construction, and maintenance costs. With the substitution of an HeS 011 turbojet, Dornier officials hoped to prevent an out-right cancellation of their 335. None of the Do 335s were modified into a P.254, in fact, the 335 program was canceled near war's end in favor of the Ta 152.

Specifications: Project 254

Engine	1xDamiler Benz DB 603A piston engine mounted in the nose and 1xHeinkel-Hirth HeS 011 turbojet with 2,866 lb (1,300 kg) thrust mounted in the tail
Wingspan	NA
Wing Area	NA
Length	NA
Height	NA
Weight, Empty	NA
Weight, Takeoff	NA
Crew	NA
Speed, Cruise	NA
Speed, Landing	NA
Speed, Top	NA
Rate of Climb	NA
Radius of Operation	NA
Service Ceiling	NA
Armament	NA
Bomb Load	NA
Flight Duration	NA

Dornier built the world's first turbojet-powered vertical takeoff and landing transport plane, the Do 31, shown here flying overhead.

In the 1960s, Dornier manufactured the Lockheed F-104 fighter under license for the Luftwaffe. Shown here is the F-104G. Its razor-blade thin wings extend only 1.3 meters (7.5 ft) from the fuselage and are so sharp that felt covers are placed over them when the aircraft is on the ground to protect the ground crew from cuts.

Dornier Werke GmbH - Friedrichshafen, Germany

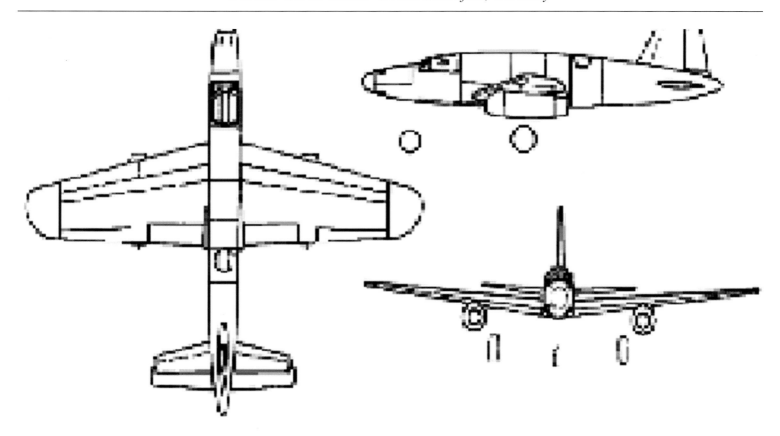

PROJECT 256

In February 1945 the Chef der Technischen Luftrüstung (CTL), or Chief for Technical Air Equipment, announced a design competition for a twin-turbojet-powered night and bad-weather fighter. Although Dornier had been restricted from building fighters and interceptors by the Air Ministry (they had demanded that Dornier concentrate entirely on bombers), but Dornier submitted a design based in large part on thier unique twin-engine Do 335 bomber design. Other designs entered in the competition were Arado's P.NJ-1 and P.NJ-2, Blohm and Voss' P.215, Focke-Wulf's P.2 and P.3, and Gothaer's P.60C.

Dornier's P.256 design was characterized by placement of twin HeS 011 turbojets in nacelles under the wings. It also featured an unusually large tail rudder, which Dornier designers believed was necessary because of the wide separation of its turbines. They felt that if one turbine had to be shut down, a large amount of rudder surface would be needed to counteract the other turbine's thrust and turning the aircraft either to starboard or port. In addition, the P.256 was to have a short, stubby fuselage, and this too, Dornier designers believed, would require a tall vertical stabilizer.

Aside from the twin HeS 011s mounted under its wings, the P.256 was to be a typical low-wing monoplane similar in many ways to the Do 335. However its wings had a slight backsweep, but for all practical purposes they were unswept. Although the P.256 did borrow a number of design concepts from the Do 335 and would have used some of the Do 335's hardware (for example, its complete tricycle undercarriage), it still can be considered a unique design. As an offensive night fighter, it was to be fitted with four 30 mm MK 108 cannon, each having 120 rounds of ammunition, set in a fixed position and housed in the nose. Two additional MK 108 cannon were to be mounted in the fuselage, in a position to fire upward into the belly of an Allied bomber as the P.256 passed beneath it in the night. Although the P.256 was designed specifically as a night fighter, bomb racks were built in so it could carry one or two 1,102 lb (500 kg) bombs under its wings.

The P.256's three-man crew was to sit inside a pressurized cabin. The pilot and radar operator were to sit side by side, while the navigation/radio operator was to sit behind the two and face the rear bullet-proof glass would be provided, protecting the crew from 20 mm cannon fire aimed into the cockpit. Since the night fighter's speed was projected at 516 mph (830 km/h), attack from the rear was not anticipated, and there was no bullet-proofing from rear attack.

The P.256 did not fare well in design competition. Although none of the entries met all of the requirements set forth by the Chef der Technischen Luftrüstrung, Dornier's P.256 was judged the least desirable. Officials disliked the underwing location of its twin HeS 011s; they felt the engine nacelles would lead to a "low Mach number, bad flight behavior, and high operating interference (low turbine efficiency) for the turbines." Its unswept wings were also cited as an undesirable feature. No redesign work was completed on the P.256, and no prototypes were constructed.

Specifications: Project 256

Engine	2xHeinkel-Hirth HeS 011 turbojets each having 2,866 lb (1,300 kg) of thrust
Wingspan	50.9 ft (15.5 m)
Wing Area	441 sq ft (41 sq m)
Length	45.3 ft (13.8 m)
Height	NA
Weight, Empty	21,605 lb (9,800 kg)
Weight, Takeoff	26,896 lb (12,200 kg)
Crew	3
Speed, Cruise	336 mph (540 km/h)
Speed, Landing	106 mph (170 km/h)
Speed, Top	516 mph (830 km/h)
Rate of Climb	44.6 ft/min (13.6 m/min)
Radius of Operation	874 mi (1,404 km)
Service Ceiling	41,013 ft (12.5 km)
Armament	6x30 mm MK 108 cannon
Bomb Load	2x1,102 lb (2x500 kg) bombs
Flight Duration	2-1/2 hours at 68% thrust and at 19,686 ft (6 km) altitude

7

Focke-Wulf Flugzeugbau GmbH - Bremen, Germany

Secret Projects and Intended Purpose:

Project 183 - This single seat, single turbojet fighter by Multhopp is considered to be Germany's best advanced fighter design to come out of Germany during World War II. The Soviet's thought so too, and called their shameless copy the MiG-15.

Project "Triebflügel" - A vertical take off and landing craft powered by two or three ramjet engines mounted on the tips of helicopter-like blades.

Project 283 - A Multhopp designed airframe to be powered by twin ramjet engines despite the fact that ramjet engines had never satisfactorily powered any manned aircraft in World War II.

Project 3x1000C - A delta-wing aircraft capable of 1000 km/h speed, 1000 kg bomb, 1,000 km.

Project "Kamikaze" - A self-sacrificing aircraft carrying a high-explosive bomb in its nose. This aircraft, when released from its mother-carrier would dive into military targets, exploding on impact.

Project "Kamikaze Carrier" - A very large-size transporter of approximately six "kamikaze" self-scrificing aircraft to the battle front.

Project "Five" - A three HeS 011 powered fighter-escort for the long-range He 177 bomber.

Project "Six" - An HeS 011 powered fighter-interceptor with twin tail booms.

History:

In terms of airframe design philosophy, Focke-Wulf Flugzeugbau GmbH was an enigma throughout the 1930s and 1940s. The firm's general

manager, Kurt Tank, was renowned in design circles for his long-standing conviction that safety must come before performance, whether in designing planes or flying them. Becoming Focke-Wulf's general manager in 1933 at the age of 35, Tank had a well-deserved reputation as a "damn good" engineer - but his approach was conservative. He believed that progress should never become an aim in itself, and he accepted

Left, an unknown highly decorated officer and right Kurt Tank, standing in front of a Focke-Wulf Fw 58.

Hans Multhopp with Professor Adolf Becker, physicist and fellow member of the German Academy for Aeronautical Sciences, during a meeting at Bad Ellsen in 1944.

110

Hans Multhopp's Fw Ta 183 proposed fighter of World War II which President Juan Péron wanted for his Argentine Air Force. Scale model by Günter Sengfelder.

Known the world over as the "Father of German Aeronautics," Professor Ludwig Prandtl considered Hans Multhopp the brightest student of aeronautics he'd ever taught.

design innovations only when they promised greater air safety through time-tested research, were based on solid test flight results, and had received favorable evaluations by other engineers and designers.

In contrast was Tank's design assistant, a brilliant young "whiz kid" in the field of aerodynamics named Hans Multhopp. A dedicated family man, author of numerous scientific articles (several of which are required reading even today in courses on aeronautical engineering), holder of aircraft patents, and at age 25 the youngest person ever elected to membership in the German Academy of Aeronautical Sciences, Multhopp was chief designer of the state of the art proposed sweptwing, T-tailed Ta P.183. He was of the new generation of engineers trained in modern mathematics and aerodynamics who had great faith in the results of wind tunnel research. Wind tunnel test data, this group believed, were so accurate and reliable that they could dispense with the old step-by-step evolution of design and jump immediately to fast turbojet-powered flight from the slow piston-powered world of Tank and his coworkers.

Hans Multhopp believed that standard designs like Bell Aircraft's X-1, which was first to exceed (officially) the speed of sound in 1947, was already obsolete by 1940. Had it not been for its brute power, Multhopp said, the X-1, lacking sweptback wings and tail surfaces, would have gone no where near the speed of sound.

Heinkel-Hirth HeS 011, turbojet engine. At war's end only 20 units had been produced for field testing.

Left to right: second row, Martin Winter, and Hans Multhopp, post-war England when they were working for the Royal Aeronautical Establishment, London 1945/1946.

Academicians, he thought, could theorize over the proper aerodynamic shape of future aircraft in relative comfort. Yet at the same time, in academia it was next to impossible to see ideas translated into metal held together by rivets. Therefore, Multhopp left his studies at Göttingen University for a new job in private industry, where his ideas could take shape before his eyes. Joining the Focke-Wulf organization in 1938, he was filled with great anticipation. Here, he felt, he could have a more powerful voice in forthcoming decisions about the aerodynamic shape of future high-speed fighters and bombers.

If Multhopp thought it was next to impossible for a desk-bound professor of aerodynamics to get his theories and concepts translated into metal, it was equally difficult for a young theoretician to bring about any rapid change in the way the highly successful and conservative Focke-Wulf organization run by Tank carried on its aircraft design and construction activities. Nevertheless, Multhopp never appeared to regret his decision to leave academia before finishing his Ph.D., thus becoming ineligible ever to return as a teacher. Nor did his doleful years at Focke-Wulf, where he so often was expected to think and act like everybody else, dampen his pioneering spirit, which had been welcomed and nourished by his mentor, Ludwig Prandtl. Prandtl described Multhopp as the most gifted student of aeronautics he had ever taught and kept reminding him that what could be conceived could be fashioned - that nothing was impossible.

Thinkers like Prandtl who believed that nothing is impossible, frequently succeed in accomplishing the impossible, and people who knew Prandtl saw in Multhopp a Prandtl-like approach to solving aeronautical problems. Multhopp had that coveted inner strength - and it got tested at Focke-Wulf and later throughout his life. Multhopp couldn't move his co-designers at Focke-Wulf away from a commitment to conservative, albeit time-tested, aerodynamic shapes of the past, even in view of the speed-of-sound performance soon to be made possible by the forthcoming turbojet. They believed that their old-style aircraft designs, with their unswept wings and standard tails, would be adaptable even beyond the speed of sound - and as it turned out they weren't entirely wrong. (American test pilot Chuck Yeager exceeded the speed of sound on 14 October 1947, in a rocket-powered plane of the old style, the Bell X-1, however due entirely to pure brute power). But in the early 1940s Multhopp felt his fellow workers were dead wrong. He believed they knew nothing - and to know nothing, he fumed, was to imagine everything to be all right.

Throughout Multhopp's employment at Focke-Wulf, his co-workers found the generally acknowledged genius in aerodynamics a brash, often immature, upstart and a very difficult team player. His determination to dispense with the step-by-step transition from Focke-Wulf's very fast piston-powered aircraft to the superfast turbojet-powered aircraft, they thought, reflected not a sound approach to aircraft development, but merely an enormous self-confidence in the correctness of his ideas and a possibly unjustifiable belief that everything would work out as his wind tunnel test data indicated it would.

Multhopp never viewed himself as any kind of eccentric. He appeared to be a man who enjoyed a good laugh, and he remained unresponsive to peer pressure to conform. "I've always had my own internal

Typical gesture of President Juan Péron in the 1950s.

The Argentine Pulqui Dos a virtual copy of Multhopp's Ta 183.

Seated with one of Focke-Wulf's test pilots, Kurt Tank, right, checks out an early Fw 189A-1 short-range reconnaissance aircraft in the summer of 1938.

Focke-Wulf Fw 200 "Condor" designed by Kurt Tank and serving with Lufthansa.

gyroscope," he once remarked, "and I choose the design path based on what my wind tunnel test data thought of my theories." His unshaken belief in himself and his grumbling good humor with fellow workers enabled Multhopp to survive the private sector's slow embrace of change. It was his self-confidence that also enabled him to overcome many of the internal conflicts he must have felt when his ideas for achieving speed-of-sound performance were watered down time and again in peer group review.

Eventually, Multhopp did get a chance to try his ideas. In late 1944, after Tank became convinced that the new Heinkel-Hirth HeS 011 turbojet had sufficient thrust and reliability to propel a single-engine, high-altitude fighter, Multhopp was allowed to form his own design team to pursue a fighter aircraft design according to that internal gyroscope of his. As the design drawings gradually came together, based on the theories of aerodynamics that had occupied many hours with Prandtl and the information gathered during countless hours at the wind tunnel test facilities at the Aerodynamische Versuchs-Anstalt (AVA; Aerodynamic Research Establishment) at Göttingen University, Multhopp fashioned what came to be called the Ta P.183, a turbojet-powered aircraft that design experts around the world considered the most significantly advanced fighter design to come out of Germany during World War II.

Although the design work for the P.183 was completed by the end of the war, the aircraft was never built. At war's end, Hans Multhopp was the most sought after of the German aerodynamicists. He worked in England for the RAE, then came to the United States and ended his career in relative obscurity, still frustrated by rules imposed by others.

Focke-Wulf Fw 190A-8 shown in 1944. The open space to the right of the "5" is the baggage compartment. This Fw 190A is shown carrying a light weight drop fuel tank.

Kurt Tank responded to an invitation from President Juan Perón and left Germany secretly (and illegally) in 1946 for Argentina where he built five aircraft resembling the P.183. The first two prototypes crashed killing their pilots and the remaining three eventually were grounded due

Professor August von Parseval, builder of airships and teacher of Kurt Tank. Tank believed that further work on airships was a waste of time. "No one ever saw birds or insects flying around like gas balloons," he said.

The airship LZ-127 "Graf Zeppelin" first flight 18 September 1928, dismantled Spring 1940 at Frankfurt.

to their vicious low-speed stalling characteristics on account of the changes Tank had made in Multhopp's original wing profile. For without Multhopp, Tank was lost because being "damn good" engineer and pilot was nowhere good enough in the emerging era of swept back wings and supersonic flight.

Kurt Tank

Among the many distinguished aircraft designers who flourished in Germany between 1930 and 1945, Kurt Tank was not a household name. Only a few people outside the aircraft industry knew who he was, for his creations did not bear his name until late in the war. Yet the initiated knew him as the brains of the Focke-Wulf Flugzeugbau GmbH. Sport enthusiasts knew him as the creator of the Focke-Wulf Fw 44 Stieglitz (Goldfinch) and the Fw 56 Stösser (Hawk), both produced in the thousands for aerobatic flying. The Fw 56 gained a unique place in aviation history because it gave Ernst Udet the idea for the dive bombing Stuka as well as the Curtiss F11C-2 Hawk 11. Commercial pilots knew the Fw 200 Condor, the plane that flew throughout the world in 1938 and 1939 setting new speed and distance records; and airmen knew Tank as the designer of the Fw 190, one of the fastest fighters produced during World War II. The Fw 190 was the only completely successful new aircraft

Airplane builder Adolf Rohrbach in the early 1930s.

introduced on any scale by the Luftwaffe after the start of the war. Despite Tank's design talents, Focke-Wulf was a late-comer to turbojet-powered aircraft. Only a few Focke-Wulf turbojet-powered designs were completed as prototypes - but one design did win worldwide attention. The American and British considered Focke-Wulf's P.183 fighter to be the best fighter design to come out of Germany. The Russians apparently thought so, too. They found a complete set of plans when they sacked the Air Ministry offices in Berlin, modified and adopted them - and called the product the MiG-15.

Dr. Kurt Waldemar Tank was, until May 1945, in charge of all technical matters at Focke-Wulf Flugzeugbau. The company's main depart-

A stern-faced young Kurt Tank (in flight suit third from right) stands with his boss, Dr.Ing. Adolf Rohrbach equally stern-faced (fourth from right), some time in the mid-1930s at Rohrbach's flight facilities near Berlin.

The Rohrbach flying boat "Romar."

The Rohrbach "Roland" ten-passenger airliner.

Popular aerobatic pilots Paul Bäumer and Theo Rasche at Hamburg. Date unknown.

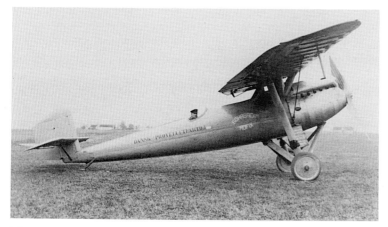

The Kurt Tank designed Rohrbach "Rofix" prototype aerobatic plane in which Paul Bäumer lost his life.

ments were at Bremen, but subsidiary factories were dispersed all over Germany. Like his contempories such as Ernst Heinkel, Willy Messerschmitt, Alexander Lippisch, and Walter Blume, Tank was actively involved in the design, production, and sales of commercial, sport, and military planes. Yet Tank was different. He was the only German designer who personally test-piloted every one of his own creations before anyone else flew it. "The ordinary type of test pilot is far too poorly equipped from a technical point of view and does not speak the same language as the designer," he believed. "I regard it as fundamental that an aircraft designer should be able to fly his own machine." Few designers did, however.

Kurt Tank was born at Bromberg-Schwedenhole on 24 February 1898. He left high school at 17 to volunteer for the Army Infantry and experienced his first combat on the Eastern Front. After the Armistice in 1918 Tank gained admission to the Technical College in Berlin. He had long been fascinated by engineering, especially by the rules governing the behavior of bodies in liquids and gases and the motion of a solid body in a direction contrary to that of the medium. This was an infant

Professor Heinrich Karl Focke in the early 1920s.

Georg Wulf prior to his death in 1927

Hans Multhopp at war's end May 1945.

Wind tunnel model of Multhopp's Ta 183 fighter.

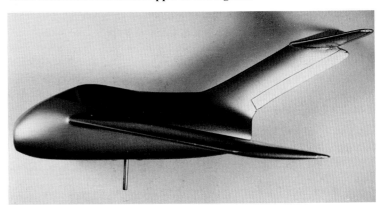
Wind tunnel model of Multhopp's Ta 183 fighter.

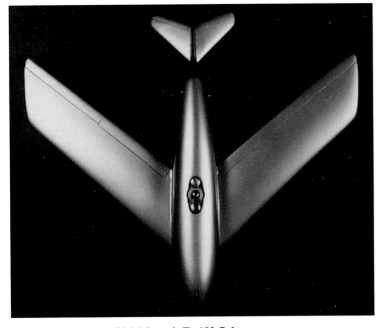
Wind tunnel model of Multhopp's Ta 183 fighter.

Multhopp's proposed Ta 183 fighter. Scale model by Günter Sangfelder.

Hans Multhopp showing a scale model of his proposed Ta 183 World War II fighter to fellow engineers at Martin Aircraft Company April 1952. With Multhopp are George Trimble, Jr., Manager of design development, and William Bergen, Martin Aircraft Vice president and chief engineer.

A dramatic view of the Multhopp T-tail on a Martin XB-51 bomber.

science in the early 1920s - even the word "aerodynamics" was only then being introduced into college textbooks. Tank was also attracted to the new field of electrical engineering. Electricity was already being put to various uses - lighting, telephone, tramways, and railroads. Believing that statics and dynamics would be useful in the developing electric industry, he decided to major in electrical engineering. He had no interest in aircraft, flying, or the budding field of aeronautics - but that would change when he entered the Technical College.

One of Tank's lecturers in physics at the Technical College was August von Parseval, who, earlier in his career during the period 1900-1919, had been Count Ferdinand von Zeppelin's rival in building airships. Because Germany had used airships for bombing London and other British cities during World War I, the Treaty of Versailles ending World War I banned their construction in Germany until after 1926. Von Parseval spoke frequently about the good ole days of the airship, i.e. a steerable lighter-than-air craft, sometimes called a dirigible from the French word meaning "to steer." Tank himself never had gone anywhere near a steerable lighter-than-air craft, but he had some strong feelings

The Martin XB-51, the first Martin aircraft with a T-tail.

A C-141 military transport with a Multhopp T-tail.

about them. Aside from regarding them as nothing more than "flying crematoriums" (initially they used highly flammable hydrogen gas to achieve lift), Tank thought it nonsense to believe in a flying craft that had so much wind resistance. "You want a surface which helps flight," Tank argued with von Parseval, "not one which puts the brake on." Tank frequently taunted von Parseval by reminding him that he didn't see birds or insects flying around like gas balloons. "We shall fly better if we follow nature and keep our eyes on how birds do their flying."

To prove his point, Tank, along with three friends from the Technical College, founded the Akaflieg, or Berlin Academicals Flying Group. Inspired by Otto Lilienthal's pioneering work in gliding and the example of Ferdinand Schulze's "broomstick glider" from East Prussia, Tank felt he could produce a more efficient aircraft. He and his friends built a workshop on the roof of the Technical College and began sketching their ideas and building scale models. About the same time one of Tank's professors began a course on the mechanics of flight. Tank enrolled in the course, and afterward, in 1923, he and his friends built their first manned aircraft - a glider. It didn't perform as well as they had hoped and was partially damaged during its maiden flight when it came in for a hard landing. It was never rebuilt due to lack of funds and a waning interest in aviation on the part of Tank's three friends. Tank had made the maiden flight, and the others did not care to fly after witnessing Tank's close encounter. Alone with the damaged glider, Tank was forced to put aside temporarily his thoughts of building a more stable, more efficient flying machine.

But he did not have to wait long to get back into aircraft construction. Although Tank, the practical engineer, did not believe in fate on any significant scale, an incident that seemed guided by the hand of fate occurred shortly after his final examinations at the Technical College. Something that would influence him and his career throughout his life. Upon boarding a train from Berlin to Potsdam, Tank met his mathematics teacher, Professor Moritz Weber. Sitting together, they pondered the uncertain future for recent graduates of the Technical College in depression-ridden 1920's Germany. Within a few days Weber was going to visit his former assistant at the Technical College who was now Director of Rohrbach Metallflugzeugbau in Berlin. Tank was invited to meet Weber in Berlin and go with him to Rohrbach. Three days after the visit, Tank had a job in the company's Berlin planning office.

At Rohrbach Tank became immersed in charts, slide rules, compasses, drawing boards, and models. His inability to pilot aircraft due to his lack of training kept nagging at him, for he felt flying experience was essential to his design work and to his ultimate goal to be a chief engineer. He considered his work incomplete if he was unable to test the products of his designs. It was not until 1925, three years after joining Rohrbach, that Tank was able to find the time to take flying lessons. By 1936 he had earned every certificate offered in German aviation: private pilot, aerobatics, blind flying, captain of an aircrew, and seaplanes. Tank regarded his flying skills as an invaluable source of knowledge and marveled at how other designers could put into the air planes that they themselves had not first experienced in flight.

Tank's application of aerodynamics and hydrodynamics to airframe design became increasingly apparent in the models that came from the Rohrbach works. Before introducing new models Tank would spend months of work in Göttingen at the Kaiser Wilelm Institute, which was the center of experimental and theoretical aerodynamics. It was at Göttingen that Tank became acquainted with Professor Ludwig Prandtl, who was known worldwide for his mastery of aerodynamics and was the originator of the theory of winged flight. Tank also got to know Albert Betz, the man who succeeded Prandtl at Göttingen and who developed the theories that led to the sweptback wing. Tank never hid the fact that his Focke-Wulf aircraft benefitted enormously from his close association with Prandtl and Betz. Later the research at Göttingen on high-speed and supersonic wing shapes helped designers throughout the world achieve their long-sought-after faster-than-sound aircraft. Although Tank was there frequently to witness the research he and Focke-Wulf tended to stand back from the crowd in applying the superfast wing designs to their own aircraft.

Early in his career Tank designed seaplanes and commercial aircraft. Consequently, he, more than his colleagues in other firms, had to look at things that would improve a heavy airplane's performance and increase its operating speeds and payloads. Since no spectacular advances in engineering could be anticipated, designers such as Tank had to concentrate on improving basic aircraft design. The challenge was to get greater payload without increasing wing size so much that the additional weight of the wing structure canceled out the advantages of the greater lift obtained. During Tank's early years with Rohrbach, he introduced a number of technical improvements that were of great importance to the future of aviation. For example, at one time in Germany pilots were required by law to have "unimpeded (visual) contact" with the sky, and, as a safety measure in the air, no cockpit canopy could be fitted to any commercial aircraft. It was Tank who designed a well-supported glass cabin to surround the cockpit and got the representative of Lufthansa (the German state owned and operated airlines) to allow their use (although pilots at first had to purchase the cabins with their own funds and install them on the planes they flew).

Another technical improvement Tank made was to the aircraft wheel brake. Prior to 1928, an aircraft, after touching down, simply ran on until it stopped. On the Rohrbach BERO, Tank introduced wheel brakes that did not cause the aircraft to "nose over" or "ground loop" when they were applied (as some prototype wheel brakes by others had done), and this breaking system was applied to all future aircraft.

During the 1920s Tank earned increasing responsibility for the design of Rohrbach aircraft. He designed the ROMAR series of flying boats and the ROLAND commercial ten passenger, tri-motor, all-metal monoplanes which were assembled by Rohrbach's subsidiary in Copenhagen. With the design of the Rohrbach ROFIX, a fighter aircraft, Tank created a monoplane fighter, a rarity for the time. It would have tragic implications for Tank, Rohrbach, and Rohrbach's test pilot, Germany's most celebrated aerobatic pilot, Paul Bäumer. Earlier, with the help of the Günter twins, Bäumer had built his successful Sausewind (Whistling Wind) which he had flown in aerobatic shows throughout Germany. When Bäumer, recipient of the Order Pour le Mèrite for his military flying during World War I, saw the ROFIX for the first time, he became very enthusiastic about its possibilities. One morning in 1927 Tank allowed Bäumer to take his one and only ROFIX out for a test flight. Wanting to test the aircraft's spin characteristics, Bäumer took the prototype monoplane fighter up to 9,843 ft (3 km), put it into a spin,

The Multhopp family at home in the mid 1950s. Left to right: Heiko, 10; Hans Multhopp, Folker, 4; Mrs. Multhopp and Ralf, 4.

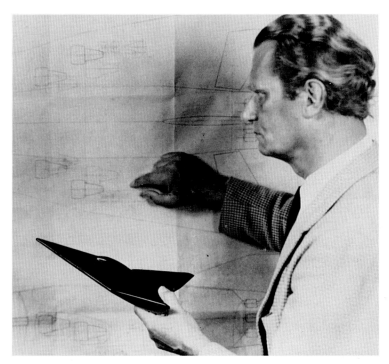

Hans Multhopp holding one of his proposed lifting body re-entry (Space Shuttle) vehicles.

recovered at 3,281 ft (1 km), flattened out, and flew away toward the sea. Then, deliberately or by accident, Bäumer suddenly went into another spin, and down went the aircraft, out of control. When the aircraft was recovered, Bäumer was still in it, drowned. Tank was blamed for the accident by a court of inquiry. The court found him to be "thoughtless" in building an aerobatic flying machine in the form of a monoplane. The crash of the ROFIX, and with it the death of much admired Paul Bäumer, ended the entire ROFIX fighter series for Rohrbach and ultimately contributed to the company's demise.

By 1929 the Great Depression was making it increasingly hard for aircraft manufacturers throughout the world to sell their products. It was especially difficult in Germany because the manufacture of military aircraft was still forbidden by the Treaty of Versailles. As Rohrbach's commercial sales declined, the firm was forced to suspend its operations, and all its employees, including Tank, went looking for work. For Tank it was easier than most, and before leaving Rohrbach he took a job with the Bayerische Flugzeugwerke (BFW; later Messerschmitt) at Augsburg. Tank's appointment there as the chief of the projects bureau came, in part, because of his reputation as a highly talented designer. It was also due, in part, to his friendship with Ernst Udet, who had founded BFW. Formerly, while Tank was at Rohrbach, he and Udet frequently had flown Rohrbach's seaplanes. Once the two of them had to ditch a flying boat over the water after the craft's wooden propellers came apart in flight. Tank's association with BFW lasted only 18 months, for that firm began experiencing financial difficulties similar to those that had driven Rohrbach out of business.

On 1 November 1931, Tank joined Focke-Wulf Flugzeugbau GmbH. Although the economic conditions that had driven Rohrbach and BFW out of business were still severe in 1931, Tank believed that things were improving slowly. Frightened by the fact that his two previous employers had lost their markets and subsequently their business, Tank made some personal stipulations with the management of Focke-Wulf. He believed that if any aircraft manufacturing firm were to hope to regain a share of the world market, it would have to design and construct aircraft on a scale comparable to the efforts in foreign countries, and he demanded that Focke-Wulf allow him to direct the design activities entirely without interference. He wanted to build the firm's aircraft entirely of metal, which Focke-Wulf had never done. He wanted the firm to break away from the notion that planes were suitable for weekend flying, to concentrate on high performance aircraft instead.

In 2 years, at 35 years of age, Tank became Technical director of Focke-Wulf. Although he had made some pretty strong demands on his new employer before he went to work, Tank at heart was a conservative. What he had proposed to Focke-Wulf, although radical to the old firm at Bremen, was what was necessary to make the company competitive throughout the world. Yet as a conservative, Tank felt strongly that progress in aircraft design must never become an aim in itself. Progress must be firmly based on the laws of logic, otherwise it was foolhardy and ultimately would bring its own retribution.

Tank's work already showed how he adhered to this principle. He would not support for long the development of new designs of radical engineering work if he thought technology was not equally advanced. Consequently, his firm, although one of the largest aircraft manufacturing companies in Germany during the 1940s (40,000 employees), was among the last to get involved in the design of turbojet-powered aircraft (and the firm's portfolio at the end of the war was one of the smallest of any aircraft firm).

The basis of Tank's conservatism was a strong belief that aircraft safety must come before performance, especially in the case of fighters which had to rely on a single engine, even a single turbojet engine, for propulsion. He considered the turbojets that were coming from the factories of BMW and Jumo neither powerful enough nor reliable enough to be used in any single-engine fighter aircraft. Later on the vastly increased power of the HeS 011, 3,527 lb or 1,600 kg, of thrust) promised for delivery in 1945, (production difficulties kept it from being introduced prior to war's end) gave him confidence that the time was nearing for a truly high-speed, high-altitude fighter to be designed around a single-turbojet engine. But to be on the safe side, Tank believed, all Germany's fighter eggs" should not be placed in the "turbojet basket;" development of other high-speed fighter aircraft powered by conventional twin piston engines such as the promising Dornier Do 335 and Focke-Wulf's own Fw 187 should continue as well. This wait-and-see attitude frustrated technology-smitten engineers such as Hans Multhopp. The old maxim, "Be not the first when the new are tried, nor be the last when the old are cast aside," fit Tank better than any other aircraft engineer-designer in Germany in the 1940s.

Tank flew every day, Sundays included, never missing a day unless his duties took him away from Bremen. He encouraged those who had important practical or scientific responsibilities to learn to pilot airplanes so they could truly "communicate" with pilots. For Tank, flying was not just a matter of vocabulary, but rather a question of "atmosphere," the way people looked at things. He believed that a designer's ability to experience the products of his designs ruled out the probability of misunderstandings. After his daily flying Tank would return to the drawing office to inspect the progress made. As he passed from board to board, viewing the shape and component parts of the latest project, agreeing, rejecting, or outlining changes, his staff felt that here was a man who knew what he was talking about. "I can't regard an aeroplane as anything else but a metal container made as skillfully as one possibly can, and provided with power to carry a given load of human beings or freight or even, unfortunately, bombs through the air at a favorable height and in the most economic way," Tank said.

A perpetual task of aviation technicians is to reduce the effects of human weakness or unforeseen weather conditions while the plane is in the air. There should be a constant, patient struggle, Tank believed, to perfect equipment so as to reduce the risks of flying. The invention and application of such aids as radio, radar, and blind-flying devices that

A Martin Aircraft SV-5 lifting body designed by Hans Multhopp.

The world's first truly operational supersonic fighter aircraft, the English Electric P.1. This aircraft incorporated many of Hans Multhopp's appropriate aerodynamics for supersonic flight when it was built in 1947.

would help pilots in bad weather, mist, or darkness or under stringent combat conditions - these were the expressions of perfection for which Tank strove, for he was one not to take risks. On the basis of his flying experience, Tank noted: "The more a man's character develops, the less likely he is to take risks. A man who takes risks is usually spiritually comatose. I have seen any number of such men," he recalled, "they find themselves in situations which they have neither the character nor the development to face it, and the results are inevitable."

As Tank's reputation for efficiency in management and design spread, he was invited to head various committees dealing with the production of war materials and supplies. Although he was not one to advocate that research continue at any price, he finally admitted that the days were coming when Germans would see over their country the giant high-altitude Boeing B-29s with their massive bomb loads - and eventually the atomic bomb (which German intelligence had known about prior to its use on Japan in August 1945). He knew something had to be done, and decided to have his designers work on a single-engine, turbojet-powered, high-altitude fighter. To make certain his firm didn't go off on some aerodynamic tangent or, even worse, a dead end, he set up two design teams. Hans Multhopp was to be in charge of one team, and Ludwig Mittelhüber the other. Each chief designer was to have a free hand to pursue the perfect aircraft for the task ahead, and the Air Ministry would decide on the winning design. In the end it was Multhopp's team that triumphed and in January 1945 began making preparations that would lead to production of the highly unusual looking P.183. Kurt Tank, still full of doubt over the apparent aerodynamic broadjump from fast piston to superfast turbojet-powered fighters, kept his word and turned Multhopp, his in-house tornado, loose.

Hans Multhopp

The son of a newspaperman, Hans Multhopp was born at Alfeld, near Hannover, Lower Saxony, on 17 May 1913. Completing his general schooling in 1932 at age 19, he entered the Technische Hochschule (Institute of Technology) at Hannover and spent the next 2 years studying engineering. In 1934 he matriculated at the University of Göttingen, where he stayed until 1937, taking advanced work in his specialty - aerodynamics. It was during these years that Multhopp studied under Dr. Ludwig Prandtl, the father of modern aerodynamics.

While at Göttingen, Multhopp designed a number of sailplanes of the general type that had become popular among students and others when civilian training in powered aircraft came under rigid regulation by the provisions of the Treaty of Versailles. Multhopp, along with several other students, built a high-performance sailplane, one with a wingspan of 53 ft (16.2 m). While working toward a Ph.D. in aerodynamics (which he never completed) Multhopp supplemented his theoretical learning with practical experience gained through a job at the Aerodynamische Versuchsanstalt (AVA). This establishment for aeronautical research corresponded to the National Advisory Committee for Aeronautic (NACA) in the United States. Like NACA, the AVA was subsidized by the government (in AVA's case by the Reich, at that time through Hermann Göring's Air Ministry). By 1937 Multhopp was in charge of one of AVA's wind tunnels. When he joined the Focke-Wulf Flugzeugbau in Bremen in 1938, at the age of 25, he had already published one of his most important papers on wing-lift theory. By 1940, the year of his marriage to a Bremen woman, he was Kurt Tank's assistant in the aerodynamics department. His responsibilities were extended 3 years later to embrace the new design department as well. In 1944 he became one of Focke-Wulf's chief engineers. Only then, 6 years after joining Focke-Wulf, was Multhopp allowed to pursue his plans for the P.183 - Germany's last single-engine design to be completely developed before the war ended in Europe.

Multhopp's P.183, intended as an interceptor against Allied bombers, was expected to have a speed of 575 mph (925 km/h) and an operational range of up to 1,150 mi (1,850 km) at 40 percent power. Multhopp boasted that it would have a 47,200 ft (14.4 km) service ceiling. The P.183's configuration was distinguished by a T-tail, pioneered by Multhopp, and by swept wings. Backsweep had been proven out in German wind tunnels as early as 1938; Multhopp's P.183 was one of the first designs actually to incorporate this feature.

Multhopp believed that a standard tail configuration was an absolute necessity for controlled flight. Although he agreed that it was possible to transfer some of the tail's usual functions to other parts of the

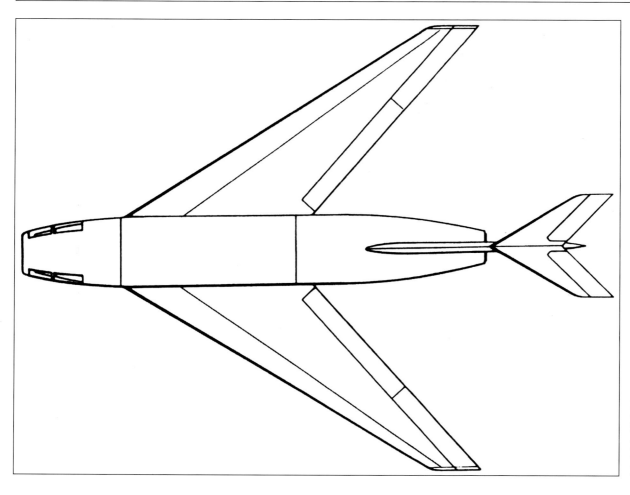

An overall view of Hans Multhopp's proposed supersonic project for the Royal Aircraft Establishment (RAE) in 1946. Multhopp believed his experimental aircraft design would reach Mach 1.24 in level flight. Later showed up as the English Electric P.1.

craft, for example to the wings of tailless aircraft, these parts could rarely take on the functions of a tail without impairing the performance of their own duties. On the other hand, Multhopp liked to point out, people rarely found the tail attractive from the standpoint of performance; its only direct contribution to performance was drag and weight. Therefore, he felt it only natural to look for the least costly way of proving the function of the tail. This could mean a tail arrangement that gave the smallest area (drag) or weight, or a combination of the two, taking into account the area and weight contributions of the fuselage or tail boom needed to support the tail.

To Multhopp, the T-tail represented the best compromise to provide for the functions of a tail unit: stability and control. Any other arrangement, such as a flying wing or even a tailless aircraft, Multhopp believed, would be less effective in the absence of some vertical tail area. Plus the further the horizontal tail was away from the fuselage, the better, as far as control effectiveness was concerned. Multhopp's exten-

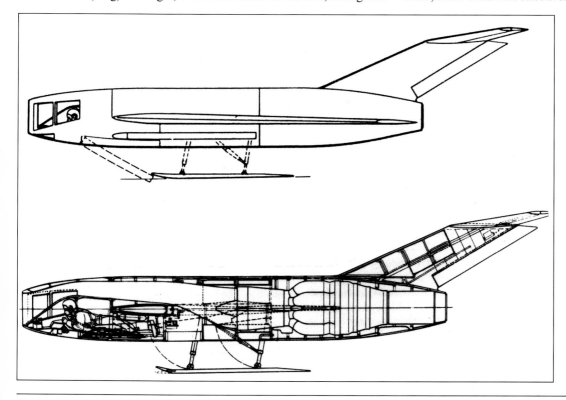

The side view of Multhopp's supersonic project for the RAE. The pilot was to fly in a prone position similar to the DFS 346, however unlike the 346, the proposed RAE aircraft would have been powered by a turbojet whereas the 346 was powered by an HWK bi-fuel rocket drive.

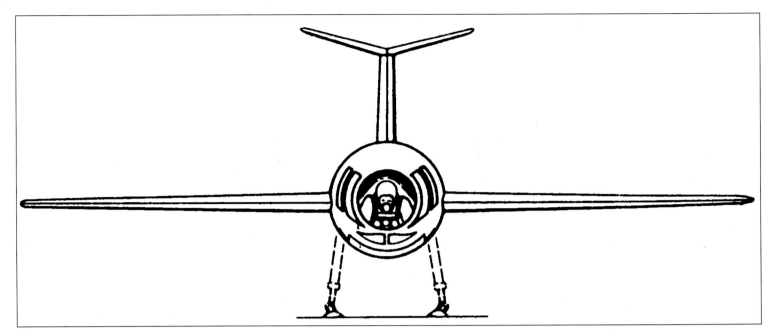

A head-on view of the RAE proposed supersonic project. Air for its turbine was ducted from the nose of the aircraft back to the turbine in the rear.

sive wind tunnel testing showed that during takeoff and landing, where control usually is most difficult, a very high tail was not too much affected by the turbulent airflow behind the wing. Multhopp also found that a horizontal tail surface on top of a sweptback fin gave more leverage than one mounted directly on a fuselage.

The only P.183-related plans and drawings Multhopp was able to bring out of Germany when the Red Army overran Focke-Wulf facilities was the scale model used in wind tunnel testing. It has been reported that Tank had a complete set of plans on microfilm - the only known copy outside the Soviet Union - and that when he escaped to Argentina in 1946 he took along the valuable microfilm, which enabled him to build an aircraft similar to the P.183, the Pulqui Dos (superdart) for Juan Perón in 1950. Multhopp was taken to England for interrogation immediately after V-E Day, then was sent back to Germany to join the collection of scientists at the AVA at Göttingen. (Multhopp was on the British list of preferred scientists; in contrast, rocket specialists were on the U.S. list). There he was required to write down in detail the state of the art of aircraft design in Germany as of August 1945. His friend and fellow designer from Focke-Wulf, Martin Winter, also joined the group. In the spring of 1946 the British offered employment to a handful of the German scientists, and in March, Multhopp and Winter left Göttingen for Royal Aircraft Establishment (RAE) facilities in Farnborough, Hants, England.

At Farnborough, Multhopp found that the English did not have any particular aircraft project in mind but expected each of the German designers to work only within his line of specialty. As frequently happened to Multhopp while he was working at Focke-Wulf, RAE officials completely underestimated him. Not long after the two began work at Farnborough, Multhopp and his friend Winter came up with a design for an experimental aircraft which they believed would achieve Mach

Hans Multhopp in the mid 1960s.

A collection of Focke-Wulf designers left to right: Käther, Stampa (builder of the Ta 183 scale model), Mittelhüber, Naumann, Otto Pabst. Shown at Focke-Wulf's relocated offices to a spa-hotel, Bad Eilson, Westfalia in 1942 after Focke-Wulf's design offices in Bremen were destroyed by Allied bombing.

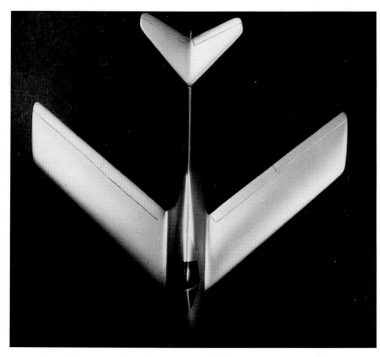

An overhead view of a wind tunnel model of the Ta 183.

1.24 in level flight. As it turned out, an American aircraft, Bell Aircraft's X-1, beat the British in reaching the speed of sound, on 14 October 1947 - and not by means of a special aerodynamic design, but by brute power from its rocket engines. With its unswept wings and tail assembly, the X-1 bore no resemblance to Multhopp's and Winter's design. Their design called for a turbojet-powered aircraft having a wingsweep of 55 degrees. The pilot was to lie prone beneath the air intake duct for its engine, an AJ65 Avon, one of Rolls-Royce's first axial-flow turbojet engines. With a projected speed of more than 800 mph (1,287 km/h), the proposed RAE supersonic aircraft would have used jettisonable wheels for takeoff and a pair of built-in skids during landings. Landing speed was estimated to be 160 mph (258 km/h). By English standards, Multhopp's RAE design was pretty radical stuff, with its mid-mounted wing 25 ft (7.6 m) long and sweptback 55 degrees at the quarter-chord. The tube-like fuselage was 26.6 ft (8 m) long and had a diameter of only 3.3 ft (1 meter), and the tail, which was swept 64.5 degrees, carried the horizontal surface at its top - a Multhopp trademark to this day.

Instead of being enthusiastic about an aerodynamic shape that Multhopp and Winter believed would easily surpass the speed of sound, British officials quickly slapped a "Top Secret" label on the design. Then a sort of "Catch-22" situation developed. Since neither Multhopp nor Winter had the citizenship necessary to gain access to secret materials, the RAE told them they could no longer work on the transonic/supersonic research project. Both would have to go.

The RAE supersonic research aircraft was destined never to be built, in part because the sound barrier had already been broken and in part because of the worsening economic conditions throughout England which made continuation of support for pure aerodynamic research impossible. Multhopp and Winter believed that the overriding issue on the cancellation of their supersonic design was the design itself. "It was because of its design that we both got pushed out of the RAE," said Winter years later. "We seemed to have a much more all-round aircraft knowledge than they had experienced or desired," Winter said, "and as a result we had to go because we and the design were too hot to handle in 1948."

Because Multhopp and Winter were considered security risks, they were placed under "house arrest" in London until such time as they could find work outside England, and were given a temporary office at the Imperial College. It was at the Imperial College the Multhopp had time to think about his theories on lifting surface for swept wings. Theories he developed during this period were put to immediate use by British aircraft manufacturing companies that did not consider his and Winter's design too hot to handle. While some British "experts" in 1947 were arguing that sustained supersonic flight was impossible, the aircraft firm known as English Electric was designing its powerful Mach 2 "Lighting" interceptor. This became the world's first truly operational supersonic aircraft, built on the basis of official specifications that grew out of the design conceived by Multhopp and Winter for the supersonic RAE project and later strongly articulated by Multhopp in is seminal theory on lift surface for swept wings.

After their RAE supersonic project had been taken away from them, Multhopp and Winter began sending out resumes. A number of countries showed interest, but both had had enough of western Europe, especially the highly conservative British aircraft community. In late 1949 both received offers of employment from three American companies: Boeing Aircraft, Goodyear Tire and Rubber, and the Glenn L. Martin Aircraft Company. In 1950 the pair joined the design staff of the Martin

Side view of the Ta 183, scale model.

company in Baltimore, Maryland, which Multhopp felt was the most advanced of the three, having recently built the sweptwing, T-tailed, B-51 jet bomber prototype which had first flown on 28 October 1949. It appeared to Multhopp that the Martin company was more eager to go after military aircraft contracts by submitting bold new design ideas. As manager of preliminary design, Multhopp came to head a team of engineers and scientists in advanced design work. A number of Martin aircraft, the XB-51 bomber prototype, the P5M-2 Seaplane, and the XP6M Seamaster (a four-turbojet-powered seaplane) among others, were designed because he was there with the Multhopp T-tail.

His decision to work in private industry in the United States was a decision Multhopp came to regret. By the early 1950s aircraft designed entirely by single individuals was almost a thing of the past and the design team approach was being widely used. Multhopp found that the design rules imposed by the Pentagon were just as restrictive as similar ones under Tank at Focke-Wulf. He remained with Martin until 1967, when the firm discontinued its aircraft design and manufacturing activities to concentrate instead on missiles and space systems, (the firm was then known as the Martin-Marietta Corporation).

After leaving the Martin company, Multhopp worked as a self-employed consultant in aerodynamics. A year later, in 1968, he joined the Re-entry and Environmental Systems Division of the General Electric Company.

Hans Multhopp, the most gifted aerodynamicist Ludwig Prandtl had ever taught, holder of aircraft patents and numerous acclaimed papers, a design genius and a true lone wolf, died on 30 October 1972, in Cincinnati, Ohio, at the untimely age of 59.

During his four years with General Electric, Multhopp developed preliminary configurations for NASA's space shuttle program and the concept of an integral, fully reusable launch and re-entry system. But it was while he was still with the Glenn L. Martin company in the mid to late 1960s that he did the work that was his most important contribution to the space shuttle program. It was his design work on the Space Vehicle 5 (SV-5), a lifting body type of re-entry vehicle having the configuration that was proved successful, on the flight of the space shuttle Columbia in 1981, through the full speed regime, from orbital (18,000 mph or 28,967 km/h), down to landing speed (200 mph or 332 km/h). Although he did not live to see the results of his research on lifting bodies or his contributions to the Martin-Bell Aircraft "Dyna Soar" avenue to the space shuttle, Multhopp's work on the Martin-Bell "Dyna Soar" has much more resemblance to the space shuttle than lifting bodies ever had.

Multhopp had predicted that aviation would continue to reap the technological advances he and others pioneered in Germany during the 1930s and 1940s. He believed that the three great revolutionary innovations in aviation occurred in the 1940s: turbojet propulsion, supersonic flight, and electronic guidance. After that, the job for aeronautical engineers such as himself, he said, was to refine and extend the useful applications of those three significant discoveries. This is a statement of remarkable modesty from the man who many in the world of aerodynamics, including Lippisch, believed to have made the most significant contributions to supersonic flight of anyone, anywhere! Although his contributions were numerous, one clearly stands out, something Claudius Dornier called a distinctively personal signature or trademark, and a feat most designers strive for but few ever achieve. One sees it frequently, on small, single-engine Piper's and large, multi engined transports such as the Lockheed C-5, and it will be a feature of aircraft for decades to come, the Multhopp T-tail.

After the war Kurt Tank was invited by Argentine President Juan Perón to build a fighter aircraft, and in 1946 he secretly and illegally left Germany for Argentina. Then came the former heads of the Focke-Wulf design offices, Wilhelm Bansemir, Paul Klages, and Karl Thalau. Ludwig Mittlehüber followed, as did theoreticians Gotthold Mathias and Herbert Wolff and, finally Otto Pabst, the specialist in gas dynamics. In all, sixty former Focke-Wulf employees and their families, a community of one hundred people, illegally left Germany to begin work on production of a prototype machine officially designated IAe 33 and named Pulqui Dos. For Tank and his faithful Focke-Wulf group, it was the beginning of business as usual again - only this time they would be able to design and construct a turbojet-propelled fighter under bomberless skies.

Although Multhopp and Tank were not on the friendliest of terms during their 7-year association at Focke-Wulf and did not socialize, their uneasy collaboration in late 1944 on what was to become the P.183 would come to be recognized as an historic encounter. The P.183 ended in one giant sweep the era of piston-powered fighter aircraft. The shape of the P.183 made every other fighter obsolete, and fighter design would never be the same. Yet Tank continued to believe that supersonic flight could be achieved by a piston-powered engine. Still, we can visualize Tank and Multhopp, both anxious for results, both hoping for a "miracle fighter" to save the Reich from the USAAF B-24 and B-17 heavy bombers, arguing over each other's concept of the appropriate fighter for that moment in time - and neither realizing that together they were tumbling over a new threshold in aircraft design. Although they had accomplished a breakthrough, neither Tank or Multhopp ever capitalized on what their collaboration had achieved during those desperate times in 1945. Others would take Multhopp's bold new design and successfully put it in flight, as did the Russians, calling their version the MiG-15. Multhopp believed that what could be conceived could be fashioned, and he kept reassuring Tank, his superior: "Trust me - I know what I'm doing." Tank, equally enthusiastic over the possibilities of a miracle aircraft yet uncertain and reluctant to jump into the engineering unknown (where Multhopp was taking him) until more research could illuminate the way, resigned himself to the coming adventure and sought to minimize Multhopp's penchant for the radical tangent by struggling to keep the design project within a few commonsense engineering boundaries. Tank never forgot either, the loss of Paul Bäumer in 1927.

Some critics believe that Multhopp achieved a sort of "beginners luck" with his P.183. Others say that what one sees is only the tip of the iceberg of what he could have achieved had he found a more receptive and appreciative place to apply his intellect. But, like many other geniuses who held inside of them their own clock, Multhopp was a difficult person to work with and had a hard time fitting, in, whether in academia or industry, and was never able to prove his detractors right or wrong. But Multhopp was never after fame or a personal design signature, and pursued what he did according to that internal gyroscope, regardless of the external environment."There is a little watchman in my

Overhead view of a new Boeing B-29 bomber.

A MiG-15 in flight with similar design features as the Ta 183 as seen from below.

heart," he would say, "who is always telling me what time it is." As far as Multhopp was concerned, it was always time to try out things that were new and different, to challenge the mediocrity he found everywhere. When Tank asked Multhopp to join him in Argentina in 1946, he said no - that he wanted to go on to other things. Although Tank eventually surrounded himself with his whole Focke-Wulf staff from Bremen, they found that without Multhopp they couldn't build the Pulqui Dos for only Multhopp knew the sophisticated aerodynamic calculations required to make the Pulqui Dos an outstanding fighter aircraft. Completely lost without Multhopp, Tank struggled with Multhopp's wing-lift distribution calculations and then decided to move Multhopp's swept back wing higher up on the fuselage. It was a tragic mistake and turned the P.183 into a design failure. Perón allowed only five Pulqui Dos' to be built, purchasing instead 100 surplus F-86s from the United States government for about what a new turbojet engine would cost. Perón then proceeded to fire Tank as his assistants. The whole bunch, suddenly unemployed were then told to leave Argentina immediately.

In 1953 Multhopp flew to Cördoba, Argentina, to hire some of Tank's associates, but Tank was cold and uncommunicative toward Multhopp and privately blamed him for the Pulqui Dos' failure. Multhopp's reaction was "Well, what the hell did you expect would happen? Look how you screwed up my perfect design. All you had to do was follow the plans, maybe a minor change here and there." Multhopp returned to the United States and the Martin aircraft company and a few of Tank's former engineers. The Pulqui Dos ended Tank's career in Argentina and the entire group disbanded. Some migrated to the United States, some back to Germany, and a handful of loyalists followed Tank on his odyssey to India and then on to Egypt. Unable to find friendly faces after his Egyptian work ended in 1965, Tank now absent from Germany for nearly twenty years, returned home again. Willy Messerschmitt offered his aging friend and one time competitor an honorary position in his factory which allowed Tank to later retire on a small pension. He died in München June 1983, never able to successfully make the transition to turbojets from pistons without the design genius of Multhopp. Multhopp had no such problem but may have found the direction he needed in his life if he had stayed with Tank and the other members of the former Focke-Wulf team after the war. Still, the most brilliant aircraft designer to come out of Germany after World War II apparently never lost faith in himself as he went on his own odyssey parallel to that of Tank's. He did continue to express one basic wish throughout his brief life - that more people would not be so timid in the face of needed change, but would dare more often to be different. But accepting change was easy for Multhopp when in fact it is resisted by most people because of the uncertainly which change can foster.

PROJECT 183

Allied intelligence officers considered it the best example of Germany's advances in turbojet-powered aircraft. Its basic configuration later was adopted by the Soviet Union for its MiG jet fighters. It was the first German fighter designed for optimum performance at 26,200 to 46,000 ft (8 to 14 km), and the first single-jet aircraft to have its jet unit installed in the fuselage, with the air intake and the exhaust tube running the length of the fuselage. It was also unconventional in that its wings had a pronounced backsweep (40 degrees). Its vertical tail and stabilizer were different from anything the world had seen before. The vertical tail, swept back at an angle of 60 degrees, contained the rudder and was attached to the upper surface of the fuselage. But it was the horizontal stabilizer that became the aircraft's signature and gave it the name, "Multhopp T-Tail." The stabilizer was swept back at 40 degrees, like its wings, but it was attached to the very top of the vertical rudder. This very high altitude, very high speed aircraft which could serve as a fighter-bomber, or an interceptor, was the brainchild of Focke-Wulf's design genius, Hans Multhopp, and the aircraft was his Project 183, later designated the Ta 183 after Kurt Tank, Focke-Wulf's managing director, who took over the design and claimed it as his own.

The P.183 was Focke-Wulf's response to a turbojet-powered fighter design specification issued by the Air Ministry in July 1944. Competing against Messerschmitt's P.1101, P.1110 and P.1111, Blohm and Voss' P.212, Heinkel's P.1078C, and Junkers' P.128, the Ta 183 was selected for test manufacture. However, the war ended before the design could be converted into a prototype for testing.

The proposed Ta 183. Scale model by Dan Johnson.

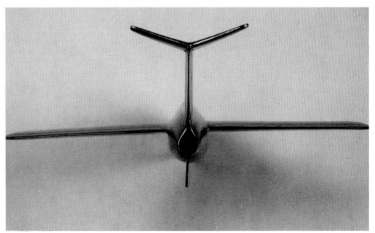

Ta 183 wind tunnel model, viewed from behind.

Ta 183 wind tunnel model, viewed from front.

The Air Ministry was in urgent need of an interceptor capable of operating reliably above 36,091 ft (11 km). The first-generation turbojet engine such as the Jumo 004 and the BMW 003 did not operate reliably above that height. Air Ministry officials were concerned that the Americans and the British would shortly introduce turbojet-powered bombers and that the Me 262, with its Jumo 004, would not be able to go up after them if they were flying higher than 36,000 ft (11 km). To pursue the powerful new American heavy bomber, the Boeing B-29 Superfortress, (although they were never used over Germany), which had a service ceiling of 33,000 ft (10 km), the Me 262 would be operating near its design limits.

By 1943 the HeS 011 turbojet of 2,866 lb (1,300 kg) thrust was thought to be soon available to aircraft manufacturers. It never was. This second-generation turbojet would be able to replace the less-powerful Jumo 004 and BMW 003, with their 1,984 lb (900 kg) of thrust. With the news that the HeS 011 was being readied for mass production, the OKL (Oberkommando der Luftwaffe), Luftwaffe High Command, quickly issued design specifications for what came to be called "Emergency Fighters." Any proposed aircraft was required to be designed

When Kurt Tank tried to build the Ta 183 for Argentinian dictator Juan Péron as the IAe 33 (Pulqui Dos), it turned out to be a design failure without Multhopp there with him. Tank fiddled with Multhopp's design, placing its wings farther back and higher on the fuselage sides. These modifications made the Pulqui Dos a brute to fly, and it had a tendency to "drop like a rock" while making its landing approach.

Kurt Tank's Pulqui Dos (Arrow Two) scale model in wind tunnel, front view.

Kurt Tank's Pulqui Dos scale model in wind tunnel, rear view.

A Pulqui Dos on permanent outdoor display at Aeroparque in Buenos Aires, with a good view of the Multhopp T-tail. 1988.

Left to right: Tank, Péron talking after Tank had made a dramatic high speed, low pass flight just above the heads of the General and his entourage with the Pulqui Dos.

around the new HeS 011 turbojet, to have a level speed of about 621 mph (1,000 km) at 22,966 ft (7 km), to operate at altitudes up to 45,934 ft (14 km), and to be armed with four 30 mm MK 108 cannon.

In mid-March 1945 the Luftwaffe High Command selected Focke-Wulf's proposal, the P.183. The aircraft was to be developed and produced at the greatest possible speed. With factories operating two shifts per day, production was expected to run at a rate of 300 per month. Preceding production were to be 16 experimental or Versuchs units powered by Jumo 004B turbojets, pending the arrival of the HeS 011s. The P.183 project never got beyond detailed engineering drawings because the following month, April 1945, all Focke-Wulf factories suitable for production were overrun by Allied troops.

Reflecting the shortage of war production material, the P.183 was designed to be as small as possible, extremely simple, and easy to build. For example, a single sheet of plywood was to be used to form the wing covering from fuselage to tip and from the top of the spar around to the bottom of the spar. In addition, the wings were to be interchangeable and were to be constructed of wood, except for a steel main spar. The upper surfaces of the fuselage were to be of steel sheeting (if aluminum were available, it was to be used throughout the fuselage). The engine cowling was to be formed from steel, too. The intake duct for the HeS 011 turbojet as well as the vertical stabilizer were to be fashioned from aluminum, while the horizontal tail and all control surfaces were to be wood.

Kurt Tank in retirement back home in Germany after a long career in avaition.

A proposed variant of Multhopp's Focke-Wulf Fw Ta 183 (Design II) single-seat fighter. This proposed design was considered as an alterative to the Ta 183 and referred to as Design III. January 1944.

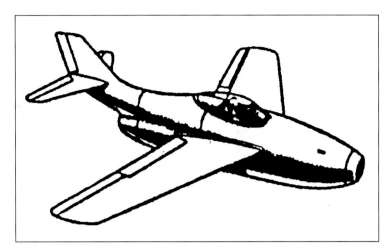

A pen and ink drawing of the Ta 183 variant referred to as Design III. January 1944.

Multhopp gave the aircraft, including pilot, a large measure of armored protection. The fuselage tanks were to be protected by 15 mm steel armor plate at the rear and by 3 mm steel deflector plating (up to 10 degrees incidence of hits) along the sides and the top. There was to be armor plate at the front and the back to protect the pilot from up to 20 mm cannon fire.

The tricycle landing gear was to be operated hydraulically. The nose wheel was to retract backward into the fuselage, and the main wheels were to retract forward and inward into the fuselage. When the gear was fully retracted each wheel was to be covered by two doors for protection, the doors lying flush with the fuselage.

Armament for the P.183 were to be two 30 mm MK 108 cannon, each with 120 rounds of ammunition, to be located in the front fuselage. Two additional MK 108s with sixty rounds each could be installed, but this would affect climb, ceiling, and endurance adversely.

Multhopp designed the aircraft to be adaptable for special uses. If it were to be used as an interceptor (designated P.183R), a HWK 509 bi-fuel liquid-rocket drive of 2,205 lb (1,000 kg) thrust could easily be installed beneath its regular HeS 011 turbojet, thereby greatly enhancing its interceptor capability. The HWK 509 rocket drive was designed to burn for 3 minutes. The two fuels (C Stoff and T Stoff) were to be carried in external wing tanks. When the craft reached its desired altitude, or when the rocket fuel had been expended, these underwing drop tanks would be jettisoned.

Although it was not produced in Germany during the war, versions of Hans Multhopp's P.183 were built later. (It is the most copied German turbojet aircraft design of World War II). After the war, in 1946, Kurt Tank, Focke-Wulf's former general manager, migrated to Córdoba, Argentina, and built an aircraft similar to the P.183 for Argentine dictator, Juan Perón. It first flew in 1950 - but it failed to live up to its name, "Pulqui Dos" or super dart, for Tank had to build the aircraft without Multhopp's help and in the process changed with wing lift distribution characteristics. Tank's modifications turned the Pulqui Dos into a brute of a flying machine and Perón allowed only five prototypes to be constructed before canceling the entire project.

The MiG-15, which began making headlines early in 1950 as foil for the US Air Force's F-86 Sabrejet, was the Soviet Union's first adaption of the P.183. It incorporated a number of changes from the original as Multhopp himself later suggested. The MiG-19 was a later - and hence, a battle-refined version of the MiG-15. It had more pronounced T-tail, swept wings, and an overall configuration that was even more closely related to Multhopp's original design.

Specifications: Project 183

Engine	1xHeinkel-Hirth HeS 011 turbojet having 2,866 lb (1,300 kg) of thrust (standard fighter), plus a Walter HWK 509 bi-fuel liquid-fuel rocket engine of 4,409 lb (2,000 kg) thrust for the interceptor.
Wingspan	32,8 ft (10 m)
Wing Area	242 sq ft (22.5 sq m)
Length	20.6 ft (6.3 m)
Height	NA
Weight, Empty	6,570 lb (2,980 kg)
Weight, Takeoff	9,480 lb (4,300 kg)
Crew	1
Speed, Cruise	391 mph (629 km/h) at about 68% throttle
Speed, Landing	103 mph (165 km/h)
Speed, Top	575 mph (925 km/h) at 9,843 ft (3 km)
Radius of Operation	348 mi (560 km) at 100% throttle at sea level with standard internal fuel tanks, or 1,150 mi (1,850 km) at 40% power at 22,967 ft (7 km), with 1,764 lb (800 kg) of additional fuel carried in drop tanks
Service Ceiling	47,200 ft (14.4 km)
Armament	4x30 mm MK 108 cannon with 120 rounds of ammunition each
Rate of Climb	79.5 ft/sec (24.2 m/sec) at sea level, or 53.5 ft/sec (16.3 m/sec) at 19,700 ft (6 km)
Bomb Load	1x1,102 lb (500 kg) bomb and/or 1 torpedo
Flight Duration	30 min at 100% throttle and 2,646 lb (1,200 kg) fuel, or 4.5 hr at 40% power and 4,409 lb (2,000 kg) of fuel

PROJECT "Triebflügel"

In the colorful language of modern aeronautics, Focke-Wulf's Triebflügel (driving-wing) project would be called a vertical takeoff-and-landing (VTOL) aircraft. Designed in 1944 but never built, it was the forerunner of all vertical-takeoff airplanes.

The concept of vertical takeoff aircraft in Germany at least, began with the studies of Dr. Erich von Holst, a Göttingen University zoologist, in the 1930s. Von Holst was curious about the mechanics of bird and insect flight, and he hoped that he might be able to apply some of his research findings to future aircraft design. Of particular interest to

The evolution of Dr. E. von Holst's Focke-Wulf Fw "Triebflügel" came about in an attempt to convert the dragonfly's ability to land and take-off vertically to an aircraft.

Focke-Wulf Flugzeugbau - Bremen, Germany

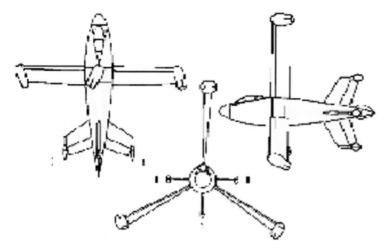

An artist's airbrush of a proposed Focke-Wulf Fw "Triebflügel." Shown is the proposed three ramjet version, which became the preferred model.

An artist's airbrush of a proposed Focke-Wulf Fw "Triebflügel." Shown is the proposed two ramjet version.

The proposed Focke-Wulf Fw "Triebflügel," three ramjet version. Scale model by Günter Sengfelder.

The proposed Focke-Wulf Fw "Triebflügel." Scale model by Dan Johnson.

VTOL - In post World War II several countries experimented with vertical take-off and landing vehicles (VTOL). The Convair XFY-1 and Lockheed XFV-1 were radical experiment single seat shipboard fighters evolved as part of a program to examine the feasibility of operation VTOL aircraft from small platforms on ships as an alternative to carrier launching. The principal mission of the fighter was the protection of convoys from aerial attack, and both the XFY-1 and XFV-1 were designed to take off vertically from an erect position. They were designed to meet the requirements of a specification prepared by the US Navy in 1950. Prototypes of the two designs were ordered in March 1951. The first to fly was the Lockheed XFV-1 which began normal horizontal trials in March 1954 with the aid of a temporary fixed undercarriage. A 5,850 hp Allison T40-A-6 turboprop was installed for initial tests pending the delivery of the Allison YT40-A-14. This engine was expected to provide 7,100 hp for take-off driving broad bladed contra-rotating propellers. Full performance trials with the XFV-1 were not completed, however, it was calculated as follows: test weight of 15,002, maximum speed of 580 mph with a rate of climb to 20,000 feet of just under three minutes

The Convair XFY-1 had a similar turboprop installed for its first vertical flight on 1 August 1954. Although both the XFV-1 and XFY-1 made a number of successful transitions, they presented severe piloting difficulties, and the program was abandoned by the US Navy.

The French SNECMA experimented with a VTOL aircraft in the late 1950s. Known as the C-450 "Coleopter" it was essentially a tail-sitter with an annular wing intended to operate in a similar fashion to the Lockheed XFV-1 and Convair XY-1. The "Coleopter" had a take-off weight of 6,615 lbs with an estimated performance of 497 mph. It was powered by an 8,150 lb static thrust SNECMA Atar E5V turbojet. Directional control was attained by means of jet deviation. Directional control during horizontal flight was pro-

Vertical take-off and landing has intrigued aircraft designers for years, Lockheed's XFV-1 shown here is about ready for a vertical take-off test.

The Convair XFY-1 "Pogo" is shown hovering a few feet off the ground during a test.

vided by four swivelling fins. Initial vertical take-offs and landings were made on 17 April 1959. Shortly afterward the "Coleopter's" test pilot, Auguste Morel, was forced to eject from the C-450 at an altitude of less than 230 feet after completing several inclination maneuvers, the aircraft being almost totally destroyed. The result of this crash brought an end to the "Coleopter" program.

The C-450 is interesting for another reason. It was designed by the high-ranking SS member Dip.Ing.Helmut von Zborowski. During World War II, Zborowski worked at BMW-Spandau and was responsible for the development of the bi-fuel liquid rocket engine attached to a standard BMW 003. This take-off assist rocket was intended for the Arado Ar 234 and the Heinkel He 162 "Volksjäger". The 003R rocket engine as it came to be called, was an originial design by Zborowski and known as the BMW 718 bi-fuel rocket engine. Unlike the HWK 509, the pump required to pressurize the C and T Stoff its two fuels was driven by the 003 over a drive shaft from its accessory gear box. Zborowski had been working on a VTOL aircraft when the war ended with Germany's surrender in May 1945. How Zborowski managed to find work in post-war France is unknown because he belonged to that group of high-ranking SS who were routinely prosecuted as war criminals. Nor did he serve in the French Foreign Legion in Indo-China, as many other SS did to escape prosecution. Zborowski also had problems with Ernst Heinkel post-war. Heinkel and other former employees held patents on their proposed "Lerche" VTOL aircraft of 1944/45. None of the Heinkel group would give Zborowksi the needed patents when he sought to build his "Coleopter" in the mid 1950s. When Zborowski did build his "Coleopter" for the French it did not have mechanical rods or cables going to the controlling surfaces. Instead it had "wires" only which connected with small electric motors. This VTOL could claim the distinction of being the first aircraft to "fly by wire" of which we are hearing so much about today as the new technology in modern aircraft.

him was a bird's ability to create enough lift for flight regardless of how fast it is going to fly once it gets airborne. Birds, he observed, achieved lift through the flapping motion of their wings. This was not a technically desirable solution in aircraft because of the very high force developed by such slow up-and-down motion, but there was one flying creature that achieved lift without flapping its wings - the well known dragonfly. This large insect, he noted, used two sets of wings, one behind the other, with the sets swinging 180 degrees of phase, that is, one set going up while the other is going down.

When von Holst built a model of a dragonfly he found that he could imitate its flight characteristics. But he also found the two complete sets of movable wings a bit too complex, and he began to simplify his artificial dragonfly. First, he changed to two rigid wings, one behind the other, which moved about hinges parallel to the body axis, again 180 degrees out of phase. His next step in simplifying the dragonfly's flight mechanics was to replace the two wings with the technically much simpler complete rotation about the same fuselage axis. After some experimentation with the size and location of tail surfaces, von Holst came up with a

The SNECMA C-450 "Coleopter" by Zborowski.

A proposed Heinkel VTOL code-named "Lerche-2" with two DB-605-D piston engines driving contra-rotating propellers.

model plane that flew any direction in space - vertical or horizontal - or at any odd flight path angle, with a contra-rotating airscrew acting as propeller and lifting device.

When von Holst's scale model was tested at the Aerodynamische Versuchanstalt Göttingen (AVA), it was found that the model did not compare unfavorably with orthodox airplanes of the same period. However, significant technical problems remained, particularly the mechanics of the large contra-rotating propeller and the placement of the engine. AVA engineers estimated that such an aircraft weighing about 10 tons would require 7,500 horsepower, which, at that time, was beyond the state of the art in Germany in the early 1940s. Thus a third step - finding a suitable power plant - was absolutely necessary. The solution came from a different direction. Focke-Wulf, where von Holst worked, had been interested in developing turbojet engines for its own use. However, the Air Ministry, under pressure from Junkers, Heinkel, and BMW, had forbidden Focke-Wulf from manufacturing their own turbojets, with the exception of ramjets.

When Focke-Wulf's Dr. Otto Pabst was working on ramjets in the early 1940s, practical application for their use were not readily apparent. However, it looked as if Dr. Pabst's ramjets could be put on tips of the three blades, or wings, of von Holst's P.Triebflügel, resulting in a combination that avoided the problems encountered earlier: the complex reduction gear, excessive weight, and an inability to supply power to handle the contra-rotating propeller. The ramjets appeared to be a "natural" for the vertical takeoff-and-landing P.Triebflügel, and by 1944 the need for an aircraft with VTOL capabilities was becoming obvious. With the Luftwaffe's loss of air superiority, operations from conventional bases, which were under attack around the clock, became very expensive and exhausting in terms of military manpower and material. A ramjet-powered Triebflügel, it was believed, could be placed throughout Germany and would not have to rely on airstrips in order to carry out its mission as an interceptor.

The P.Triebflügel interceptor was to stand vertically on the ground, supported by four tail fins, each of which had a small outrigger wheel at its tip. The main landing load was to be taken on a single main wheel at the base of the fuselage. During flight all wheels were to be enclosed by streamlined, tulip-shaped pods. The P.Triebflügel pilot was to be accommodated conventionally in a nose cockpit. Armament which consisted of two 30 mm MK 103 cannons with one-hundred rounds of ammunition each, plus two 30 mm MG 151 cannon with 250 rounds each was to be placed in the nose.

Control of the aircraft was to be accomplished by means of control surfaces at the trailing edges of the tail fins. For flying in a horizontal position the tail would be depressed slightly to direct part of the thrust force into a lift force. In addition, the blades, or wings, could be rotated. During lift-off they would provide the lift needed to get the aircraft off the ground. As it gained altitude, each of the blades would be rotated about its axis and take on more of the function of a fixed wing, while the ramjets at each blade tip would provide the thrust for forward flight. Output of each ramjet was estimated at 3,400 horsepower, more than enough power to lift the 11,385 lb (51,642 kg) interceptor and propel it through the air on the order of Mach .09. Additional versions were planned: a twin-bladed and a four-bladed versions were considered. It was believed that a four-blade arrangement would create excessive drag

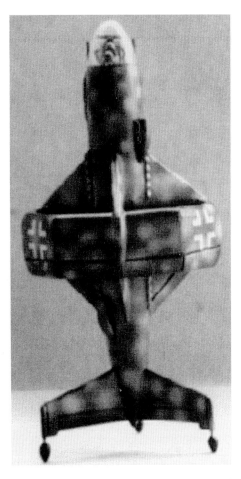

Another view of the proposed Heinkel VTOL "Lerche-2."

A view of the proposed Focke-Wulf FwTa 283 as it may have looked in operational status. Scale model by Günter Sengfelder.

and, consequently, unsatisfactory performance. Nevertheless, no one had the slightest idea if the concept would even fly and if so how well.

Specifications: "Triebflügel"

Engine	2 or 3 ramjets designed by Focke-Wulf, each having 1,852 lb (840 kg) thrust, brought up to ramjet operating speed by 2 or 3 solid fuel rockets having 661 lb (300 kg) thrust.
Wingspan	37.7 ft (11.5 m)
Wing Area	177.6 sq ft (16.5 sq m)
Length	30.2 ft (9.2 m)
Height	NA
Weight, Empty	7,055 lb (3,200 kg)
Weight, Takeoff	10,560 lb (4,790 kg)
Number of Crew	1
Speed, Cruise	419 mph (675 km/h)
Speed, Landing	NA
Speed, Top	621 mph (1,000 km/h) at sea level
Radius of Operation	1,491 mi (2,400 km)
Service Ceiling	45,934 ft (14,000 km)
Armament	2x30 mm MK 103 cannon and 2x30 mm MK 151 cannon
Landing Speed	NA
Rate of Climb	24,608 ft/min (7,500 m/min)
Bomb Load	NA
Flight Duration	2.5 hr

PROJECT 283

One of the several pioneers in ramjet power units was Dr. Otto Pabst. Head of Focke-Wulf's gas dynamics department, Pabst had conducted significant research for the superchargers used on many Focke-Wulf piston-powered aircraft. Although Focke-Wulf took no part in the initial development of the turbojet engine, Kurt Tank had been interested in the ramjet engine for many years and had been financing Pabst's gas dynamics group to investigate the possibilities of using the aero-thermo dynamic duct (or ramjet) for high-speed flying above and below the sound barrier. Some of Pabst's experiments had advanced considerably, and in 1945 he had the Luftfahrt Forschungs Anstalt (LFA) or Aircraft Research Establishment, test his prototype ramjets in its high-speed wind tunnels. (The only other experimental ramjet tested by the LFA in the late 1940s was a prototype built by Hellmuth Walter, manufacturer of the famed HWK 509A bi-fuel liquid rocket drive used in the Me 163 and the Bachem Ba 349A Natter. Wind tunnel tests proved encouraging relative to the possibilities of using the ramjet as a propulsion system for fighters and interceptor aircraft. Pabst had intended to use his ramjet on his P.Triebflügel vertical takeoff-and-landing aircraft, but the Air Ministry's decision to investigate the ramjet as a suitable propulsion unit for subsonic winged aircraft intervened. The Air Ministry's decision surprised many people in the aircraft industry because no aircraft had flown solely on ramjet power. Moreover, the ramjet was not a self-sufficient engine. It had to be boosted up to its self-operating speed of 149 mph (240 km/h) by a solid-fuel rocket drive or a HWK 509A bi-fuel liquid rocket drive.

In November 1944 the Luftwaffe's Chief of Technical Air Equipment issued a design request for a ramjet-powered fighter. Kurt Tank responded with his P.283 design which was to be powered by two of Pabst's ramjet engines. Tank had asked Hans Multhopp to handle the airframe design. Multhopp developed a very pleasing aerodynamic design on which to place Pabst's untried in flight twin propulsion units. To minimize airflow disturbance, Multhopp placed the ramjet units at the tips of the sharply sweptback rudder-elevator tailplane. The wings were

Focke-Wulf Flugzeugbau - Bremen, Germany

A side view of the proposed Focke-Wulf Ta 283. Scale model by Günter Sengfelder.

The proposed Focke-Wulf Fw Ta 283. Scale model by Günter Sengfelder.

to sweep back in characteristic Multhopp style, with a leading edge sweep of 45 degrees. The P.283's vertical rudder was to be massive, beginning at the rear of the cockpit canopy and extending in a long sweep beyond the fuselage. With the cockpit set well back, the P.283 was to have a long slim nose in which its twin 30 mm MK 108 cannon were to be installed. One HWK 509A bi-fuel liquid-fuel rocket drive of 2,646 lb (1,200 kg) thrust would propel the P.283 up to its ramjet's operating speed of 149 mph (240 km/h).

No prototype was constructed.

Specifications: Project 283

Engine	2xFocke-Wulf Pabst ramjets (thrust unknown), and 1xHWK 509A bi-fuel liquid rocket drive having 2,646 lb (1,200 kg) thrust to boost the aircraft up to ramjet operating speeds
Wingspan	26.2 ft (8 m)
Wing Area	NA
Length	38.7 ft (11.8 m)
Height	NA
Weight, Empty	NA
Weight, Takeoff	11,861 ft (5,380 kg)
Number of Crew	1
Speed, Cruise Speed	NA
Speed, Landing	NA
Speed, Top	699 mph (1,115 km/j)
Radius of Operation	429 mi (690 km)
Service Ceiling	32,810 ft (10 km)
Armament	2x30 mm MK 108 cannon
Rate of Climb	2.5 min to 9,840 ft (3,000 m)
Bomb Load	NA
Flight Duration	NA

An HWK bi-fuel rocket engine of the type to be used in the proposed Focke-Wulf Fw Ta 283.

The proposed Focke-Wulf Fw Ta 283. Scale model by Günter Sengfelder.

The proposed Focke-Wulf Fw Ta 283. Scale model by Günter Sengfelder.

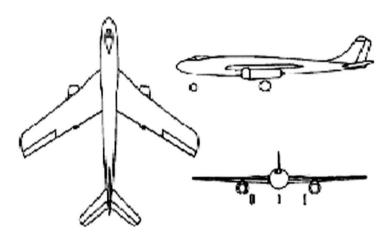

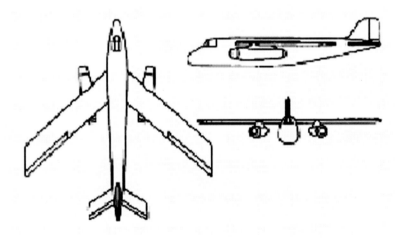

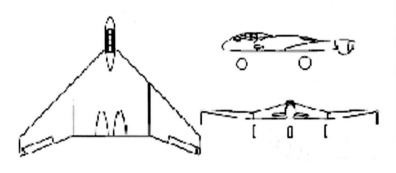

PROJECT 3x1000C

By 1943 nearly everyone in the Air Ministry, including Luftwaffe officials, had come to the conclusion that Heinkel's He 177 long-range heavy bomber, though a design success, was an engineering failure. Although the Air Ministry would not officially give up on large bombers until July 1944, a large number of firms started thinking about turbojet-powered bombers in 1943 when the OKL (Oberkommando der Lutwaffe or Air Force Luftwaffe High Command) publicly announced it wanted a replacement for the He 177 (which by then was known as "Udet's Folly). Even then, many people wondered about the wisdom of pursuing construction of large bombers. By 1943 the Wehrmach was on the retreat, and Germany was experiencing a growing shortage of the very materials, metals and fuels that would be required to construct and operate a large bomber fleet. Nevertheless, a number of companies such as Focke-Wulf gave considerable time to designing new turbojet-powered, long-range heavy bombers to meet the Air Ministry's 3x1000 bomber requirements. The Air Ministry wanted an aircraft that could carry a 2,205 lb (1,000 kg) bomb at 621 mph (1,000 km/h), with a range of 621 mi (1,000 km) thus the 3x1000 designation.

Focke-Wulf proposed three different bomber designs: sub-types A and B were conventional aircraft with sweptback wings and a standard tail assembly. Their twin turbojets were carried beneath the wings. Sub-type C was a delta-wing design with twin turbojets housed inside the wing's center section. The delta-wing sub-type (P.3x1000C) was thought by Fw to be their only version able to meet the 3x1000 requirements for speed, bomb load, and distance and do so with only two turbojets for propulsion. Fw's sub-type C bomber was based on a design concept suggested by Alexander Lippisch, who served Focke-Wulf from time to time as a consultant during the 1940s. Typical Lippisch delta-wing design features, such as the P.3x1000C's pressurized cockpit pod extending forward from the delta apex and the downward-bending wing tip rudders are apparent in this Fw design. Lippisch's practice of fully enclosing the propulsion system inside the wing's center section also was followed. The center section was to be sufficiently deep to accommodate the fuel, the bomb load, and the twin HeS 011 turbojets with 2,866 lb (1,300 kg) thrust which were to be mounted side-by-side aft of the cockpit pod. Air intakes for the turbojets were to be built into the two leading-edge wing roots.

To stretch as far as possible the dwindling supply of pilots, the OKL required that the fast bomber be a single-seater; the pilot would also serve as the navigator and bombardier. The only armor protection was to be some light armor in the front of the cockpit pod to protect the pilot from flak. The OKL felt that at 621 mph, (1,000 km/h), this fast bomber's speed would be its best protection.

The proposed Focke-Wulf Fw 3x1000A. Medium bomber with two HeS 011 2,866 lb thrust turbojet units. Anticipated speed was 621 mph at 32,000 ft.

The proposed all-wing Focke-Wulf Fw 3x1000C. Medium bomber with two HeS 011 2,866 lb thrust turbojet units. Scale model by Hans Peter Dabrowski.

Fw's P.3x1000C did not progress beyond the design stage. With the cancellation of large bombers by the Air Ministry in 1944, Focke-Wulf abandoned the project.

Specifications: Project 3x1000C

Engine	2xHeinkel-Hirth HeS 011 turbojets, each having 2,866 lb (1,300 kg) of thrust
Wingspan	45.9 ft (14 m)
Wing Area	592 sq ft (55 sq m)
Length	19 ft (5.8 m)
Height	NA
Weight, Empty	9,259 lb (4,200 kg)
Weight, Takeoff	17,857 lb (8,100 kg)
Number in Crew	1
Speed, Cruise	404 mph (650 km/h) at 65% throttle
Speed, Landing	NA
Speed, Top	621 mph (1,000 km/h)
Radius of Operation	1,554 mi (2,500 km)
Service Ceiling	49,212 ft (15 km)
Armament	4x30 mm MK 108 fixed-forward cannon
Rate of Climb	NA
Bomb Load	2,205 lb (1,000 kg)
Flight Duration	NA

PROJECT "Kamikaze"

At a top secret meeting of the German Academy of Aeronautical Sciences in January 1944, it was concluded that the Air Ministry should establish a policy calling for Luftwaffe pilots to systematically crash-dive bomb-laden planes "kamikaze" style into vital enemy targets including electric power generating plants, water reservoirs, important industrial sites, and supply ships. The idea of self-sacrifice, although practiced successfully by the Japanese, did not correspond with European mentality; nevertheless, a few people believed the action could be justified if the suicide of Luftwaffe pilots meant saving Germany from total ruin by bringing about an end to the war through an early negotiated settlement. They argued for example, that though a Luftwaffe pilot

A Fuji Hikoki MXY-7 "OHKA" suicide bomb released by its mother airplane a Japanese "Betty" bomber in the vicinity of its target. Gliding to its target and establishing a favorable position, the pilot fires the "OHKA" rocket engine and dives at high speed into its victim, perishing in the resultant explosion.

This example of the "OHKA" was found on Okinawa on 6 April 1945. Range at time of release was about 55 miles from its target at 27,000 ft. Maximum speed obtained in its power dive was 650 mph.

A Fieseler Fi 103 "Reichenburg" hung under the starboard wing of a Heinkel He 111 and then launched dropped. It had to be this way because when an Fi 103 was launched on a ramp the pilot would have experienced an acceleration of about 16 times gravity.

The Fieseler Fi 103 "Reichenburg."

would lose his life by crashing into and destroying a 15,000-ton Allied aviation fuel tanker, the loss of fuel would mean the enemy would be unable to carry out three major bombing attacks against the Reich.

Despite the impressive losses being inflicted on American naval ships by Japanese Kamikaze pilots, there remained widespread resistance in Germany to any form of self-sacrifice. However, some advocates felt that if a very fast aircraft were used, one that could deliver a large bomb to a vital enemy target regardless of fighter attack or anti-aircraft guns, then more people would support the idea. The Japanese Kamikaze pilot did not believe that they were committing suicide, but rather were only doing their job as pilots by inflicting mortal damage on the enemy. Hanna Reitsch wanted more German pilots to take hold of this attitude.

The initial search for fast and accurate kamikaze aircraft started with the Messerschmitt Me 328, a single-seat fighter. Powered by two Argus 1,323 lb (600 kg) thrust pulsejets, it had an estimated top speed of 572 mph (920 km/h) and a range of 478 mi (770 km). Hanna Reitsch, Germany's famed female test pilot with the Deutsches Forschungsinstitut für Segelflug (DFS), found the Me P.328's flight characteristics impressive and adequate for use as a kamikaze aircraft. The Air Ministry placed a production order. However, just as production was gearing up, the Messerschmitt facilities at Thuring where the Me P.328 was to be manufactured, was hit by a substantial Allied bombing raid. All assemblies, facilities, equipment, jigs, and engineering drawings were destroyed, and the Air Ministry had to look for another suicide aircraft.

The second aircraft to be considered was the piloted version of the V-1 and known as the Fieseler Fi 103. In May 1944 workers from the Gerhard Fieseler Werke, along with designers from DFS, started making changes on the V-1 to convert it to a piloted suicide attack airplane. By October 1944 the first of several prototypes were ready for flight testing. The new model was designated Fi 103 and given the code name "Reichenberg." The Reichenberg could not be launched in the same manner as the unmanned V-1 because pilots could not withstand the tremendously high acceleration of about sixteen times gravity provided by its engine; thus, it was to be hung under the wing of a Heinkel He 111 and then "launched-dropped." Once separation from the He 111 had been achieved, the Reichenberg's pilot was to ignite the single Argus 014 with its 772 lb (350 kg) thrust pulsejet.

Although the Reichenberg was expected to be fast 404 mph, (650 km/h), it was feared that the Argus pulsejet, firing away at 47 cycles per second, would create very destructive acoustical problems for the craft's airframe, tail unit, and wing structure, as it had on the original V-1. The

A "Kamikaze" aircraft nose-loaded with high explosives with its ferrying plane attached. Disengagement would come about with the use of explosive bolts designed by Dip.Ing Butter of Heinkel AG.

sound frequency of the pulsejet engine had been found literally to shake the V-1 flying bomb apart. This explains why half of the estimated 9,000 V-1's which carried 1,000 lb (2,200 kg) of Amatol or TNT plus Ammonium Nitrate, may have never made it to London from their launch sites, but instead self-destructed in the air. Since it was to be powered by the same Argus pulsejet engine, not much hope was held out for the manned version of the (Fi 103) V-1, however these being desperate times, the idea was pursued. Flight tests confirmed that the pulsejet engine did indeed create harmful vibrations, causing glue in the wooden wings to shake loose. With no quick solution available, the idea of using a variation of the V-1 as a piloted kamikaze aircraft was abandoned. Some thought was given to replacing the Argus pulsejet with a single Jumo 003 or BMW 004 turbojet; however, this idea was dropped because of the significant amount of design rework required.

The third aircraft considered for the kamikaze program was a single-turbojet aircraft wich would be carried aloft in a very large carrier or "mother" plane. Focke-Wulf designers suggested that a multi-engined turboprop-powered carrier plane with up to 28,660 lb (13,000 kg) of thrust could carry five kamikaze-style turbojet-powered planes to the general location of the enemy targets. It was thought that the mother plane with its kamikaze "flock," would be unattackable because of its very high speeds. A Focke-wulf kamikaze aircraft powered by a single HeS 011 turbojet having 2,866 lb (1,300 kg) of thrust, was expected to achieve 576 mph (927 km/h) at 19,686 ft (6 km). The kamikaze would be constructed out of wood. It was to carry a 2.5-ton explosive charge in its nose and only a few instruments - a gyro compass, an altimeter, an air speed indicator, a clock, and a turn and bank indicator. Flight controls were to be the conventional stick and rudder foot pedal. Personal equipment was to include a parachute, a life preserver, a helmet with headphones and throat microphones, sunglasses, a safety belt, shoulder straps, and a plywood bucket seat. The plan was to have the pilot direct the self-sacrifice airplane down to the target, bail out just before collision and parachute down to safety. However, given the kamikaze aircraft's projected high speed, especially when it was making a high-speed dive, it is doubtful that the pilot could have escaped safely from the aircraft before ramming its target and exploding.

Neither Focke-Wulf carrier aircraft nor the kamikaze-type airplanes progressed further than the drawing board.

A ramp used to propel the unmanned version of the Fieseler Fi 103 up to its ramjet-operating speed. Partially destroyed near war's end.

Secret Aircraft Designs of the Third Reich

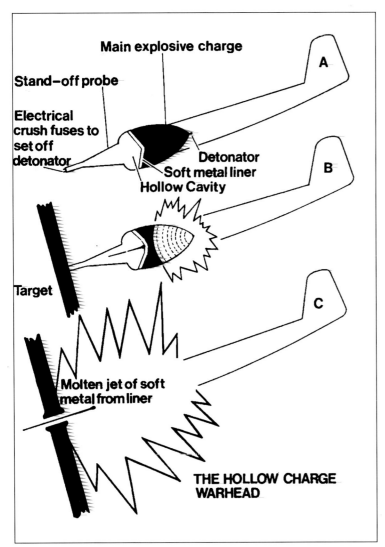

The principle used in the war head of the Focke-Wulf Fw "Kamikaze" and its attack on shipping.

Specifications: Project "Kamikaze"

Engine	1xHeinkel-Hirth HeS 011 turbojet having 2,866 lb (1,300 kg) of thrust
Wingspan	27.9 ft (8.5 m)
Wing Area	161.2 sq ft (15 sq m)
Length	30.2 ft (9.2 m)
Height	10.5 ft (3.2 m)
Weight, Empty	3,968 lb (1,800 kg)
Weight, Takeoff	9,921 lb (4,500 kg)
Crew	1
Speed, Cruise Speed	NA
Speed, Landing	None, self destructing on impact
Speed, Top	576 mph (927 km/h) at 19,686 ft (6 km)
Radius of Operation	1,068 mi (1,720 km)
Service Ceiling	34,450 ft (10.5 km)
Armament	None
Rate of Climb	11 min to 6.0 km and 34 min to 10.5 km
Bomb Load	2.5 ton
Flight Duration	2 hr

PROJECT "Kamikaze Carrier"

The kamikaze, or self-sacrifice, aircraft being considered in Germany in the early 1940s was envisioned as an aircraft that would carry little equipment yet be able to fly up to 2 hours before crash-diving at high speed into its target. Since kamikaze operations were to be based on land, at a considerable distance from any intended targets, a critical ques-

The proposed Focke-Wulf Fw "Kamikaze" mother carrier. Scale model by Reinhard Roeser.

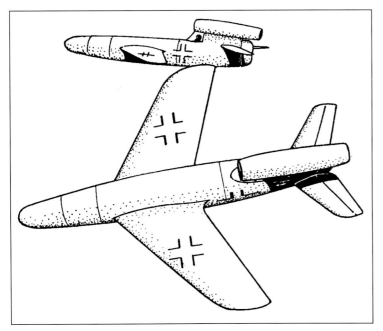

Another version of the "Kamikaze" self-propelled aircraft in flight after being released from its mother carrier. Illustration by Hugh W. Cowin.

The proposed Focke-Wulf Fw "Kamikaze" mother carrier. Scale model by Reinhard Roeser.

Front view of a proposed Focke-Wulf Fw mother carrier with two "Kamikaze" aircraft attached to its outer wings. Scale model by Reinhard Roeser.

tion arose: How could a sparsely equipped aircraft defend itself and find its way to its target? Focke-Wulf proposed that to minimize the amount of fuel, navigation equipment, and defensive armament and cannons required, the self-sacrifice planes and their pilots be carried within striking distance of their targets by some sort of mother ship. What Focke-Wulf engineers had in mind was a mammoth piston-powered aircraft capable of carrying up to five turbojet-powered kamikaze aircraft beneath its wings.

The "carrier" plane idea was preferred over the "pick-a-back" concept developed by Deutsches Forschungsinstitut für Segelflug (DFS) as early as 1940 when they began studying the problem of coupling one aircraft to another in flight. DFS engineers were working on an idea for long-range bombers to carry their own fighter protection; the fighter would detach when the bombers came under attack and re-attach after the action was completed. This work had an unsuccessful outcome in Germany, but the idea was resurrected in 1947 by the US Air Force with the help of Richard Vogt, former head of Blohm and Voss' aircraft division.

It was in 1942 that DFS scientists found that by coupling a Bf 109 with a Ju 88 twin-engine bomber, the resulting combination could be used in an attack role. The plan was for the Ju 88 to be loaded with highly explosive materials, flown to a target, and guided down, to explode on impact. By 1944 the Operations Staff of the Luftwaffe had found the idea worthwhile and had launched a program, code-named Mistel (mistletoe), that led to the modification of at least 100 Ju 88 bombers for use as flying bombs.

It was thought that the carrier airplane proposed by Focke-Wulf could transport a wide variety of aircraft from any land base having a runway as short as 1,640 ft (500 m). It could be also used as a large container aircraft to supply troops in the field.

Overall, the Focke-Wulf carrier plane was to be a mid-wing monoplane with gull-shaped wings, a fixed landing gear, and two long, slender tail booms connected at the rear by a horizontal stabilizer. The two-man crew was to be housed in a glazed-over cockpit at the very front of the fuselage. Aft of the fuselage was to be the equipment used in starting the turbojets of the aircraft's piloted kamikaze aircraft.

The wing assembly of the kamikaze carrier was to have only a moderate backsweep, about 32.8 ft (10 m); otherwise it would have a rectangular contour, with a span of 177 ft (54 m). For power, the kamikaze carrier would have used six Daimler-Benz DB 603N, twelve cylinder, 1,900-horsepower engines, the same engines used in the Focke-Wulf Ta 152, the Messerschmitt Me 410, the Heinkel He 219, and the Dornier Do 335 night fighter. Estimated maximum speed of the 120-ton kamikaze carrier was 348 mph (560 km/h). Landing the giant carrier was not expected to be a problem since the wing load of the empty aircraft would be light and the landing weight would be only about 70 tons. A tricycle landing gear was proposed, with each of the two main landing gear assemblies having six tires and held in a fixed position during flight. The nose wheel was to contained two tires with the gear retracting into the fuselage.

As Focke-Wulf engineers viewed the entire kamikaze program, their carrier airplane would play only one small part. Its sole function would be to carry the "self-sacrifice" aircraft from its operations base flying through Luftwaffe-controlled airspace, where it would be then met by a long-range turbojet-powered reconnaissance airplane such as an Arado Ar 234B. At this time the kamikaze aircraft would start their turbojet engines, release themselves from the carrier plane, and follow the reconnaissance airplane to the intended target. The kamikaze airplane was to have a flight duration of about 2 hours and a range of 1,068 mi (1,720 km). After releasing its load, the kamikaze carrier would return to the base of operations, where it would be reloaded with another flock of kamikaze aircraft.

A proposed Focke-Wulf Fw mother aircraft releasing her brood of "Kamikaze" piloted suicide aircraft. Airbrush by Gert W. Heumann.

Side view of the proposed Focke-Wulf Fw mother carrier with its "Kamikaze" aircraft hanging from its wing tips. Scale model by Reinhard Roeser.

Specifications: Project "Kamikaze Carrier"

Engine	6xDaimler-Benz DB 603N, 12-cylinder piston engines each having 1,900 horsepower.
Wingspan	177 ft (54 m)
Wing Area	5,382 sq ft (500 sq m)
Length	114.8 ft (35 m)
Height	39.4 ft (12 m)
Weight, Empty	113,095 lb (51,300 kg)
Weight, Takeoff	268,961 lb (122,000 kg)
Number of Crew	2
Speed, Cruise	NA
Speed, Landing	NA
Speed, Top	348 mph (560 km/h)
Radius of Operations	1068 mi (1,720 km)
Service Ceiling	NA
Armament	None
Rate of Climb	1,496 ft/min (456 m/min)
Bomb Load	None
Flight Duration	2 hrs

PROJECT "Five"

During the time the Air Ministry was seriously interested in development of a tactical and strategic bomber fleet (1943-1944), officials were also thinking about building a long-range fighter escort. They believed the trouble-plagued, dual-engine (two Daimler-Benz DB 606s coupled to form one engine) bomber, the Heinkel He 177, would eventually be debugged and put into service. It never was debugged, and the Air Ministry dropped the entire bomber concept in July 1944 as unworkable. But Air Ministry planners knew that the relatively slow-moving He 177 would require a special type of fighter escort. The escort would have to

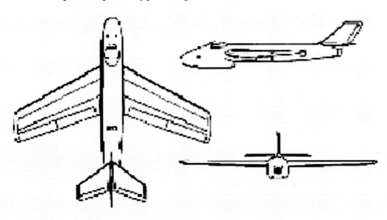

be an aircraft that could watch over the He 177 throughout its 3,417 mi (5,500 km) range and match its slow 258 mph (415 km/h) cruise speed, yet could quickly attack enemy fighters and bring them down regardless of their altitude or speed if it had to. Although no fighter escort competition was ever established, Air Ministry officials, along with designers from Focke-Wulf, did draw up a proposed project for an extended long-range fighter-escort. Focke-Wulf called this consulting work its "Project Five" or shorten to P.5 activities.

The result of this cooperative "brainstorming" was a fairly standard-looking aircraft powered by three HeS 011 turbojets, each having 3,307 lb (1,500 kg) of thrust. Early in the project planning sessions single and twin turbojet designs were eliminated from consideration; an extended long-range escort would have to carry a considerable amount of fuel in order to match the He 177's 3,417 mi (5,500 km) range, and one or even two turbojets would not supply sufficient power to guarantee good performance. It was thought unlikely that a fuel-laden long-range fighter escort could even be lifted off the runway by only a single HeS 011 turbojet. A twin-engine fighter escort could probably takeoff, but it would not have sufficient power to engage fast, high-flying enemy fighters. A three-engine version was finally settled on. With three HeS 011 turbojets it could easily lift off with a heavy load of fuel and ammunition. By cutting one of the three turbojets entirely and throttling back the other two, the aircraft could be slowed to speeds of only 249 to 311 mph (400 to 500 km/h), the typical cruising speed of an He 177 bomber. When enemy fighters were sighted on their radar screens, the third turbojet could be started, giving the fighter escort superior performance characteristics, in terms of altitude and speed, in its role as a fighter.

The Focke-Wulf P.5 was to have what its designers called a "free-carrying middle wing" (a wing attached midway up the fuselage). All three HeS 011 turbojets were to be mounted inside the fuselage - two in the nose and one in the tail. For the tail-mounted unit, a border-layer suction unit was to be attached to guarantee flawless air-inlet performance. The wings were to be swept back at 25 degrees and would be constructed and covered with wood, with two steel spars for strength and tightly glued fuel areas between the spars. The craft was to carry 23,810 lb (10,800 kg) of fuel, 13,228 lb (6,000 kg) inside the fuselage and 10,582 lb (4,800 kg) in wing tanks. The three-man crew (pilot, FM-radio operator, and navigator radio-operator) was to be housed inside a pressurized cabin. The main landing gear was to retract hydraulically into the fuselage, with the nose wheel retracting under the cockpit between two of the turbojets. Attack armament was to be substantial, including four forward-firing 30 mm MK 108 cannons firing 80 degrees up, each with 200 rounds. For protection from attack by 20 mm cannon

Focke-Wulf Flugzeugbau - Bremen, Germany

The proposed Focke-Wulf Fw Project 5 fighter escort for Heinkel He 177 bombers. The Project 5 was to be powered by three HeS 011 turbines in the fuselage: two forward and one aft. *Courtesy of Lufwaffe Secret Projects: Fighters 1939-1945.*

fire of enemy fighters, the crew was to be surrounded by armor steel plate. Other equipment was to include de-icers on the rudder, elevators, and wings, and fire extinguishers on each of the three HeS 011 turbojets.

Specifications: Project "Five"

Engine	3xHeinkel-Hirth HeS 011 turbojets, each having 3,307 lb (1,500 kg) of thrust
Wingspan	63.3 ft (19.3 m)
Wing Area	805.1 sq ft (74.8 sq m)
Length	58.4 ft (17.8 m)
Height	NA
Weight, Empty	NA
Weight, Takeoff	42,613 lb (19,392 kg)
Number of Crew	3
Speed, Cruise	249 to 311 mph (400 to 500 km/h)
Speed, Landing	NA
Speed, Top	NA
Radius of Operation	3,417 mi (5,500 km)
Service Ceiling	NA
Armament	4x30 mm MK 108, forward-firing cannon with 240 rounds of ammunition each, and 2x30 mm MK 108 cannon firing 80 degrees up with 200 rounds each
Rate of Climb	NA
Bomb Load	NA
Flight Duration	NA

PROJECT "Six"

About the time Hans Multhopp was working out the design details on what would eventually become the Project183 (1943), Ludwig Mittelhüber, one of Kurt Tank's close personal colleagues was leading another design team at Focke-Wulf's Bremen headquarters also working on a single-seat, single-turbojet interceptor. Mittelhüber came with Tank to Focke-Wulf when they met in 1930 while working together at Messerschmitt AG. He would accompany Tank, Otto Pabst and others from Focke-Wulf to Argentina in 1946.

Multhopp and Mittelhüber were competing to come up with a "Flitzer" (streaker) aircraft that could intercept the USAAF's B-24 and B-17 bombers that were flying into Germany from England and Italy. Mittelhüber's design called for a central fuselage nacelle and twin tailbooms. Twin HeS 011 turbojets, each having 2,866 lb (1,300 kg) of thrust, would be attached beneath the wing, below the fuselage nacelle. In addition, a single HWK 509A bi-fuel liquid-rocket propulsion unit could be attached to give even quicker short-term thrust than that provided by the powerful HeS 011 turbojets. It was estimated that if all three propulsion units had been used, the P.6 "Flitzer" would have climbed to 38,060 ft (11.6 km) in 1.9 minutes from a standing start. If only the twin HeS 011s had been used for power, estimated range would have been 800 mi (1,287 km) and estimated climb to 36,091 ft (11 km) in only 2.5 minutes. Estimated maximum speed on turbojet-power was a modest 516 mph (830 km/h) at 20,014 ft (6.1 km).

Considerable doubts arose over Mittelhüber's placement of the P.6's air intakes and this probably led to its rejection by the Air Ministry. Although a wooden mockup was completed, no order was placed and the entire project was abandoned when the Air Ministry selected

Close up of the slow-flying Heinkel He 177 bomber which needed round-trip fighter protection to complete its intended bombing plan.

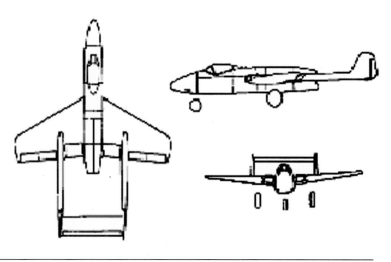

Secret Aircraft Designs of the Third Reich

Paper and wood scale model of the proposed twin-boom Focke-Wulf Fw turbojet fighter interceptor "Flitzer" meaning dasher or whizzer. Its designer Ludwig Mittelhüber is seen on the far left in 1944. This particular "Flitzer" version is known, too, as "Project 7," an improved model (sharply swept wing and tail surfaces) from the previous so-called "Project 5" of 1943.

The De Havilland "Vampire" shown in this September 1951 photograph of the "Vampire's" maiden flight bore a strong resemblance to the Fw P.6 "Flitzer." The "Vampire" was fitted with two Rolls-Royce Avon turbojets and combined the De Havilland twin-boom with sharply swept back wings.

Multhopp's P.183 "Huckebein" (a cartoon raven who always got others into trouble) for immediate construction.

Specifications: Project "Six"

Engine	2xHeinkel-Hirth HeS 011 turbojets each having 2,866 lbs (1,300 kg) thrust plus 1xWalter HWK 509A bi-fuel liquid rocket drive having 2,646 lb (1,200 kg) thrust.
Wingspan	NA
Wing Area	NA
Length	NA
Height	NA
Weight, Empty	NA
Weight, Takeoff	NA
Crew	1
Speed, Cruise	NA
Speed, Landing	NA
Speed, Top	516 mph (830 km/h)
Rate of Climb	38,060 ft (11.6 km) in 1.9 minutes with use of the HWK 509A bi-fuel liquid rocket drive
Radius of Operation	800 mi (1,287 km) on 2xHeS 011 turbojets
Service Ceiling	NA
Armament	NA
Bomb Load	NA
Flight Duration	NA

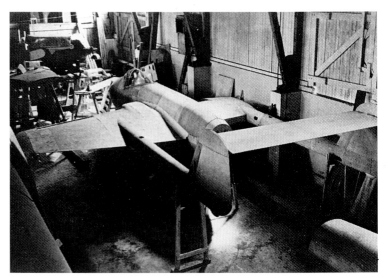

Ludwig Mittelhüber of Focke-Wulf built a wooden mockup of his proposed twin-boom fighter design, the so-called "Project 7" in order to give the RLM a better idea of his thinking. It was to no avail, however; the RLM selected fellow Focke-Wulf designer Hans Multhopp's Ta 183 proposal instead.

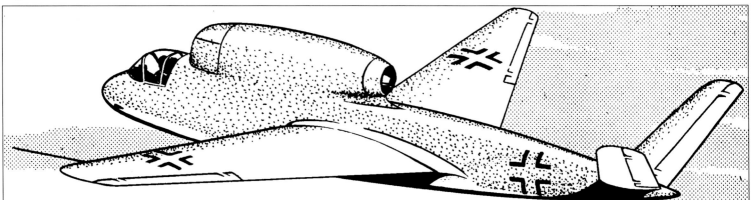

One of Kurt Tank's early concept designs for a proposed turbojet-powered fighter. This design stems from December 1942 and was known within Focke-Wulf as "Project 1." It looked surprisingly similar to the "Volksjäger" fighters of late 1994-45. This concept came with a swept-forward wing and was to be powered by a dorsal-mounted Jumo 004B turbojet. Pen and ink illustration by Hugh W. Cowin.

142

Focke-Wulf Flugzeugbau - Bremen, Germany

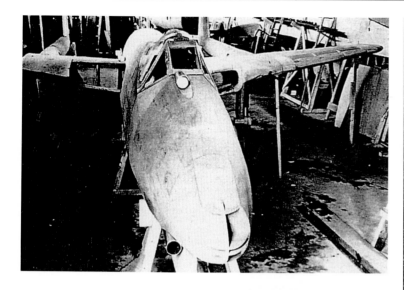

The proposed Blohm & Voss Bv P.209.02. Scale model by Dan Johnson.

Front view of the Focke-Wulf Fw P.7 "Flitzer." Shown is the full scale mock-up out of wood, perhaps to show it to the RLM and convince them to build the real thing. 1944/45.

The proposed Focke-Wulf Fw Project II. 1942. Scale model by Steve Malikoff.

The proposed Focke-Wulf Fw Project II. Scale model by Steve Malikoff.

8

Gothaer Waggonfabrik AG - Gotha, Germany

Secret Projects and Intended Purpose:
PROJECT 60B- twin turbine day fighter with a prime position pilot
PROJECT 60C- twin turbine night-fighter

History:
By April 1945 thirty-some companies were manufacturing aircraft for the Third Reich. Most of them produced aircraft under license for Heinkel, Messerschmitt, and Focke-Wulf, while others produced various sub-assemblies and components. Only three of these thirty companies, Aktien Gesellschaft Otto, Junkers Flugzeug, and Gothaer Waggonfabrik (Gotha) had a record of producing complete flying machines dating back to World War I. Junkers was by far the best known of the three, its aircraft having been flown by every major airline throughout the world since the 1920s. Gotha was famous, too, but for entirely different reasons. During World War I Gotha had manufactured those giant bi-winged twin-engine Gotha "G" bombers that Germany used to attack the Great Britain during the German offensive against England in the summer of 1917. Although the physical damage as measured by destroying property and lost lives was slight, the effect of these aerial bombing raids on British morale was devastating. People had never before witnessed an aircraft used to drop bombs on civilians and urban populations, and the word "Gotha" came to mean all large enemy bombers, regardless of manufacturer. The British came to regard Gothaer Waggonfabrik as a sinister company that had no right to exist post-war and needed to be eliminated. Gotha was the only German aircraft manufacturer specified in the Treaty of Versailles to suffer complete destruction of all its manufacturing facilities post-war. The dismantling of Gotha's factories was completed in 1919, and 15 years were to elapse before aircraft manufacturing was resumed at Gothaer. In the interven-

These Gotha "G" bombers were known as the "England Fliers."

ing years, the firm occupied itself by manufacturing large locomotives, diesel engines, railway cars, freight trailers, fire engines, farm tractors - even furniture.

In 1933 Gotha re-entered the aircraft industry, manufacturing a number of aircraft under license for Arado and Heinkel as well as a variety of training, sports, and touring aircraft of its own design. How-

The Gotha "G" bomber of World War I. Notice the man near the nose wheels of the fuselage, this aircraft was huge!

144

In World War I the bi-plane ruled. However Igor Etich, of Gotha, came out with a revolutionary model plane called the "Taube" or "Dove" because of the distinctive shape of its wing.

This Gotha bomber was brought down by the French near Soissons, France. The large number of French soldiers standing in front of the aircraft gives some indication of its huge size.

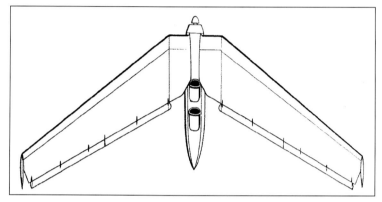

A top-view drawing of August Kupper's Go 147B air observation prototype aircraft.

This Kupper-designed Go 147B, an all-wing short-range air observation aircraft prototype, was abandoned by Gotha after August Kuppers's death in 1937.

ever, throughout the 1930s and into the 1940s the Air Ministry looked to Gotha to build Messerschmitt aircraft for the Luftwaffe. Thus, although Gotha did produce some designs for assault gliders in the 1940s, its primary activity was producing the Messerschmitt Bf 110. It built some 3,000 of these widely-used twin-engine fighter bombers.

When Gotha re-entered aircraft manufacturing, one of the more experienced and talented designers hired in was Dr.Ing. August Kupper, a pioneer in all-wing airplanes during the 1920s and early 1930s. It was Kupper, along with Alexander Lippisch, G.T.R. Hill of England, Alexander Soldenhoff of Switzerland, and others, who developed,

A pioneering builder of all-wing gliders, August Kupper designed a series of prototype all-wing aircraft for Gotha in the mid 1930s. He died in 1937 aged 32, in a crash of one of his own glider designs.

through constant trial and error, a number of high-performance all-wing sailplanes and powered tailless airplanes.

In 1929 Kupper built and successfully flew his famous Uhu (owl) all-wing glider. A giant aircraft for its time, the Uhu had a wingspan of 55.8 ft (17 m), the largest span of any sailplane built up to that time. Only the Horten brothers produced a sailplane with a greater wingspan - their popular Ho 3 series which had a wingspan of 65.5 ft (20 m) and first flown in 1938.

Kupper's all-wing designs were characterized by their normal rectangular wing and wing-tip rudders. Unlike Lippisch, who was beginning to favor sweptback wings that had a triangular shape, Kupper was able to obtain satisfactory results with rectangular wings. Much of Kupper's design theory had been tested at the Deutsche Versuchsanstalt für Luftfahrt (DVL), where early in the 1930s he had been manager of a group involved in studying flight mechanics. Upon joining Gotha in 1933, he continued experimentation on a series of all-wing sailplanes and powered aircraft. In 1934 he designed Gotha's first tailless fighter prototype, the Go 147, which had a wing of semi-gull form and swept back at an angle of 38 degrees on the leading edges. Similar to Kupper's sailplanes in the 1920s, the Go 147 had wing-tip rudders for directional control. A Siemens Sh 14A radial, air-cooled engine of 240 horsepower was mounted in the front of the fuselage. The fuselage was relatively short 21 ft (6.4 m), in length and provided accommodation for a pilot and a gunner who faced rearward. Kupper believed that since the Go 147 had no tail mounted on the fuselage, a rearward-facing gunner, with his 7.9 mm MG 15 machine gun on a flexible mount, would have an excellent field of fire.

The flight characteristics of the Go 147 were disappointing and the Air Ministry rejected it in favor of a more conventional design from another firm. After August Kupper's death in 1937 in the crash of one of his all-wing sailplanes, Gotha discontinued development work on the Go 147 and related designs. Instead, the firm's design activity remained

After the fall of France, Gotha engineers converted the Go 242A-1 transport glider into a twin-engine version by using captured French Gnome/Rhone 14 m fourteen-cylinder radial engines. Gotha called this powered version the Go 244B-1.

Dr.Ing.Rudolf Göthert, chief designer at Gotha and a recognized authority on aeronautical stability and control, wanted the RLM to scrap the Horten Ho 229 in favor of his proposed all-wing Gotha Go P. 60 turbojet-powered fighter design.

confined largely to light aircraft, and its rapidly increasing assembly capacity was devoted almost exclusively to the Messerschmitt Bf 110. But company officials continued to harbor an interest in all-wing aircraft, which they felt was the design of the future. In 1942 they persuaded Dr.Ing. Rudolf Göthert of DVL in Berlin-Adlershof to join Gotha as chief aerodynamicist and head of its new design group for all-wing and tailless aircraft. This came after a photograph appeared of American Jack Northrop's experimental all-wing N-1M design in the October 1941 issue of "Inter Avia," a highly respected international aviation magazine. Gotha officials believed that the Americans had evolved some secret principles to make the all-wing directionally stable in flight and thus would soon be on their way to putting advanced military aircraft into the war effort. Gotha officials were eager to become leaders in the all-wing field of Germany.

Rudolf Göthert, born 30 July 1912, in Hannover, was raised by his brothers and sisters after their parents' early death. After completing high school in 1931, he studied mathematics and natural sciences at the Technical University of Hannover, from which he graduated in 1940 with a Ph.D. in aerodynamics. As part of his work-study program between 1935 and 1936, he was employed by the DVL at its wind tunnel facilities at the University of Braunschweig.

After the founding of the Deutsche Forschungsanstal für Luftfahrt (DFL), or German Research Institute for Aviation in Braunschweig-Volkenrode, Göthert became in 1937 manager of DFL's newly built low-speed wind tunnels. There he took the measurements on wing-flaps that had been made at the University's wind tunnel and studied them under different conditions. The results of these thorough tests provided the basis for his Ph.D. dissertation, which was titled "Systematical Testing of Wings With Flaps and Auxiliary Flaps." After 1940 he wrote numerous technical articles about aircraft controlling surfaces and their different wind tunnel measurements. In addition, he started collecting, for the first time in Germany, information on aircraft controlling surfaces (rudders, elevators, landing flaps, and so on) from German aviation companies and then placed it all in a catalog so that everyone could refer to the data. By 1942, Göthert was considered Germany's ranking expert on stability and control and held the position (at DVL) of Fachsbearleiter (Specialist) for elevator and rudder control surface design.

When Göthert joined Gotha in 1943, his task was to design a turbine-powered all-wing fighter airplane. But the Horten Ho 9 V2 all-wing turbojet-powered prototype fighter was nearing completion in mid 1944, both the Hortens and the Air Ministry began investigating alternatives for mass production of the bat-shaped aircraft. Lacking any se-

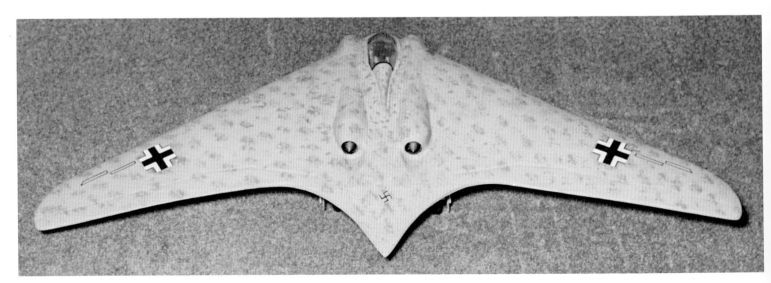

A Gotha manufactured Horten Ho 229. Scale model by Reinhard Roeser.

The Horten Ho 229 V3 discovered by US Army troops at Gotha facilities in Freidrichroda. The RLM had contracted with Gotha to build in series the Horten Ho 229 turbojet-powered all-wing fighter. Only a handful were built before the war ended. Shown here (a, b, c,) is the first Horten Ho 229 Gotha built. Known as the Horten Ho 229 V3, it was captured by US Army troops and brought to the United States after the war with the thought that it might be completed and flown. It never was and now is in storage at the Smithsonian Institution's National Air and Space Museum-Washington, DC.

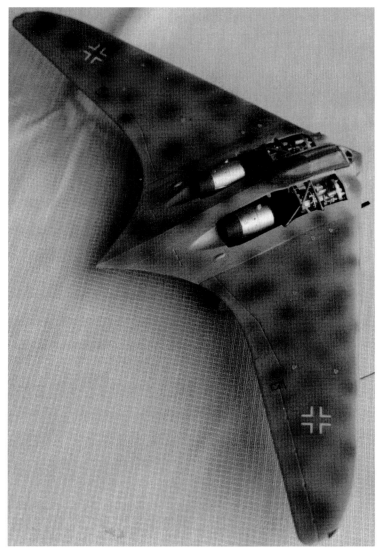

A Horten Ho 9 V2 with engine cowling removed from both Jumo 004B's. Scale model by Reinhard Roeser.

ries production facilities/abilities themselves, the Air Ministry ordered Gotha in 1944 to build initially twenty Horten Ho 9 V2s which now had the RLM designation of Ho 229 (not the Go 229 as it is sometimes mistakenly written). Gotha and Göthert admired the Horten brothers dedication and respected their record, with more than 10 years of experimentation with the all-wing planform. However the people at Gotha believed that they could design a better all-wing fighter than the Ho 229. Better in terms of superior directional control. Gotha had become very reluctant to continue mass-production of the Ho 229 beyond the twenty the Air Ministry had ordered. In mid 1944 Göthert and his staff

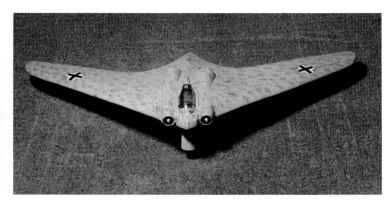

An overhead view of a Gotha manufactured Horten Ho 229 V3. Scale model by Reinhard Roeser.

of engineers began their own all-wing fighter design project known as the Project 60. Their goal was to design a "fresh" all-wing aircraft to replace the Ho 229 which Göthert had come to really dislike. This dislike was based on his wind tunnel tests at DVL during which a scale model of the Ho 229 had been tested. Later, however, Göthert learned to his embarrassment that his wind tunnel test activities on the scale model Ho 229 were inaccurate and wrong. He went on to pattern his own new P.60 after the Ho 229.

Dr. Göthert designed three models of the P.60. The P.60A, to be powered by twin BMW 003 turbine engines, was to have a crew of two, a pilot and observer, both in prone positions inside the aircraft. The P.60B, an improved and slightly larger version of the P.60A, was to be powered by twin HeS 011 turbines, and again was to have a crew of two in a prone position. The P.60C was a proposed night fighter version of the P.60B, also to be powered by twin HeS 011 turbines. It was found necessary to enlarge the center fuselage-wing section to house the radar scanner (Spiegel), and this allowed the crew of three to sit upright in a normal flying position. The P.60A and B were all-wing aircraft, but the extended nose for the radar unit on the P.60C required Göthert to place two vertical tail surfaces on the aircraft. The engines for all three models were to be mounted externally at the rear of the fuselage-wing center section, one above and one below in the plane of symmetry.

The P.60C was Gotha's only model to be entered in a design competition. In December 1944 the Air Ministry called for a night fighter. Five firms (Arado, Blohm and Voss, Dornier, Focke-Wulf, and Gotha) responded with seven proposed aircraft. During the Air Ministry's review in March 1945, Gotha's P.60C was judged to have the best estimated performance, and also to be the least costly to build and operate. It is believed that Gotha would have gone on to win the design competition had not the war ended the following month. However, the Air Ministry apparently asked Blohm and Voss to construct a prototype of its entry, the P.215. The details of this decision are not known.

After the war Göthert was interned by the Americans at St. Germain near Paris for 4 months. Returning to Germany, he received an assignment from the Royal Air Station in Braunschweig-Volkenrode to write several monographs about German research concerning the aerodynamics of controlling surfaces. In 1946 he went to France to become department head for aerodynamics and flight mechanics at the Society

A view of the Horten Ho 229 V3's center section looking forward. The center section had plywood covering as did its outer wings.

Nationale des Construction Aeronautiques du Sud-Est (SNCASE), Usiness de Marignane. During 1946 he worked on the all-wing large French airline, the SE 200.

Dr. Göthert returned to Germany in 1949 and found a position as a department head at Rollei (the camera company), which was a subsidiary of Franke and Heidecke in charge of development of new camera types and production methods. In 1955, with the aircraft manufacturing ban imposed at the end of World War II lifted, Göthert worked as a part-time consultant for Heinkel AG and other aviation manufacturers. When Franke and Heidecke changed hands in 1964 and new management came in, Göthert returned to aviation full time, becoming manager of the department for low-speed aerodynamics and wind tunnel activities for the research institute he had worked for 25 years earlier - Deutsche Versuchsanstalt für Luftfahrt in Braunschweig-Waggum. His main activity was research on a program connected with the high-lifting power of wing and auxiliary air inlets for turbine engines. He also devoted himself to the improvement and enlargement of testing facilities of the Braunschweig low-speed wind tunnels. He died on 16 October 1973, at the age of 61.

Although Gothaer Waggonfabrik survived World War I despite the dismantling of its facilities in 1919, the firm was not so fortunate after World War II. Its major manufacturing facilities were located in the city of Gotha in the region that became East Germany, and the Soviet Union forbade manufacture of all types of aircraft. With the ban, a famed producer of multi-engine heavy bombers during World War I and some proposed exotic all-wing fighter designs late in World War II ceased aircraft manufacture.

PROJECT 60B

In the summer of 1944, Gothaer Waggonfabrik received a purchase order from the Air Ministry to produce at least ten possibly twenty all-wing Horten Ho 9's with the RLM designation Ho 229. It was no secret that Gotha and its chief designer, Dr. Göthert, were unhappy with the Ho 229 as an aerial fighting machine. Its predicted lack of stability as an effective gun platform was borne out in flight tests, and Göthert was hoping to replace some of the twenty Ho 229s with Gotha's own Go P.60B. Instead, Gotha was told by the Air Ministry in early 1945 to construct twenty more for a total of forty Ho 229s.

Designed for high speeds, the P.60B was to be a two-man day fighter and interceptor with an unusual pilot cockpit arrangement. To minimize aerodynamic drag, Göthert designed the aircraft to carry its two-man crew in a pressurized cabin in a prone position. Gotha believed that the pilot would experience no discomfort in flights up to 3 hours with a second pilot then taking the controls. Prone piloting had already been tested by the DVL and the DFS. DVL had built an aircraft called the Berlin B-9 for the purpose of conducting research on prone-pilot cockpits with regard to vision, instrument layout, and ability of the pilot to remain in the cockpit for extended periods. DFS had built a prone piloted aircraft called the DFS 228 and powered by a single HWK 509A bi-fuel rocket drive. Numerous gliding tests were carried out with the prototypes but never under power with the HWK 509A rocket drive. DFS' supersonic research aircraft, the DFS 346, was also designed for prone piloting.

The unorthodox nature of the prone pilot position concerned many Air Ministry officials, they thought the P.60B would be vulnerable from rear attack due to the pilot's poor rearward visibility. Göthert told the RLM that he had taken the visibility factor into account when he placed the pilot(s) in the prone flying position, and he felt that the benefit of less aerodynamic drag outweighed the disadvantage of poor rearward vision. Although Göthert conceded that rear vision would be a constant problem for the pilots, he thought it unlikely that the P.60B, with speeds up to 590 mph (950 km/h), would never be surprised by an enemy aircraft sneaking up from behind.

Overall, the estimated straight line performance of the proposed P.60B did not differ much from the aircraft it was designed to replace. Wind tunnel tests showed that the P.60B to have considerably less drag than the Ho 229, but that its HeS 011 turbines had slightly less thrust than the Jumo 004E's scheduled to power the Ho 229, making the two aircraft about equal in performance. However, Göthert claimed that wind tunnel testing indicated that the P.60B would be far more stable at high speed and have better control characteristics than the Ho 229. He felt, too, that an all-wing required some type of a vertical stabilizer. Göthert placed a pair of movable wing-end vertical plates or rudders, near the tip of each wing to help retard the tendency of most all-wing aircraft to suffer what pilots call the "Dutch Roll." This is a kind of oscillating or turning from side to side while the all-wing is in straight, level flight, Göthert felt that "Dutch Roll" problem experienced by an all-wing would not be completely solved until engineers were able to develop an automatic gyrostabilizer.

The Go P.60B did not progress beyond the design stage. Plans called for the aircraft to be built in three sections like the Ho 229: two outer wings and one center section fuselage. The tubular steel center section would have contained the two-man crew. Undercarriage, armament, and twin turbines. The outer wings would have been constructed of wood and the entire aircraft covered with a plywood skin. Although Göthert

Side view of the Horten Ho 229 V3 taken at Wright-Patterson Air Force Base, Ohio. 1945

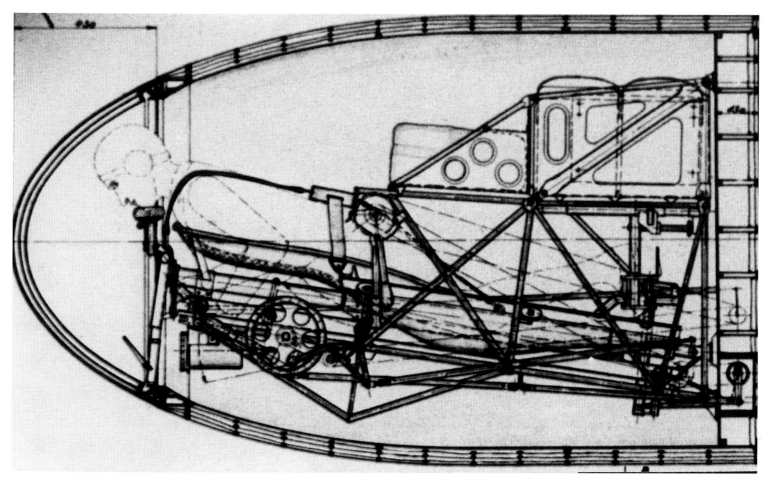

The Gotha Go P.60A and P.60B would have had unique cockpit arrangements. Designed for high speeds, the Go P.60 series (except the "C" subtype) was to have carried its crew in a pressurized cabin in a prone position. Tests by the DVL, the Luftwaffe's research laboratories, had shown that pilots wold not experience discomfort in flights of up to 3 hours. For a longer period of time, it was considered desirable for two pilots be aboard. It was thought that the lack of rearward vision would be compensated for by the Go.P 60's high speed. The cockpit in both the Go P.60A and the P.60B had conventional controls except that the stick and rudder pedals were of the hanging type. Both side and top view of the Gotha P.60A and P.60B are shown.

believed that his P.60B would have replaced the Ho 229 as a more stable gun platform, this is very doubtful. By 1945 the Air Ministry was interested mostly in single-engine fighter designs due to fuel shortages and was also starting to favor the delta-wing planforms pioneered by Alexander Lippisch. With the scarcity of pilots, fuel, and strategic material, the twin-engine, dual piloted Go P.60B fighter was a luxury Germany could no longer afford.

Specifications: Project 60B

Engine	2xHeinkel-Hirth HeS 011 turbojets, each having 2,866 lb (1,300 kg) of thrust, plus 4xHWK solid-fuel, 4,409 lb (2,000 kg) thrust rocket boosters for takeoff.
Wingspan	44.3 ft (13.5 m)
Wing Area	587 sq ft (54.5 sq m)
Length	36 ft (11 m)
Height	NA
Weight, Empty	NA
Weight, Takeoff	16,535 lb (7,500 kg)
Crew	2
Speed, Cruise	384 mph (618 km/h) at 65% throttle
Speed, Landing	93 mph (150 km/h)
Speed, Top	590 mph (950 km/h)
Rate of Climb	3,240 ft/min (990 m/min)
Radius of Operation	1,044 miles (1,615 km) at 100% throttle
Service Ceiling	42,653 ft (13 km)
Armament	4x30 M 108 cannon with 170 rounds of ammunition each
Bomb Load	1x1,102 lb (500 kg) bomb
Flight Duration	1.7 hours

The proposed Blohm & Voss Bv P.215. Scale model by Reinhard Roeser.

Dr.Ing. Rudolf Göthert also proposed a night-fighter version of his Go P.60 series. Pictured here is a two-man Go P.60C modified to carry a radar dish in its nose. Unlike the GoP.60A, both crew men would sit upright in the cockpit. Scale model by Reinhard Roeser.

Rear side view of the proposed Gotha Go P.60C. Scale model by Reinhard Roeser.

PROJECT 60C

On 27 February 1945, the Chef der Technischen Luftrüstung in the Air Ministry (Chief of Technical Air Equipment) issued a request for proposals for a twin turbine-powered night and bad weather fighter. Five manufacturers submitted a total of seven proposed aircraft: Arado with their NJ-1 and NJ-2, Blohm and Voss with their P.215, Dornier with their P.256, Focke-Wulf with their P.1 and P.2, and Gotha with their P.60C. Although no firm officially was awarded a contract before the war ended in May 1945, Richard Vogt of Blohm and Voss claimed his P.215 entry won the competition. Perhaps, but all available evidence suggests that Gotha's P.60C may have been the winner of the night and bad weather fighter competition. In the spring of 1945, the Air Ministry review group found that the P.60C would have the fastest estimated maximum speed of 606 mph (975 km/h). This speed exceeded their original requirement that the proposed night fighter reach at least 559 mph (900 km/h). The Air Ministry found that the proposed rate of climb and endurance was about the same for all entries except for the P.60C. Gotha estimated that its P.60C would climb at a rate of 3,480 ft/min (1,062 m/min) compared to their competition of only 2,460 ft/min (750 m/min) and thanks to the thrust provided by a BMW 003R the combined turbine/HWK bi-fuel liquid rocket drive.

On the negative side, the Air Ministry's Chef der Technischen Luftrüstun disliked the P.60C's external, rear mounted, twin turbines located one above and one below the center section. They felt that the turbine efficiency would suffer due to the long distance between the wing's leading edge and the rear turbine's air intake. Dr. Göthert responded saying that drag would in fact be lower and critical Mach numbers higher with his external rather than buried engines like the Ho 229, which they were building under licence, with its wing leading-edge air ducts. The reason said Göthert was because long ducts tended to disturb the air flow over that entire section of the wing. Göthert believed that external engines permitted quick access and easy maintenance. It appears that the Air Ministry was swayed by Göthert's arguments and despite the criticism, their P.60C received the highest overall rating in the competition. Göthert claimed that a prototype would have been ordered by the Air Ministry had time not run out.

Project 60C was really an enlarged night fighter version of Gotha's P.60B. The center fuselage-wing section of the P.60B was to be enlarged to house a 3.3 ft (1 m) Spiegel or radar scanner, and this would have allowed the crew of three to sit upright in a conventional flying position. The extended nose on the P.60C, plus the longer cockpit and canopy for the crew, Göthert believed, would decrease its directional stability.

Nose side view of the proposed Gotha Go P.60C. Scale model by Reinhard Roeser.

A rear view of the proposed Gotha Go P.60C. One turbojet was mounted on the aft portion of the center section, and the second mounted directly beneath on the underside. Scale model by Reinhard Roeser.

Gotha P-60C.

To counteract, this two vertical plates were to be placed outboard on the wing between the two controlling surfaces on the wing's trailing edge. As with the P.60A and P.60B subtypes, the turbines on the P.60C were to be mounted externally at the extreme rear of the center section, one above and one below in the plane of symmetry. The P.60C was to be powered by twin HeS 011 turbines, each having 2,866 lb (1,300 kg) of thrust, the same as the P.60B. In addition, an HWK solid-fuel 4,409 lb (2,000 kg) thrust rocket booster units could also be fixed to aid takeoff and increase climbing speed.

Gotha's P.60C did progress beyond the design stage, although no prototypes were constructed. It appears that the Air Ministry was about to announce that Gotha had won the design competition when the war ended.

Specifications: Project 60C

Engine	2xHeinkel-Hirth HeS 011 turbines, each having 2,866 lb (1,300 kg) of thrust, plus 1xWalter HWK solid-fuel 4,409 lb (2,000 kg) thrust rocket booster.
Wingspan	44.3 ft (13.5 m)
Wing Area	587 sq ft (54.5 sq m)
Length	NA
Height	NA
Weight, Empty	18,519 lb (8,400 kg)
Weight, Takeoff	25,132 lb (11,400 kg) with 3,950 lb (1,793 kg) of C-Stoff and T-Stoff rocket fuel.
Crew	3
Speed, Cruise	394 mph (634 km/h) at 65% throttle
Speed, Landing	106 mph (170 km/h)
Speed, Top	606 mph (974 km/h) at 16,500 ft (5.0 km)
Rate of Climb	3,480 ft/min (1,062 m/min)
Radius of Operation	2,668 miles (1,658 km)
Service Ceiling	43,963 ft (13.4 km)
Armament	4x30 mm MK 108 cannon fixed forward and 3x30 mm MK 108 cannon fixed oblique that is firing upward firing
Bomb Load	1x1,102 lb (1x500 kg) bomb
Flight Duration	1.7 hours at 32,810 ft (10 km)

The proposed Gotha Go P.60C shown here is the two-man day-fighter version. Unlike the night fighter, the day fighter would not to use a radar set, so it would have been equipped with a projectile-style nose.

Gotha P-60C.

9

Ernst Heinkel AG - Rostock, Germany

Secret Projects and Intended Use:
Project 162 - Operational single turbojet-powered fighter introduced near war's end. This "People's Fighter" was Heinkel's only operational turbojet-powered aircraft of World War II.
Project 176 - The world's first airplane to fly powered only by a bi-fuel liquid rocket drive.
Project 178 - The world's first airplane to fly powered only by a turbo-jet-engine.
Project 280 - World's operational turbojet-powered fighter prototype.
Project 343 - A four-turbojet-powered bomber. A prototype had been built near war's end but was destroyed in an Allied bombing raid.
Project 1077 -Known as the "Julia," this proposal was a HWK 509 bi-fuel rocket drive- powered target defense interceptor.
Project 1078A - A proposed single HeS 011 powered swept-back wings standard tailed fighter.
Project 1078B - A proposed single HeS 011 powered twin fuselage single-seat fighter.
Project 1078C - A proposed single HeS 011 powered tailless fighter.
Project 1079 - A twin HeS 011 powered night and bad-weather fighter.
Project 1080 - A proposed twin ramjet-powered fighter.

History:
As he lay on the grass, seriously injured form the crash of his home built bi-wing Farman in 1911, the 23-year-old student of engineering at Stuttgart Technical College kept thinking about what famed German engineer and aeronautical pioneer Otto Lilienthal had said before he died under similar circumstances. Lilienthal had been killed while testing his latest single-surface glider near Rhinow on 10 August 1896.

Otto Lilienthal shortly before his death. He believed that in the name of science some individual sacrifices should be made.

Ernst Heinkel, front and at the controls of his home-built bi-wing Farman in 1911.

Heinkel ready to take off in his home built bi-wing Farman. When he crashed this aircraft he was badly injured and nearly lost his life. Stuttgart. 1911.

The famed Albatross, B.1 of World War I. This design came about through the combined talents of Ernst Heinkel and Hellmuth Hirth, 1913.

In addition to the Günter twins, one of Heinkel's fortunate discoveries was Dr.Ing. Joachim Pabst von Ohain, fresh out of college with a background in engineering and a novel idea about building a continuous combustion engine, the turbojet. This picture was taken in 1936. Ohain is 27 years old.

Ernst Heinkel is toasting Joachim Von Ohain's success in achieving the world's first turbo-jet powered flight in the summer of 1939.

After being pulled from the wreckage, he had uttered what was to become his epitaph: "Opfer mussen gebracht" (sacrifices must be made). But the young Ernst Heinkel was not ready just then to make any such sacrifice. Somehow he survived, but had to rely on a walking cane throughout his life.

Ernst Heinkel was born on 24 January 1888, at Grunbach, Germany, and he was one of the great personalities of German aviation. Unlike Messerschmitt, who detractors claimed dreamed only of war and never of peace, Heinkel abhorred war and resisted Nazism. He was a man obsessed with engineering problems yet keenly interested in fast commercial planes, fast cars, beautiful women, and choice wines. He liked to call himself "a regular obstinate fellow;" and until late in his life he was a thorn in the side of production engineers because, like many creative people, he could never finalize designs.

Heinkel first became known in aviation circles after he teamed up with Hellmuth Hirth in 1913. Combining Hirth's flying skills (he was considered Germany's master pilot at the time) and Heinkel's engineering interests, they produced advanced monoplanes through the Albatros Flugzeug Werke. Their most famous aircraft, one of the mainstays of German equipment until after World War I, was the Albatros B.1. It was estimated that nearly three-fourths of all aircraft used in Austro-Hungary from 1914 to 1918 were designed by Ernst Heinkel.

As with other aeronautical engineers, hard times came for Heinkel following the collapse of Germany in 1918 and the dictated peace of Versailles. Yet in 1922 he was able to establish his own firm, Ernst Heinkel AG at Warnemünde, through a series of orders for naval air-

Dr.Ing. Joachim Von Ohain standing next to a cut away model of his first operational turbojet engine, the HeS 1. Standing along side Ohain is Erich Warsitz, the Heinkel test pilot who first flew the Heinkel He 178 which was powered by Ohain's turbojet engine.

A rare post-war meeting of Sir Frank Whittle, developer of England's first turbojet engine, Dr. Joachim Pabst von Ohain, and Paul Garber, the legendary curator of the Smithsonian Institute's National Air and Space Museum.

Charles Lindbergh at the head of the table upper left, Ernst Heinkel, to Lindbergh's left and other members of his organization entertaining Lindbergh and his American friends.

craft and catapults from the Japanese government. Additional orders came from the Reichwehr (German Defense Ministry) and from the Soviet Union. However, the German home market offered no long-term demand for aircraft, and Heinkel, in order to survive, developed a series of pioneering new designs, particularly hydroplanes, and sold them worldwide.

The great hit of the Ernst Heinkel AG, which later moved to Marienehe near Rostock, was a high-speed commercial plane, the He 70. In 1933, in a military version called the Heinkel "Blitz" (Lightning), it secured a number of speed records for Ernst Heinkel AG. Designed by the Günter twins, Siegfried and Walter, the He 70 owed its famed performance not so much to increased power as to improved aerodynamic qualities.

With the build-up of the German Air Corps from 1935 on, Heinkel went on to build what would become his world-famous He 111 bomber. Considered Germany's greatest striking weapon in the early German offensive, the He 111 bore the brunt of Germany's offensive bombing and, indeed, many other front-line and secondary tasks. It was the He 111, with its all-glazed, smooth contour, forward fuselage in place of the conventional raised cockpit window arrangement, that took part in the bombing raids on England in 1940. And it was because of the He 111, that Heinkel was awarded the German National Prize for Art and Science in 1938. But Heinkel's fortunes changed considerably in the early 1940s due to his constant criticism of Hitler and the Air Ministry for their combined failure to appreciate his turbojet work and his turbojet-powered aircraft designs. The consequence of this open criticism, he found in due course, was the restriction of his aeronautical activities to the development of his He 111 bombers and of the hopeless proposition of debugging his He 177 four-engine dive bomber ("Udet's Folly," it came to be called). It was not until late in 1944 that Heinkel was allowed to assert himself again within his own factories. His first undertaking was participation in the design development of the He 162 Volksjäger turbojet-powered home-defense interceptor. It was a major personal tragedy for him that, as a pioneer of rocket and jet propulsion, he had never been permitted to carry his ideas fully into aeronautical applications. For this he blamed the Air Ministry.

"The tragedy of the Luftwaffe," Ernst Heinkel said after the war, "had many interlocking causes, all stemming from Hitler's overestimation of the Luftwaffe's scare value in keeping the British neutral." equally disastrous, he believed, was the blind faith that dive bombers obviated the need for long-range heavy bombers. A major blunder, Heinkel bitterly recalled, was the freezing of most fighter development after the

Heinkel He 111 carrying a Fieseler Fi 103 "Reichenburg" piloted flying bomb for a kamikaze or suicide mission.

A close up of the Heinkel He 177's source of problems, two Daimler Benz engines coupled together and driving a huge four-bladed propeller resulting in an enormous amount of technical difficulties in the field.

Heinkel He 111 in England forced down during the Battle of Britain.

awesome Bf 109, on the assumption that the war would be over soon. But Heinkel's greatest personal blow, and ultimately a blow for the entire nation, was the official amusement that greeted his creation, the He 178, in 1939. It was the world's first jet-powered aircraft to fly, 22 months ahead of the British and a full 3 years ahead of the United States. By the time the "mammoth bureaucracy of mediocrity," Heinkel's term for the German Air Ministry, had stopped its laughing and demanded jet-propelled fighters, it was too late for anyone, including Heinkel, Messerschmitt, Junkers, Focke-Wulf, Hortens, and others to make any difference in the fortunes of war. Only one of Heinkel's turbojet-powered fighters went from prototype to production status in a matter of weeks. Ragged, bug-ridden, flown by mere boys, many of the He 162's were destroyed on the ground by American bombers. Nonetheless, in terms of production competence and imaginative designing, Ernst Heinkel AG was perhaps Germany's best.

Owner of one of Germany's largest independent aircraft manufacturers, with more than 50,000 employees during World War II, Ernst Heinkel was shamelessly interested in speed records, neck-jerking performances, and profits. His firm generally achieved all three. His was one of the first companies to invest in development of a non-reciprocating engine, the turbojet, that is, an engine in which combustion is continuous, thus eliminating the need for pistons. And one of his designs, the experimental He 178, was the world's first plane to fly powered entirely by turbojet propulsion. This milestone in aviation history was

Erhard Milch. Heinkel considered most officials at the RLM a bunch of "mediocritics" for their lack of interest in his turbojet engines. This included Göring and Milch.

The HeS 011 Heinkel-Hirth turbojet which Messerschmitt had intended to use in his Me P. 1101 and then Horten in their Ho 18B "Amerika Bomber."

The genius of the Günter twins is evident in the lines of their Heinkel He 70 when it appeared in late 1932, it was the fastest commercial aircraft in the skies. Between 14 March and 28 April 1933 the Heinkel He 70 had established eight new world speed records.

The only Heinkel He 70 sold outside of Germany. This went to Rolls Royce Limited to be fitted with their new Rolls Royce Kestral engine a forerunner of the Spitfire's Merlin engine. Critics say that the British Spitfire was patterned after the Heinkel He 70.

reached on 27 August 1939 - but Heinkel had achieved non-piston-powered flight even earlier, when his He 176, with its early Walter KG rocket drive, flew for a brief time during taxiing tests in the summer of 1938. A year later, on 30 June 1939, the aircraft was far enough advanced to make its first real flight.

According to its test pilot, Erich Warsitz, the historic flight lasted 50 seconds, having a perfectly smooth start and landing. But perhaps most gratifying to Heinkel was beating his archrival, Willy Messerschmitt, into the air with "nonconventional" power plants.

Ernst Heinkel believed strongly in rewarding talent, which he had a knack for discovering, and frequently told his engineers and designers, "When you have success you will be rich men." Impatient and continually asking individuals in his design group, "What's new today?" Ernst Heinkel was not selfish. But he did have a mania for beating Messerschmitt, for being front-and-center on the stage of aviation technology. And though becoming first in technology, out-performing his competitors, was important, while getting there, Heinkel reasoned, why not make some money, too? He believed jet-propelled aircraft would help him achieve his goals of fame and fortune, and he set about getting what he wanted earlier, and with more determination than any other manufacturer, with the possible exception of Willy Messerschmitt. Ironically, Heinkel's very independence was his downfall. Too much independence in the highly politically charged German military bureaucracy

Walter Günter, design genius of the He 176 in 1936. He died in 1937 of complications following an auto accident. He lost one lung as a soldier in World War I. He badly damaged the remaining lung in that auto accident.

The Günter twins in 1937, Siegfried left and Walter right.

Paul Bäumer, widely popular aerobatic pilot and winner of the Order Pour le Mérit in World War I.

Siegfried Günter late 1950s. A capable designer in his own right but unable to achieve the kind of success he and his brother obtained together.

Siegfried Günter, somewhat eccentric, never went anywhere without a shirt and tie; wading in a lake in Russia in the early 1950s.

The Heinkel AG Aircraft Manufacturing Facilities.

of the 1940s was not good business. Although his HeS 011 turbojet was even more powerful as units from Jumo and BMW, he found himself excluded more and more, his technology and design work ignored, and his factories run by government commissioners. It wasn't always that way for Ernst Heinkel.

In pre-war Germany the firm that held the world's speed record was in a favored position for financial support and military sales. Heinkel believed that 500 mph (805 km/h) was about the maximum a piston-driven aircraft could ever hope to achieve, and he set out to develop an aircraft propelled by a non-piston engine. Such a craft, capable of unlimited speed, held out tremendous hope for a profit-oriented manufacturer such as Heinkel. Always on the lookout for talent, ideas, and designs, Heinkel spotted two technical geniuses - Wernher von Braun and Joachim von Ohain.

In 1935 the young von Braun was looking for technical assistance and airframes in which to place his liquid-rocket engines. Liquid oxygen and alcohol were to be injected into a simple combustion chamber that was shaped somewhat like an inflated balloon with a nozzle at its end. The thrust produced, von Braun contended, would be sufficient to propel an aircraft at unbelievable rates of speed. Heinkel agreed to provide the necessary technical aid and the fuselage of a new He 112. Although much was accomplished, Heinkel lost von Braun in 1937 when the young scientist was assigned full time to develop the V-2 rocket program. Heinkel discovered von Ohain in 1936, who at the age of 27, had just graduated from the University of Göttingen with a Ph.D. in engineering and who had an exciting idea about applying gas turbines to propel aircraft. His goal was to build a continuous combustion engine. This more elegant, more powerful, and smoother form of propulsion, von Ohain believed, could be achieved by drawing air in by a centrifugal compressor and tunneling it into a ramjet type of burner. Once started, the turbine would rotate by the jet of high-velocity exhaust air, and this, in turn, would turn the compressor. Thus the engine would operate as a self-supporting system.

Perhaps the most fortuitous talent discovery Heinkel made was the Günter brothers, Siegfried, the timid one, and his congenial twin, Walter. Siegfried and Walter Günter, were born 8 December 1899, in Kaula, Thuringen. Becoming interested in aviation at an early age, they, together with a few friends who were also students of Paul Bäumer, built the Pour le Mérite-Flyer of the Boelcke-Staffel (flying unit) during World War I. Later, in the early 1920s, the Günters designed the famous, record-setting Sausewind (Whistling Wind). Flown by Bäumer during the 1930s from one record-setting performance to another, this cantilever-wing monoplane reached the then astonishing speed of 155 mph (249 km/h) with only a 60 horsepower engine. At that time the performance of this unusual aircraft was credited to the engine, which was believed to perform better than indicated. This was not the reason, of course. Instead, its record-setting performance came about through the Günters' care-

The bombed out Heinkel AG Facilities at Rostock.

fully planned aerodynamic design, the aircraft's good profile, and its super-smooth surface. In more ways than one, the "Sausewind" was a remarkable design. Built entirely of wood and polished to perfection, the aircraft had a long, slender fuselage attached to a pair of low-set, evenly tapered wings and rounded tail surfaces of the type later to become a hallmark of the Günter brothers' Heinkel AG designs.

Heinkel was most impressed feeling that the man who had designed the "Sausewind," with its careful attention to design and it smooth surface finish, betrayed an unusually gifted hand. The person who designed this machine would certainly fit in with his own plans for a series of fast planes. Heinkel soon learned that two men responsible for the "Sausewind" were twin brothers, certainly the most remarkable pair of twins ever to play a role in aviation. Even though the Günters were not pilots, this didn't matter to Heinkel who hired Siegfried Günter in 1931 because he had "that special something," and he shared Heinkel's own obsession for speed. Walter Günter joined the Heinkel AG team a few weeks later.

The Günter brothers' first assignment was to design an aircraft for the 1932 "Round-Europe" air race. The result was the He 64, which was constructed entirely of wood and polished to perfection. As anticipated, the He 64 gave an outstanding performance. Painted red and christened the "Red Devil" the He 64 was the plane that brought fame to Hans Seidemann, later a general of the Luftwaffe. For the 4,700 mi (7,564 km) race, scheduled to take 6 days, Seidemann required only 3. While the He 64 was winning fame as the first really efficient sports plane, the Günters' remained in Warnemünde working on what was to become the He 70.

When it first appeared late in 1932, the He 70 displayed truly remarkable advances in aerodynamic design and finishing. Destined to be a trend-setter for Heinkel AG designs in the years to come, with its elliptical wooden wings, small rounded tail surfaces, and all-metal ovoid, semi-monocoque fuselage, the He 70 had a mirror-like finish. With countersunk riveting throughout, its main undercarriage bay flush with the wing surface, and a low-profile windscreen to cut drag to a minimum, was the fastest aircraft then available, reaching speeds in excess of 220 mph (354 km/h). A number of aircraft firms throughout the world were interested in licensing arrangements, including Rolls-Royce of England who were looking for a suitable airframe to demonstrate its prized engine to the world and eventually to mass-produce a version of the He 70 powered by its new V-12 "Kestrel" engine.

Heinkel saw a windfall in the making. He wanted Rolls-Royce to have the copy of the He 70, which he gave them. He also wanted to grant Rolls-Royce a license to build the He 70. In return, Heinkel would obtain a license to build the coveted V-12 "Kestrel" engine in Germany.

In the mid 1930s, with Germany's engine industry far behind in the development of high-performance engines such as those being built in France, England, and America, Heinkel's engineers recognized that if the He 70 were fitted with the new French-Gnôme-Rhône 900 horsepower engine, it would reach a speed of 275 mph (443 km/h) considerably faster than any existing fighter. Shortly after Heinkel made the proposal to Rolls-Royce, London agreed to the exchange. But the new German Air Ministry said absolutely not. They told Heinkel that given the present stage of aircraft development, it was out of the question for Germany to deliver its fastest plane into the hands of the British for mass production. As for the generally technologically weak German aircraft engine industry, the Air Ministry maintained it would experience such "breakthroughs" that within 2 years, German industrialists would surpass the achievements of all foreign engine builders combined. This would not come to pass - but the episode turned out to be only the first skirmish in Ernst Heinkel's lifelong battle with the Air Ministry.

The single copy of the He 70 which Heinkel gave to Rolls-Royce, when fitted with the "Kestrel" V-12 engine, cruised comfortably at 260 mph (418 km/h). It is thought to have contributed to the development of the Supermarine Spitfire because, coincidentally or otherwise, the Spitfire had the same graceful elliptical wings and other features common to the generation of single-engine combat aircraft that came after the He 70. The Spitfire was powered by the Rolls-Royce V-12 Marlin, whose ancestry included the pioneering Kestrel V-12.

Considering Heinkel's massive ego and strong self-motivation, coupled with a stable of design talent, Ernst Heinkel AG was capable of substantially more than the Air Ministry allowed. Heinkel was a methodical person. He formulated a goal and pursued it until he had accomplished what he set out to do, flirting with this design, that engine, and so on, to achieve his goal. He appeared to be more interested in advancing the state of the art than in providing what the Air Ministry asked for. Heinkel frequently boasted that his company initiated rather than followed trends in aircraft design. There is little doubt of the truth of this statement, and over time that approach did little to popularize the Heinkel AG with the Air Ministry. On numerous occasions when developing aircraft to meet official specifications, Heinkel elected to ignore the recommendations of the Air Ministry. The old doctor of physics' acknowledged talent lay in racing planes that set speed records. When he crossed a new threshold in 1939 by flying a craft powered entirely by jet propulsion, he lost interest in propeller-driven aircraft. Although his factories were producing some of the best propeller driven military aircraft in the world, he was unable to sell the turbojet aircraft he had developed with his own money. His rocket-powered plane, the He 176, was packed in crates and sent to the Air Ministry Museum in Berlin, later to be destroyed by an Allied bombing raid.

When the success of his rocket-powered He 176 was greeted with indifference, Heinkel persisted, trying to impress on Hitler, Göring, and Udet the idea that a revolution in flight propulsion was within their grasp. But in 1939 Hitler saw no immediate need for rocket or jet-powered aircraft. Convinced that France and England would not aid Poland when he invaded, he did not anticipate war. The Luftwaffe clamped down on any new attempts to test jet aircraft without its permission.

This ban, coupled with Hitler's disinterest, dispirited Heinkel, but he continued to pursue development of the turbojet engine. His second experimental turbojet-powered plane, the He 178, was flight-tested every day at dawn while the factory was deserted so that there would be no observers.

When the He 178 was also given a cold reception, Heinkel determined to build the world's first operational jet fighter and force the Air Ministry to take it. The outcome of Heinkel's stubbornness was the He P.280, a twin-jet fighter. Again, Heinkel was criticized, and Erhard Milch,

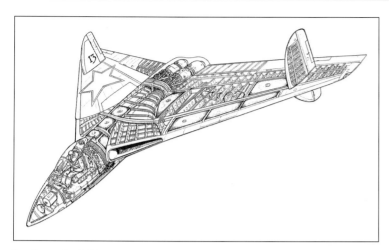

In the early 1950s the Soviets were interested in a delta-wing planform. Günter, arrested by the Soviets in 1946 and taken to Russia, offered the Soviets this proposed delta design.

the former managing director of Lufthansa and senior Field Marshall of the Luftwaffe, second in command after Göring, banned the fighter. (Messerschmitt's Me 262 suffered a similar fate about the same time.) Heinkel was beginning to have other problems, particularly a growing loss of influence. It was this loss that eventually kept Heinkel from producing more jet-powered aircraft than he did in the 1940s.

Heinkel's biggest loss of influence, indeed his near "lockout" from his own plant, resulted from difficulties his firm went through in developing for the Air Ministry the 30-ton Heinkel He 177 heavy bomber and long-range reconnaissance aircraft. A considerable amount of delay was involved, much of it the result of the ridiculous requirement that the He 177 be able to dive-bomb. The coupled-engine bomber, that is two engines geared to turn one large propeller, had first flow in 1939. By 1942 as many as 102 of the trouble-plagued aircraft had been constructed, but only thirty-three had been accepted for squadron service, and by September 1942 only two remained operational.

During this time Milch and others were openly questioning whether to leave Heinkel in charge of his own factory. The Air Ministry felt Heinkel was devoting all his attention to his profitable aircraft such as the He 111 and to the pursuit of his jet-propelled airplanes. Although this apparently was not the case, still Heinkel's factories and aircraft assembly facilities, as well as his engine development factory at Zufenhausen, eventually were turned over to commissioners. Heinkel was virtually cut off from is work between 1943 and 1945. Had it not been for his friend and chief designer Siegfried Günter, Heinkel might well have been excluded entirely from turbojet design activities. As it was, his firm did manage to put the He 162 fighter into the air, although it was full of fatal design flaws.

The remarkable thing is that, in spite of his tremendous problems with the Air Ministry, Heinkel was able to put into the air one of the three jet aircraft that were flying on any regular basis in Germany in 1945, the He 162. In terms of technology too, Heinkel AG was a standout. His HeS 011 turbojet would have powered many of his competitor's airframes, Messerschmitt's P.1101, for one, the Hortens' Ho 18B "Amerika Bomber" for another. Looking back, it is difficult to imagine just what Ernst Heinkel, builder of the first flyable jet-powered aircraft, might have accomplished had the Air Ministry's goals been similar to his own.

The five turbojet projects, the Air Ministry did assign to Heinkel AG were pretty conservative in terms of design. At the time of Germany's defeat, Heinkel was producing only one turbojet-powered plane, the fatally flawed He 162 "Peoples' Fighter." Heinkel AG's designers had reportedly completed a prototype of a four-engine bomber, the P.343, but it was destroyed during a bombing raid, and the entire project was abandoned in late 1944 when it became evident that there was no place for heavy aircraft in German war strategy.

Other Heinkel design projects included the P.1077 "Julia," a solid-propellant rocket interceptor; the P.1078A, a night fighter looking surprisingly similar to the Me P.1101; the P. 1079, a twin-jet, long-range, two-man night fighter featuring a "V" tail; and the P.1080, a twin-ramjet fighter that was tailless but retained a large rudder. These five aircraft, all pretty conventional in terms of turbojet designs (the P.1080 was to be ramjet-powered), were being planned to take Germany up to the 1950s. For the 1950s and beyond, Heinkel's design genius, Siegfried Günter, had two designs already on the drawing boards: the tailless P.1078B and the twin-nosed, tailless P.1078C. None of these projects ever went beyond the design stages, but Siegfried Günter continued to make aviation history.

Following the war between 1946 and 1954, Siegfried Günter worked for the Russians in the USSR. He was not fond of, nor did he support, the Communist cause, but he was out of work and had no job prospects. The Soviets asked him if he wanted a job? He said yes. Siegfried Günter was timid, almost shy. According to Heinkel, you could have guessed him to be anything, a priest, or high school teacher, a librarian, but under no circumstances an aircraft designer. In his efforts to find employment afer the war, he ventured to England in search of job prospects in the West. US Army intelligence officers whom he first came in contact with dismissed the old-fashioned, conservatively dressed man with the frameless glasses. They told him kindly that they were not interested, for, it seems, they did not believe that the person standing in front of their desks was the person he claimed to be - chief designer at the former Ernst Heinkel AG. Chief designer indeed! Didn't this poor chap know that Heinkel himself was the design brains behind every aircraft that came out of the Marienehe works? In fact, he wasn't - but few people outside Germany knew the difference. The timid Günter did not press the point, and retreated instead.

Upon his return to Berlin and the home of his parents, Günter was contacted by the Russians. They knew of his design activities at Marienehe and offered him the opportunity to continue his design work in Moscow. Lacking any job prospects, either in the West or from within Germany, Günter accepted the Russian offer and (together with a Heinkel colleague, Wilhelm Benz (lead designer on the He 176 and the P.1077

For a time after World War II, Heinkel built motor scooters to remain in business. Pictured here on a Heinkel built motor scooter is Heinkel's son.

159

"Julia"), he left for Moscow. Günter returned to Germany in 1954, and Ernst Heinkel welcomed him back and invited him to join the team he was forming in anticipation of 1955, when aircraft could once again be produced by Germany. Siegfried Günter died in 1968 at the age of 69. Although twin brother Walter had less impact on Heinkel designs, having died in 1937 after an automobile accident, Heinkel always believed that his Günter twins were the brightest airframe designers Germany had ever produced. This may or may not be true - but it is virtually certain that the Günters dominated airframe design at Heinkel between 1931 and 1945, all of the planes produced by Ernst Heinkel AG after 1931 might well have been designated "Gu," for Günter.

The post war years were not kind to Heinkel and his aircraft works. He was interned in London during June and July 1945, during which time he, along with other German professional men of industry, science, and finance, were interrogated about their work during the war. When he returned to Germany in late 1945, Heinkel couldn't believe what he found. His main airplane manufacturing plant at Marienehe, with its sprawling lawns and gardens, the place where most of his firm's turbojet plane designs had taken shape, had been seized and dismantled by the Soviets. His engine works at Rostock had been blown up, and the other Heinkel factory in the same town had been taken over by the Soviets. In addition, his flight-test center in Oranienburg had been completely dismantled and taken piecemeal back to the USSR, as was his turbojet engine works at Zuffenhausen. But perhaps his greatest personal blow came when he heard that his close friend and chief designer, Siegfried Günter had voluntarily gone to work for the Soviet Communists.

Heinkel's Jenbach manufacturing factory in Austria was seized by the French occupation forces. Most of the planes were carried away, and the rest were turned over to the new Austrian government, in trust until November 1948. It wasn't until 1950 that Heinkel's Zuffenhausen factory was returned to him. But it was too late to make any difference in a successful Heinkel comeback. Currency reforms throughout Germany had practically wiped out the value of large claims of many industrialists against the former Third Reich for unpaid deliveries of aircraft for which the Federal Republic was thought to be legally responsible. On the other hand, the obligations Heinkel's factories had incurred - the loans from banks and insurance companies he had obtained while he was building planes for the Reich - were not allowed to lapse. The banks and insurance companies pressed heavily for a share of what remained of the vast factories owned by Heinkel, because Heinkel AG had little in the way of cash to give them, being forbidden until 5 May 1955, from producing aircraft in Germany once more.

By 1955 financing an aeronautical industry had become a major undertaking for any firm, especially for one that had been out of the aircraft business for 10 years. Nevertheless, Heinkel did start over again, at first producing under license aircraft needed by Germany and other European countries. In the 1960s Heinkel's firm carried out some valu-

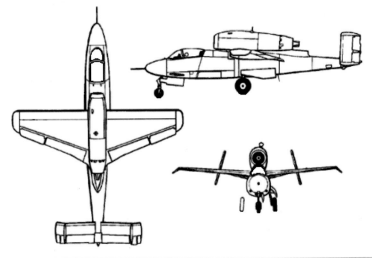

Heinkel He 162.

able work on vertical-take-off-and- landing aircraft (VTOL) and took up its turbojet engine work again. Yet by the mid 1960s, there was not enough work to keep the Heinkel factories open. In 1965 Heinkel joined the Vereignite Flugtechnische Werke (VFW), a consortium of several German aviation firms that were seeking to keep German aviation alive. With this decision, a firm that had been prominent in airplane manufacturing for more than 42 years ceased to exist. Ernst Heinkel died not long after, on 30 January 1968, at the age of 70 - but did live long enough to see his good name restored in the realm of German aviation.

PROJECT 162

It became evident in the latter half of 1944 that the Me 262 turbojet fighter and the Me 163 HWK 509A bi-fuel rocket-propelled interceptor could not combat the Allied air offensive against Germany. Due to the difficulties in obtaining enough fuel for the twin-engine Me 262 and the long manufacturing times required for both aircraft, the Air Ministry turned to a new concept: mass production of a cheaply produced, single-engine, lightweight fighter-interceptor. Believing that Germany's only hope lay in overwhelming Allied bomber squadrons with thousands of such high-speed fighter-interceptors, the Air Ministry on 8 September 1944, issued basic project requirements to six companies: Arado, Blohm and Voss, Focke-Wulf, Junkers, Heinkel, Horten, and Messerschmitt. Proposed designs and other material relating to the Air Ministry's request were to be submitted by 14 September 1944 and the aircraft itself was to be ready in less than 4 months, by 1 January 1945.

Heinkel He 162.

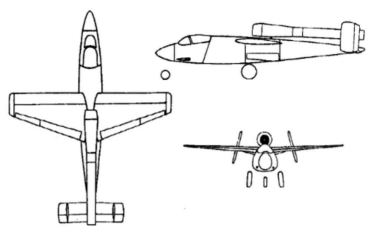

Heinkel He 162.

What the Air Ministry wanted was a light-weight fighter that would use a minimum of strategic materials, be suitable for rapid mass production, and still have a performance superior to contemporary piston-engined fighters. In addition, the new fighter was to weigh no more than 4,409 lb (2,000 kg), have a minimum endurance of 30 minutes, take-off within 1,640 ft (500 m), and be armed with two 30 mm MK 108 cannons. For propulsion, a single BMW 003 turbojet having 1,764 lb (800 kg) of thrust was to be used. The initial production was to be ready in January 1945, it was expected that within a few months output would settle down to 4,000 airplanes per month.

On 15 September the Air Ministry selected the design proposals submitted by Blohm and Voss and by Heinkel. Messerschmitt did not submit a proposal. Initially, Blohm and Voss' P.211 was judged best. However, Heinkel had had a mockup of his P.1073, whereas Blohm and Voss did not. After 2 weeks of wrangling, the Air Ministry selected the P.1073, which was a prototype of the He 162.

The tricky problems of supplying thousands of new pilots and maintaining thousands of the new aircraft were soon solved. Pilots were to come from the Hitler Youth. Many knew how to fly. Those who did not were to receive speedy training, which would be completed by actually flying the jet fighters on missions. Regarded as a consumable item, He 162's maintenance could be ignored. Damaged or worn out aircraft would merely be replaced by the vast numbers that would be produced. The intent was that hordes of He 162s would swarm over the Allied bombers' fighter escorts, leaving the unprotected bombers at the mercy of the Me 262.

The most noteworthy features of the He 162 were its low span-to-length ratio, the mounting of its turbojet above the fuselage, the arrangement whereby its nosewheel undercarriage retracted into the fuselage, and the pronounced dihedral of its tailplane. Mounting of the engine outside the fuselage was in keeping with Heinkel's effort to avoid problems with jet intake and tailpipe ducting. Placing the engine above the fuselage in itself created no problems, but when this arrangement was combined with the high (shoulder) wing, the turbulence created left the He 162 with a design flaw that was never satisfactorily corrected.

On the other hand, the He 162 was quite a tough aircraft, having been designed for a load factor of 7.5 (similar to dive bombers) and a safety factor of 1.8. It was the first production aircraft ever to have an "ejector" seat which would, by means of a powered charge, throw the pilot clear should the aircraft be disabled.

Nearly 300 of the aircraft were constructed, but despite repeated modifications, the He 162 remained aerodynamically unstable. In the end its Mach number was a relatively low 0.75. Later, in the similarly arranged Henschel Hs P.132, some of these troubles were avoided by mounting the wing at right angles to the fuselage skin (the wing of the He 162 had a cavity beneath it where the fuselage skin began to curve inward). Although a fairly "hot" plane to fly, it was one of the most difficult even for experienced pilots. A number of test pilots died, and the fatalities continued after the war when British officials were test flying the aircraft to determine its flight characteristics.

Despite the He 162's difficult handling, considerable experimental work on the original design was planned. It was hoped that by fitting a new swept-forward wing, the low critical Mach number might be raised. In addition, a new butterfly tail unit was being planned. Of the nearly 250 fully operational and 800 partially complete examples of this series known to have been built, only seven He 162s are known to exist. They are on display in museums in Canada, Europe, and the United States.

Specifications: Project 162

Engine	1xBMW 003E turbojet having 1,764 lb (800 kg) of thrust
Wingspan	23.8 ft (7.2 m)
Wing Area	120 sq ft (11.2 sq m)
Length	29.9 ft (9 m)
Height	8.4 ft (2.5 m)
Weight, Empty	3,859 lb (1,750 kg)
Weight, Takeoff	5,490 lb (2,490 kg)
Crew	1
Speed, Cruise	NA
Speed, Landing	103 mph (165 km/h)
Speed, Top	522 mph (835 km/h) at 19,686 ft (6 km)
Radius of Operation	606 mi (975 km) at 36,091 ft (11 km)
Service Ceiling	36,091 ft (11 km)
Armament	2x20 mm MiG 151 cannon with 120 rounds of ammunition each
Rate of Climb	4,231 ft/min (1,290 m/min)
Bomb Load	None
Flight Duration	20 min at sea level, or 57 min at 36,091 ft (11 km)

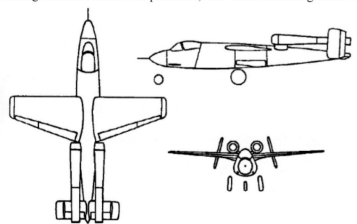

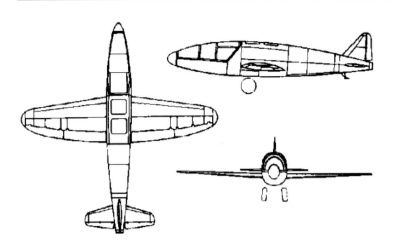

PROJECT 176

"A rocket with running boards" - is what German Air Ministry officials dubbed the He 176, the first rocket-powered aircraft to use only liquid fuel. Apparently unimpressed after viewing its first demonstration flight on 20 June 1939, Ernst Udet, Erhard Milch, and other officials deemed the aircraft and its rocket-propulsion system too dangerous and later grounded the noisy object. It received official sanction for only two or three more demonstration flights, once when Hitler was present, and Heinkel eventually gave up on rocket-propelled aircraft.

It was the use of liquid fuel that made the He 176 unique. Other rocket-powered aircraft had flown as many as 11 years earlier. Both Fritz Stamer, a German, and Fritz von Opel had flown rocket-powered gliders in 1928, and Hugo Junkers had tested float planes that took off with the assistance of underwing powder rockets.

Heinkel began testing liquid-fuel rockets in flight in the autumn of 1936. First he used an He 72 to test a simple Hellmuth Walter KG bi-fuel rocket drive which used hydrogen peroxide to give a thrust of about 300 lb (135 kg) for 45 seconds. After this success, Heinkel began to take serious interest and donated two He 112 airframes to Werhner von Braun, who later achieved about 2,200 lb (1,000 kg) of thrust for 30 seconds. By 1937 the Walter firm had improved their R1 bi-fuel liquid rocket drive, getting 2,000 lb (950 kg) of thrust for 30 seconds.

After 1937 advances came rapidly, and Heinkel became convinced that rocket-propelled aircraft could achieve dramatic performances. It was during this time that he decided to build an aircraft powered entirely by a liquid-fuel rocket. His goals were typical, reflecting his dynamism - to achieve the long-sought-after speed of 621 mph (1,000 km/h), the speed of sound, which far exceeded the existing world record, and to gain prestige from doing so.

The He 176's design was largely the work of Wilhelm Benz and Hans Regner, Heinkel wanted an airframe of the smallest feasible size

The Heinkel He 176 designed by Wilhelm Benz under Walter Günter's guidance.

Hellmuth Walter built the bi-fuel liquid rocket engine which successfully propelled the He 176 into the air on 20 June 1939.

for his attack on the world air speed record. As the He 176 came to be designed he got it: the fuselage had a diameter of only 2.7 ft (0.8 m) and was literally "made to measure" for its test pilot. A novel safety feature was a system that ejected the entire cockpit section in case of emergency. Overall, the He 176 did not look like a typical Heinkel-built aircraft. Instead of the usual shoulder-level wings, it had low-set elliptical wings which were fitted with wing-tip skids as a precaution due to the narrow track of the undercarriage. Designed as a tail dragger, that is a metal rod in place of a tail wheel, to keep the fuselage off the ground, the He 176's main wheels retracted into the fuselage. Adding to its clean lines was another first - a one-piece acrylic cockpit cover which served as virtually the entire nose section.

Although the He 176 never suffered an accident during flight testing, it failed Heinkel in two ways. First, it was unable to fly faster than 434 mph, somewhat below the world speed record in 1939, apparently because it was simply too heavy for the output of its rocket drive. Second, it did not really impress Hitler and other Air Ministry officials. In fact, it damaged Heinkel's reputation due to a miscalculation of the size of the wing (surface) needed to get the rocket-powered aircraft off the ground. The He 176 simply did not have sufficient power to keep the aircraft airborne and maintain satisfactory performance. A constant-thrust power plant such as the rocket motor used in the He 176 offered considerably less equivalent power for takeoff than did an orthodox piston-type engine. Had Heinkel experimented more with gliders, Alexander Lippisch later said critically, he would have known that a rocket-propelled aircraft requires a pretty low wing loading. Instead, Heinkel's team of Benz and Regner with input from Walter Günter doomed the He 176 by placing short, stubby wings on the aircraft and expecting it to perform. It flew, but poorly.

Heinkel believed that his problems with the Air Ministry in connection with their refusal to enter into a deal with Rolls-Royce over the He 70 really solidified when the He 176 design miscalculations became known. However, the Air Ministry was not disinterested in liquid-fueled aircraft, as Heinkel believed. In fact, Air Ministry officials approached Alexander Lippisch shortly after Heinkel's disappointing performance and asked him to develop a suitable airframe for a liquid-fueled aircraft. This request resulted in the Messerschmitt Me 163 Komet interceptor.

With the start of World War II, Heinkel abandoned plans to build a second prototype of the He 176. His own disappointment with the craft's lack of performance and official interest contributed to his decision. But perhaps the biggest reason was his growing interest in the turbojet

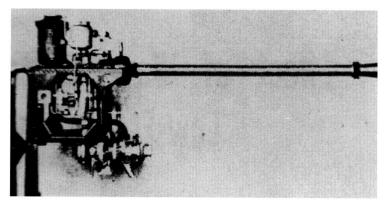

The Walter bi-fuel rocket engine of the type used in the He 176.

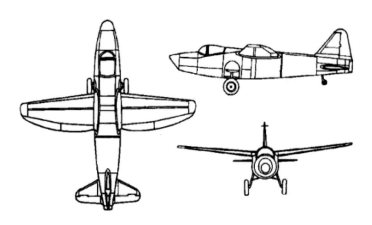

as a means of propelling an aircraft at even higher rates of speed, and with a greater margin of pilot safety.

The He 176 prototype did not survive World War II. In 1942 Heinkel shipped the entire aircraft to the Air Museum in Berlin where it was destroyed while still in its shipping crates. The following year the Air Ministry would be rapidly pursuing jet and rocket-powered fighter-interceptors to bring down the very bombers that had destroyed the He 176.

Specifications: Project 176

Engine	1xWalter HWK-R1 rocket engine having 1,323 lb (600 kg) of thrust
Wingspan	13.2 ft (4 m)
Wing Area	53.8 sq ft (5 sq m)
Length	16.5 ft (5 m)
Height	4.9 ft (1.5 m)
Weight, Empty	3,455 lb (1,570 kg)
Weight, Takeoff	4,409 lb (2,000 kg)
Crew	1
Speed, Cruise Speed	NA
Speed, Landing	NA
Speed, Top	435 mph (700 km/h)
Radius of Operation	68 mi (109 km)
Service Ceiling	29,529 ft (9.0 km)
Armament	None
Rate of Climb	NA
Bomb Load	None
Flight Duration	10 min

PROJECT 178

The He 178 was the world's first turbojet-powered aircraft to fly. Although severely underpowered and lacking sufficient wing area, it was still able to achieve 373 mph (600 km/h) during its first flight on 24 August 1939. A low power-to-weight ratio, defects in its airframe, and a retractable under-carriage that refused to operate despite repeated attempts to correct it, kept the He 178's top speed under 400 mph (644 km/h) throughout its flight demonstration. One of its major problems was the placement of its HeS 3B turbojet engine. Lengthy, power-robbing ducts were required to provide air to the internally mounted engine and to exhaust it. Dr. von Ohain, head of Heinkel's turbojet engine development, believed that if these duct work defects alone could have been corrected, the aircraft might have attained 534 mph (700 km/h).

Development of the He 178 had two long-term effects. Most immediate was that attempt to mount engines internally were abandoned temporarily and manufacturers such as Messerschmitt, Junkers, Arado, and even Heinkel placed all turbojet engines either under the wings or attached to the fuselage. Further, although it served a useful purpose as a test bed, the He 178 damaged Heinkel's reputation and acceptance in the German aviation industry for years. Prior to that time Heinkel was recognized throughout Germany, indeed throughout the world as a superior airframe manufacturer. While Air Ministry officials were highly interested in turbojet development and the military advantages it offered, they were critical of Heinkel for diverting his energies from the job his company did best - designing airframes. His pursuit of engine development represented an immediate drain on the company's financial and personnel resources. Further, past experience suggested that

Side view of the Heinkel He 178.

Wilhelm Benz was also a guest of the Soviets 1946-1954. 1986. Germany.

Adolf Jensen, the third member of the design team on the Heinkel He 176, also guest of the Soviets 1946-1954. 1986. DDR.

Heinkel, as a company having little or no background in engine design and production, was destined to fail - and failure of such a highly successful airframe manufacturer would cause the entire country to suffer.

Heinkel's innovations were given some positive attention when he lifted some of the secrecy from is project in 1938 by revealing the partially completed airframe with von Ohain's turbojet installed. Air Ministry officials were amazed. But, with the country edging toward war, they looked askance at Heinkel's private venture into jet engine development. Experimental work was discouraged and manufacturers were expected to concentrate on doing what they did best, in Heinkel's case, building efficient aircraft. Undaunted by criticism of his He 178, Heinkel went on to construct his twin-engine He P.280.

As for the He 178, with its small, nearly round duralumin fuselage and its short, typically Heinkel shoulder-level wooden wings, it was pretty much abandoned after demonstration flights. Neither of the two prototypes survives. The first-built, the history making first turbojet to fly, was disassembled, packed in crates, and sent to the Berlin Air Museum, where it was destroyed, along with Heinkel's rocket-propelled He 176, in a 1943 bombing raid. The second He 178, kept in storage at Rostock, was destroyed in a 1945 air raid.

Specifications: Project 178

Engine	1xHeinkel HeS 3B turbojet having 926 lb (420 kg) of thrust
Wingspan	23.8 ft (7.2 m)
Wing Area	98 sq ft (9.1 sq m)
Length	24.7 ft (7.5 m)
Height	6.10 ft (2.1 m)
Weight, Empty	3,565 lb (1,620 kg)
Weight, Takeoff	4,396 lb (1,998 kg)
Crew	1
Speed, Cruise	360 mph (580 km/h)
Speed, Landing	103 mph (165 km/h)
Speed, Top	435 mph (700 km/h)
Radius of Operation	NA
Service Ceiling	NA
Armament	None
Rate of Climb	NA
Bomb Load	None
Flight Duration	NA

The Heinkel He 178A open cockpit model later a full plexiglass canopy added and test flown.

The Heinkel He 178B.

The Heinkel He 178B.

The Heinkel He 178B.

PROJECT 280

The He 280 was another record-breaker for Ernst Heinkel, the Günter twins and Heinkel's engineers. On 2 April 1941, it became the first fighter aircraft to fly powered solely by turbojet. Design studies had begun in 1939.

The He 280 was a beautifully proportioned, metal-skinned aircraft, though its overall design was not really advanced. It was powered by two HeS 8 turbojets, each having 1,650 lb (750 kg) of thrust hung to the undersides of the wings. The wings had a straight leading edge, a curved trailing edge, and a slight dihedral. The tail unit consisted of a high-mounted tailplane carrying a fin and rudder at each tip. Flight controls were conventional and included ailerons, landing flaps, and a one-piece elevator placed between the twin rudders.

In the He 280's slender fuselage, the pilot sat about midships in a stepped cockpit that had a rearward-sliding canopy. Armament was mounted in the nose section, and the fuel tanks were positioned aft of the cockpit.

The undercarriage consisted of a tricycle landing gear. The two main wheels were attached to the wing just inboard the engine and retracted inward, partly into the wing and partly into the fuselage. The nose wheel was placed well forward into the nose and retracted rearward into the fuselage. This gear arrangement was another first, because fighter aircraft had always been tail draggers; that is, the third wheel was at the tail.

Few photographs of this pioneering aircraft exist. It is frequently shown as it appeared during its early trials, with the engines left uncowled or only partially cowled. This was done to minimize the fire risk result-

The Heinkel He 280 V1.

The Heinkel He 280 V3 (GJ+CB) making a landing approach.

The Heinkel He 280 V3 (GJ+CB) making a banking turn.

Secret Aircraft Designs of the Third Reich

Artist's impression of the layout of the HeS 8A turbojet engine as mounted on the He 280.

The Heinkel HeS 8A of 1,650 lbs (759 kg) of thrust which powered the first He 280.

ing from fuel accumulation when the engine nacelles were fitted. Later the fire risk problem was corrected.

Eight prototypes of the He 280 were completed and flown, the second and sixth model being equipped with armament. The final two prototypes were fitted with a "butterfly" tail arrangement. Overall, the world's first jet fighter performed surprisingly well. During a demonstration on 5 April 1940, for the Air Ministry, an He 280 flew in competition with the fastest mass-produced Luftwaffe plane of the time, the Focke-Wulf Fw 190. With its extraordinary maneuverability and speed, the He 280 completed four circles before the Fw 190 had made three.

The He 280 was never accepted by Erhard Milch and the Air Ministry. Heinkel believed their reluctance was due mainly to the He 280's use of a tricycle landing gear. The Air Ministry accepted Messerschmitt's "half-ready" Me 262 instead, because it was being designed as a "tail dragger." Later, the tail-dragging arrangement of the Me 262 was scrapped to keep the hot exhaust trail horizontal, thereby reducing fire damage to the plane as well as to the runway itself.

As for the He 280, Heinkel continued to experiment with it, installing different types and models of jet engines such as the Jumo 004. The various prototypes continued up through July 1943 for use as test flights for engine, performance, and other investigations. The use of the eight prototypes as flying test beds was discontinued in mid 1944. No examples of the He 280 survive.

The Heinkel He 280 V3 between test flights with its engine cowling removed.

A Heinkel He 280 lifting off for a test flight with all its front engine cowling removed for easy adjustments.

Specifications: Project 280

Engine	2xHeinkel HeS 8A (001A) turbojets having 1,650 lb (750 kg) of thrust each
Wingspan	39.5 ft (12.2 m)
Wing Area	232 sq ft (21.5 sq m)
Length	33.6 ft (10.4 m)
Height	10 ft (3.1 m)
Weight, Empty	7,073 lb (3,215 kg)
Weight, Takeoff	9,482 lb (4,310 kg)
Crew	1
Cruise Speed	NA
Speed, Landing	87 mph (140 km/h)
Speed, Top	559 mph (900 km/h) at 19,680 ft (6 km)
Radius of Operation	404 mi (650 km) at 19,680 ft (6 km)
Service Ceiling	37,720 ft (11.5 km)
Armament	3x20 mm MG 151 cannon
Rate of Climb	3,755 ft/min (1,145 m/min)
Bomb Load	None
Flight Duration	45 min

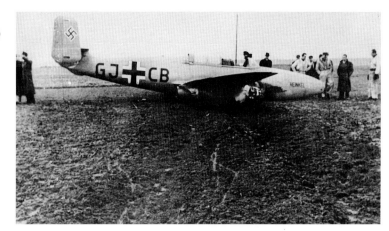

The Heinkel He 280 V3 suffering a landing mishap due to a failed engine.

PROJECT 343

After Ernst Heinkel's He P.280 lost out to the Me 262, he turned his attention to other projects. One was the He P.343, a relatively straightforward four-turbojet bomber with a crew of two. To be propelled by four HeS 011A turbojets, each having 2,866 lb (1,300 kg) of thrust, it was expected to have a cruising speed of about 540 mph (869 km/h) at sea level and a range of 1,000 mi (1,609 km). Like other Heinkel aircraft (the He 176, He 178, and He 280, for example), this bomber was a private venture. Never very enthusiastic about Heinkel AG since the days of their He 280, or about four-engined bombers for that matter, the Air Ministry reluctantly sanctioned the construction of bomber prototypes in April 1944. It was to be Heinkel AG's final wartime design to reach the construction stage.

Construction began in February 1944. By the time the first prototype was completed later that year, it was becoming obvious there was no longer a place for heavy aircraft in German military strategy.

Several distinct versions of the P.343 were envisioned. The prototype carried a design layout of a mid-set straight wing, with the four turbojet engines slung below in separate pods. The main undercarriage retracted rearward into the fuselage, and each wheel unit held only one tire. Project 343 A-1 would be a bomber, with a 6,615-pound load dis-

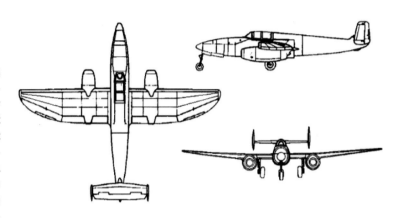

tributed between the fuselage and underside of the inner wings. A reconnaissance version and an attack aircraft were to complete the series. In each case, all functions, including armament deployment, would be handled by the two-man crew in the pressurized nose cockpit.

Several wing forms were also planned. One was a 25 degree sweptback wing, and another a 35 degree sweptback wing. The attack version would be fitted with Fowler flaps and additional nose flaps as well. Heinkel AG purposely avoided the sweptback wing on its first P.343 prototype. It was hoped that the bomber would be in service by

Scale model of the Heinkel He 343, mid 1940s. Plans drawn up May 1944 with four separate Jumo 004s shown is the reconnaissance version.

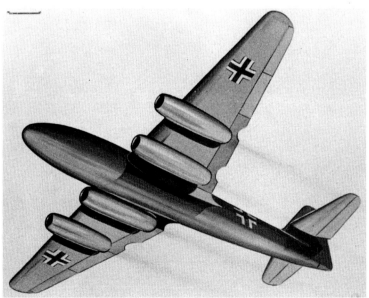

Heinkel AG artist impression of the Heinkel He 343 in flight.

The Heinkel He 343 A-1. Scale model by Dan Johnson.

The Heinkel He 343 A-1, front view. Scale model by Dan Johnson.

early 1945, and he wanted to get it into the air with a conventional mid-set straight wing. Later the P.343 could be fitted with an experimental sweptback wing arrangement. In effect, the P.343 was an enlarged version of the Arado P.234B twin-jet bomber already flying early in 1944.

The first nearly complete prototype of the P.343 was destroyed on the ground in a bombing raid near the end of the war. The Air Ministry then asked the planning office of the Wrede Gesellschaft in Freilassing to help Heinkel in the construction planning of all subsequent P.343 prototypes. While a little later a P.343 was undergoing final preparations for flight testing early in 1945, a fire at the Wrede factory destroyed the prototype as well as the reconnaissance versions, which was about 50 percent complete. Portions of the attack version were also lost in the fire along with components, jigs, material, and the series' drawings.

Specifications: Project 343

Engine	4xHeinkel-Hirth HeS 011A turbojets having 2,866 lb (1,300 kg) of thrust each
Wingspan	59 ft (17.9 m)
Wing Area	455 sq ft (42.2 sq m)
Length	44.2 ft (15.4 m)
Height	17.7 ft (5.4 m)
Weight, Empty	23,748 lb (10,772 kg)
Weight, Takeoff	43,108 lb (19,554 kg)
Number of Crew	2
Speed, Cruise	540 mph (369 km/h)
Speed, Landing	NA
Speed, Top	565 mph (909 km/h)
Radius of Operation	1,006 mi (1,619 km)
Service Ceiling	48,500 ft (14.7 km)
Armament	2x20 mm MG 151 cannon
Rate of Climb	NA
Bomb Load	4,410 lb (2,000 kg) in the bomb bay, plus 2,206 lb (1,000 kg) carried under the wings
Flight Duration	2 hr

The Heinkel He 343 A-1, front view. Scale model by Dan Johnson.

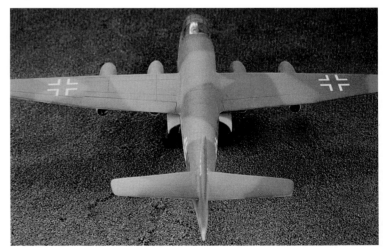

The Heinkel He 343 A-1, rear view. Scale model by Dan Johnson.

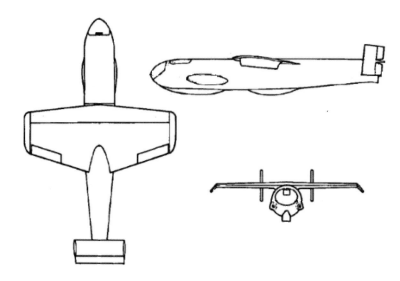

The proposed Heinkel HeP.1077, "Julia." Scale model by Jamie Davies.

PROJECT 1077 "Julia"

By 1944 it had become painfully apparent to most Germans that the Allied bomber force headed by USAAF B-24s and B-17 heavy bombers were the major factor contributing to Germany's overall war losses and suffering. Unchecked, this bombing would eventually lead to the country's complete collapse and destruction. Therefore, every effort was put into developing weapons to intercept and destroy the bombers that were beginning to range the entire length and breadth of the homeland.

The German Air Ministry hoped to come up with a turbojet-powered fighter-interceptor which could shoot down the bombers. Such an all-purpose aircraft could be used for both reconnaissance and dive bombing, as both a fighter and an interceptor, but it was also costly in terms of materials and production time. A less costly alternative was needed, at least until a fighter-interceptor could be produced in sufficient numbers. The Air Ministry decided on a weapon that had the capabilities of an interceptor, could be built easily, and could be produced economically using a minimum of strategic material.

Four manufacturers submitted designs. Heinkel's entry, the P.1077, won the competition. Designed by Wilhelm Benz, the P.1077 was to be a semi-expandable, single-seat, rocket-powered interceptor. The losing designs included the Junkers P.127 "Walli" and Messerschmitt's P.1104. The fourth design, Bachem's P.20 "Natter," was later also selected for development and was given the official designation of Ba 349A. At war's end, two versions of the P.1077, or "Julia," the cover name by which it became known, were being considered. They differed only in pilot accommodation. In one version the pilot was placed in a prone position, while in the other the pilot sat upright in the usual manner.

Heinkel engineers generally favored the prone position. This preference, as well as the call for rocket power, stemmed from the intended use of the craft against B-17 heavy bombers. The Fw 190, with its large-diameter, radial, air-cooled engine, presented B-17 gunners with some seventeen square feet of target area. By discarding the bulky piston engine and placing the pilot in a prone position, the target area could be greatly reduced.

A proposed Heinkel HeP.1077, "Julia" in flight showing its centerline wooden landing skids. Scale model by Jamie Davies.

The proposed Heinkel HeP.1077, "Julia" in flight. Scale model by Jamie Davies.

German Aircraft Exhibition at RAE Farnborough in November 1945. In the foreground is the only genuine Re IV brought to the UK, the Ju EF 126. (The Fieseler Fi 103).

The other competitor to the Heinkel He P.1077, the SS sponsored Bachem Ba 349 "Natter." It was chosen over everything else because of SS Führer Himmler's interest.

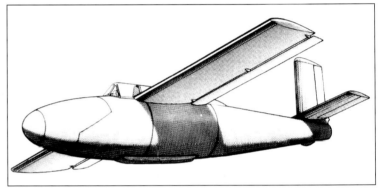

The Messerschmitt Me P.1103/1104, also a competitor to the Heinkel He P.1077.

A second reason for preferring the prone position was related to takeoff acceleration. Research had shown that the average pilot in the normal sitting posture usually blacked-out at pressures between four and five times that of gravity. While the P.1077's rate of acceleration was expected not to create even 2.1 times the pressure of gravity, pilot blackout was always a concern, and it had been found that blackout could be delayed until a pressure as much as ten or eleven times that of gravity had been reached if the pilot were lying prone.

The P.1077 was intended to take off almost vertically from guide rails and to land on skids after completion of its mission. During flight it was to be powered by a HWK 509C bi-fuel, liquid-rocket dual chambered drive having 3,750 lb (1,695 kg) of thrust. Four solid fuel Schmidding 533 booster rockets, each providing 2,650 lb (1,200 kg) of thrust for twelve seconds were to aid takeoff and then to be jettisoned automatically on burnout. The forward portions of the tandem skids beneath the cockpit were to extend to absorb the impact of touchdown. Instrumentation and equipment were to be minimal. A normal reflector sight was to be provided. Armament was to be two 30 mm MK 108 cannons mounted semi-externally on the side of the forward fuselage and providing sixty rounds apiece.

The P.1077 was to have a very simple shoulder wing spanning only 15.1 ft (4.6 m), with a surface of just 77.5 sq ft (7.2 sq m). The high wing was a typical Heinkel AG design. Overall, the P.1077's length was to be 22.5 ft (6.9 m). In the prone-pilot design, takeoff weight was expected to be 3,950 lb (1,900 kg) after the solid-fuel booster rockets had been jettisoned. Calculations indicated that the P.1077 would reach a maximum level speed of 559 mph (900 km/h) at 16,405 ft (5 km). It was estimated that initial rate of climb would be 39,405 ft/min (118,389 m/

A Soviet artist's rendering of the Heinkel He P.1077 interceptor.

The Heinkel He 162 "Volksjäger." Scale model by Jamie Davies.

min) and that it would reach 5 km in 31 seconds after takeoff and 49,000 ft (14.9 km) in 72 seconds. Based on the amount of fuel and thrust of the engine's cruising chamber, range was expected to be about 495 mph (797 km/h). Heinkel AG was unable to get a P.1077 prototype crafted before the war ended. The only project of this type to reach the flight test stage was Erich Bachem's Ba 349A Natter.

The Heinkel He 162 "Volksjäger." Scale model by Jamie Davies.

Specifications: Project 1077

Engine	1xWalter HWK 509C bi-fuel liquid-rocket drive having 3,750 lb (1,695 kg) of thrust, plus 4xsolid-fuel booster rockets having 2,650 lb (1,200 kg) thrust each.
Wingspan	15.5 ft (4.6 m)
Wing Area	77.5 sq ft (7.2 sq m)
Length	22.5 ft (6.9 m)
Height	NA
Weight, Empty	NA
Weight, Takeoff	3,950 lb (1,900 kg)
Number in Crew	1
Speed, Cruise	497 mph (800 km/h)
Speed, Landing	NA
Speed, Top	559 mph (900 km/h)
Radius of Operation	40 mi (64 km)
Service Ceiling	49,000 ft (14.9 km)
Armament	2x30 mm MK 108 cannon
Rate of Climb	39,400 ft/min (118,389 m/min)
Bomb Load	None
Flight Duration	NA

American military intelligence's impression of the proposed Heinkel He 162B, powered by a single Argus ramjet.

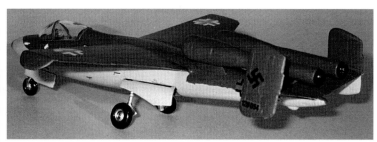

The Heinkel He162 A-10 with two Argus 014 pulse jets. Scale model by Steve Malikoff.

The Heinkel He162 A-10 with two Argus 014 pulse jets. Scale model by Steve Malikoff.

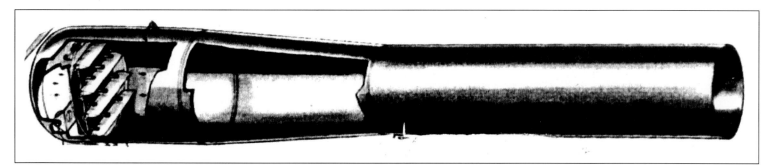

A cut away of the Argus 014 ramjet. This engine was proposed for a variety of German aircraft projects, such as the Heinkel He P.1077, Messerschmitt Me 328, and Heinkel He P.1080.

PROJECT 1078

Siegfried Günter's P.1078 series, turbojet models A, B, and C, represent Heinkel AG's most advanced and daring airframe experiment. Even after the war ended, Ernst Heinkel was particularly bitter about the P.1078B. With its twin fuselages, Heinkel called it the most freakish aircraft he had ever seen. Something as unnatural as the "B" model, he told his design team, does not last long in aircraft circles, and he chastised Günter for squandering his talents and precious time with child's play. Günter, however, felt that this was unfair criticism.

Each of the three aircraft in the P.1078 series was to have been powered by a single turbojet, Heinkel AG's HeS 011 having 2,866 lb (1,300 kg) thrust. The P.1078A was similar in appearance to Messerschmitt's P.1101 and its single-engine mounted well forward in the fuselage and the cockpit atop the air intake. Like the P. 1101, this model would have had a standard tail separated from the main fuselage by a long, slender boom. The P.1078B was a tailless design made even more exotic with its two slender noses extending out of a single fuselage. The port nose housed the cockpit, and the starboard nose held a radar dish, armament, and the nose wheel. A rectangular air intake for its single HeS 011 was set well back between the two noses. The P.1078C was tailless too; however, the design showed its wing tips bending downward for control, and its cockpit placed above the air intake duct well forward to the front of the fuselage.

Heinkel's single-engine P.1078 series interceptors, as well as other designs in mid 1945, were born out of a need to conserve and stretch the country's dwindling fuel stock. Due to a major shortfall of oil production, German aircraft engineers were being forced to design lighter single-place fighters. It was pretty well accepted that airframes powered by two turbojets, such as Heinkel's P.1079, would be the best in terms of space for the pilot and armament. But in Germany, higher thrust turbojet engines were not expected until the late 1940s, so the use of twin turbojets was ruled out on account of fuel shortages, and designers had to come up with aircraft projects utilizing only one HeS, BMW or Jumo turbojet.

As usual, Heinkel engineers were looking for ways to increase the performance of their fighter aircraft and Siegfried Günter was exploring the concept of tailless aircraft. Although several German aircraft designers believed that the fighter would one day evolve into a tailless configuration (and, indeed, the Stealth fighter designs seem to indicate that this is happening), Heinkel and Günter were not sure in the mid 1940s. Intrigued with the possibilities of good, or at least better performance from a tailless designs, they were still unsure of its potential, despite the considerable amount of work already conducted by Alexander Lippisch. They also felt that the possibilities for increasing the performance of an airplane with a tail was unlimited. In either event, Günter would design several airframes, as was Heinkel's habit - one with a tail, one without, and one tailless but with twin noses. None of these three designs got beyond the design stage. The P.1078 series would be viewed as a sort of competition between tail and tailless aircraft. Each of the three aircraft would have about the same wing area and landing speed. The two tailless versions, however, would have about 10 percent less surface, less air intake losses, and a lighter overall airframe than the one with a tail. The smaller amounts of surface on the two tailless designs was partly a function of the low specific weight of the front part of the fuselage. Another reason for the small amounts of surface was that a broad, flat fuselage cross-section below the wing gave less additional surface than a narrower or longer fuselage with a tail.

The designers at Heinkel AG expected both benefits and disadvantages in going to a tailless aircraft. A tailless aircraft's generally unstable moments about the normal axis due to the small height and length of the fuselage was a major problem, but the phenomenon could be reduced substantially by the use of cathedral wing tips. Another important disadvantage of the tailless airplane was that it gave pilots poor vision to the sides, and Günter recognized that the twin-nosed P.1078B was especially limiting to vision on the right. However, since enemy attack generally was expected to come from behind and above, poor side visibility was felt to be relatively unimportant.

Where tailless aircraft gained over aircraft having a tail was in performance. The tailless P.1078B and C had the smaller wing loading, an advantage because the fighter with the lower wing loading would have higher ceiling and would be able to turn in a smaller radius than would the multi-engined aircraft it would be attacking. Even though the tailless would be a better performer, it would consume larger quantities of fuel than the conventional aircraft because the turbojet(s) had to maintain full power during maneuvers in order to remain airborne, a problem not generally experienced by the conventional aircraft. This is the main reason why pure delta's such as Convair's F-102 and F-106 were replaced in favor of conventional designs - their high fuel consumption put them at a serious disadvantage in "dog fights." Heinkel AG engineers, quite correctly, also felt that the tailless aircraft would be faster than one with a tail, for example, the P.1078A. At sea level the tailless P.1078B and C were expected to reach 637 mph (1,025 km/h), a speed about 25 mph (40 km/h) greater than that anticipated for the P.1078A. At an altitude of 19,686 ft (6 km) the difference in expected speeds was less, but the anticipated ceiling for the tailless airplane was nearly 2,461 ft (0.75 km) higher.

PROJECT 1078A

Heinkel AG engineers, in designing the P.1078A, were eager to obtain the smallest possible fuselage cross-section. The need to use every bit of space for equipment meant that construction would have to be considered by carefully. The space in the fuselage nose held the radar aim-

The Heinkel He P.1078A. Scale model by Steve Malikoff.

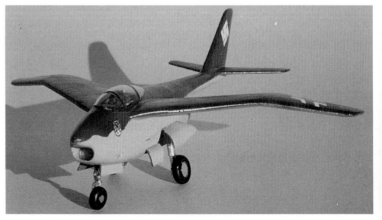

The Heinkel He P.1078A. Scale model by Steve Malikoff.

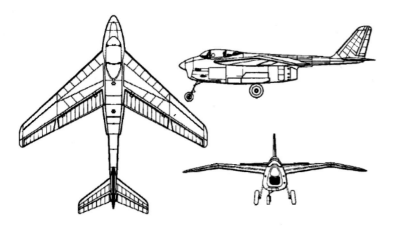

The Heinkel He P.1078A. Scale model by Steve Malikoff.

ing mirror. The main wheels were to be retracted forward to lay beside the air intake duct for the turbojet while the armament, the ammunition, the radio, and the nose wheel were to be fitted into the space left over in front of the retracted main wheels. As a result, most of the fuselage skin was detachable in order to gain access to the internal equipment.

The P.1078A's sweptback wings were to be built of wood or light alloy since the fuel was to be carried in the wings and have a dihedral angel for more efficient fuel drainage. The feasibility of carrying fuel in the wooden wing compartments had been proved on their HeP.162. Electric deicing was to have been used, and in case of fire, the pilot would escape by an ejection seat.

PROJECT 1078B

Initially, Heinkel engineers had planned to design the P.1078B as an all-wing airplane. That is, to make the ratio of wetted fuselage-to-wing as small as possible. However, Siegfried Günter felt that the P.1078B would be directionally unstable, and he sought to restore stability through the

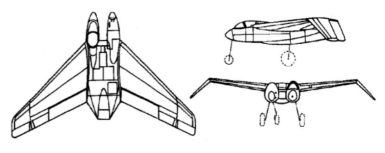

use of cathedral wingtips. The selection of twin noses was an experiment by the Günter led team to gain stability while obtaining the smallest wetted fuselage surface. It was hoped that with the two noses yaw oscillation wold be favorable. For a time Günter had flirted with asymmetrical designs, that is, placing the fuselage alongside the turbojet. Dr. Richard Vogt, chief designer/engineer at Blohm and Voss, had experienced some interesting results with asymmetrical aircraft in the 1940s.

The Heinkel He P.1078B. Scale model by Dan Johnson.

The Heinkel He P.1078B. Scale model by Dan Johnson.

The Heinkel He P.1078B. Scale model by Dan Johnson.

Günter was interested too, and would have experimented if Germany had not lost the war. However, the complications of constructing such an aircraft were overwhelming under the circumstances, and the Heinkel team compromised and chose the twin-nose alternative instead.

The P.1078B's wing was to be built in three main sections for ease of production. Later it would be riveted together in assembly, with the exception of cathedral wingtips, which were to be detachable. The two fuselage nose sections were also to be detachable, while the remaining nacelled portions were to be constructed integrally with the wing. Armament was to be carried in the right fuselage. In addition, there would be sufficient space for other equipment in the rear part of the fuselage. The landing gear was to be as simple as it could be made with the main wheels turning about a fixed axis into the sides of the fuselage and the nose wheel retracting into the right nose.

To Ernst Heinkel the P.1078B projected an unnatural impression. He believed that above all else a fighter should have a good view - and the view to the right in the P.1078B obviously was bad. Heinkel felt strongly that a limited view put a burden of insecurity on the pilot, something not to be tolerated for a fighter aircraft. He even felt that placing the pilot to the port side was unnatural and impractical. One thing that pleased him about the P.1078B was the downward bending wing tips on the aircraft's outer wing. Heinkel engineer's thought this arrangement highly interesting and should definitely be tested, although not with twin fuselages, because it just might provide the controllability absent in the Horten brothers' Ho 229 all-wing high-speed fighter/bomber aircraft.

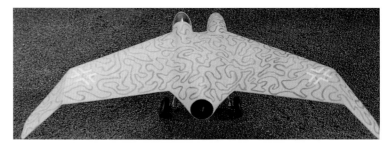

The Heinkel He P.1078B, view from the rear. Scale model by Dan Johnson.

The Heinkel He P.1078B, close-up view of its twin noses.

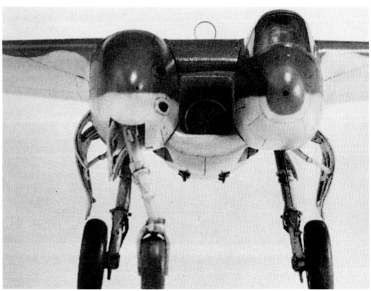

The Heinkel He P.1078B, a head on view showing the jet engines air intake between the two noses.

PROJECT 1078C

Following the Horten brothers' successful work with the all-wing and Alexander Lippisch's equally successful work with his tailless and delta planforms, designing tailless and delta-wing aircraft for all military roles was in vogue. Other leaders in the field at war's end were Blohm and Voss and Heinkel. One such Heinkel design was the P.1078C, a tailless aircraft with its wings swept back at 40 degrees. Unlike its Blohm and Voss counterpart, the P.212, and the Junkers P.128, both of which had vertical fins, the P.1078C had anhedral wingtips. It was thought that the anhedral design would cause the craft less trouble with stability as it approached the speed of sound.

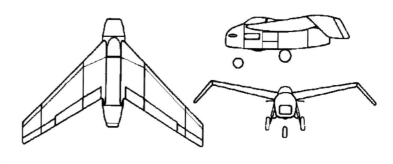

No prototypes of the P.1078C series were built, and proposed construction details are not known. It is assumed that the P.1078C would have had an all-metal structure and, like similar Heinkel AG turbojet-powered aircraft designs produced in 1945, all wood, shoulder-mounted wings. All of the aircraft's fuel was to be contained in wing tanks with a total capacity of 320 gal (1,450 l). Like the Blohm and Voss and Junkers tailless designs, it was to have a pressurized cockpit fitted with an ejector seat and armor protection. The fuselage was to be fairly long and to extend out, but stopping just short of the wings' trailing edges. Its bubble canopy was to be positioned above the curved tubular steel duct that was to lead from the nose intake to the engine and form part of the fuselage structure. Power was to be provided by a single HeS 011A turbojet having 2,866 lb (1,300 kg) of thrust. Captured engineering drawings show the air intake in the nose appearing rectangular, suggesting that the air duct itself was to be flattened to provide space above for the cockpit and below for the retracted nose wheel. The main wheels had nowhere to go except forward and inward into the side of the fuselage. The P.1078C was to have been armed with four 30 mm 108 MK cannons, two mounted on each side of the cockpit.

The P.1078C was one of Ernst Heinkel's favorite Günter-designed turbojet-powered fighter projects. The other was the P.1079 night hunter. Overall, Heinkel considered the P.1078C to be an excellent design for many reasons. He especially liked that its HeS 011 turbojet had been placed aft of the cockpit and he felt that this location would provide for a lower center of gravity and improved view for the pilot.

With the downward bending wingtips, Ernst Heinkel felt that Günter was on to something important with the P.1079. Although the wing-bend meant a more expensive construction, Heinkel believed it would provide the considerable advantage of greater maneuverability and better "rolling" ability in the air. One recommendation he offered his design team concerned safety for the aircraft during a "belly" landing. Heinkel wanted the HeS 011 protected against the danger of fire during such a landing so the P.1078C's fuselage bottom should be fortified with steel tubes in such a way that they would serve as skids similar to those found in a snow sled.

Specifications: Project 1078C

Engine	1xHeinkel-Hirth HeS 011 turbojet having 2,866 lb (1,300 kg) of thrust
Wingspan	29.6 ft (9 m)
Wing Area	NA
Length	19.8 ft (6 m)
Height	NA
Weight, Empty	5,403 lb (2,459 kg)
Weight, Takeoff	8,533 lb (3,870 kg)
Number of Crew	1
Speed, Cruise	614 mph (990 km/h)
Speed, Landing	114 mph (183 km/h)
Speed, Top	636 mph (1,025 km/h) at sea level
Radius of Operation	932 mi (1,500 km)
Service Ceiling	42,313 ft (12.9 km)
Armament	4x30 mm MK 108 cannon
Rate of Climb	4,018 ft/min (1,255 m/min)
Bomb Load	NA
Flight Duration	90 min

PROJECT 1079

Known as a Zerstörer (Destroyer) by the Heinkel AG design team lead by Siegfried Günter, the P.1079 was a night fighter design and one of Ernst Heinkel's favorite projects. Highly stressed, capable of a full-throttle takeoff with a 3 degree flight angle, armed to the teeth with six cannons (four-firing forward and two-firing rearward), radar, and enough power from its twin HeS 011 turbojets to produce a total 5,732 lb (2,600 kg) of thrust, it would be propelled through the night skies at 621 mph (1,000 km/h) for nearly three hours. The P.1079 would be a deadly, high-speed hunter in pursuit of RAF and USAAF bombers.

The Heinkel He P.1078C, an artist's version. *Courtesy of Lufwaffe Secret Projects Fighters 1939-1945.*

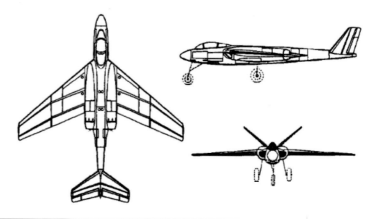

The Heinkel He P.1079A, overall view from above. Scale model by Dan Johnson.

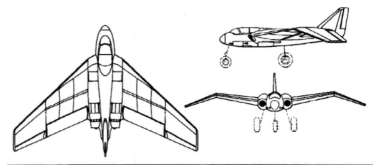

The Heinkel He P.1079A, head on view. Scale model by Dan Johnson.

The Heinkel He P.1079A look-a-like, the British Supermarine 508.

By 1945 German radar had been perfected to the point where it could be installed in an airframe, thus producing a highly effective night fighter. The P.1079 would be one of the first-generation airframes designed specifically around the radar. With a two-man crew, radar, and large-volume fueled tanks, the P.1079 would be able to range up to 1,400 mi (2,200 km) in search of enemy bomber formations.

Three versions of the P.1079 were planned. The only difference between them was the degree of the wing's backward sweep. With the preferred 45 degree sweep, the P.1079 had a designed cruising speed of 614 mph (990 km/h). Solid-fuel assist rockets were to help the P.1079 achieve an exceptionally fast and steep climb after takeoff.

In his quest to design an aircraft that would reach nearly 621 mph (1,000 km/h), Siegfried Günter incorporated the many design and finishing techniques he and his late brother Walter had used on their He 64 and He 70 high-performance aircraft from the early 1930s. Everything on his P.1079 was to be aerodynamically smooth, and the entire aircraft was to be polished to a mirror finish. It would be an all-metal aircraft except for the wing tips, which would be made of wood. Even the four 30 mm MK 108 cannon barrel holes were designed to minimize drag.

Although the P.1079 was never built, design drawings show the proposed aircraft to look surprisingly similar to the post-war British Vickers Supermarine 508 single-seat fighter. One HeS 011 turbojet would be tucked under each shoulder-level sweptback wing next to the fuselage. The "butterfly" tail unit would consist of a tailplane and elevators set at about a 32 degree dihedral angle, the elevators also acting as rudders.

The main landing gear would retract into the fuselage. With a track width of only 8.6 ft (2.6 m), the two main wheels would rotate 180 degrees by a steering control. This type of undercarriage had been designed and built for use on their P.343 four-engine turbojet bomber project. The nose wheel was to be placed virtually in the nose cone section and retract rearward into the fuselage under the pilot's seat.

The P.1079's long, slender fuselage accommodated its two pilots in a back-to-back seating position. The back-facing pilot's job was to op-

The Fouge C. 170R "Magister" fighter trainer with twin turboméca Marboré II engines. Similar to the Heinkel He P.1079A.

The Heinkel He P.1079A. Scale model by Dan Johnson.

Ernst Heinkel AG - Rostock, Germany

The Heinkel He P.1079A, bore a striking similarity to the post-war British Vickers Supermarine 508.

The Messerschmitt ME P.1110.2. Scale model by Dan Johnson.

erate the craft's electrical distance-measuring device and aiming equipment (radar) housed in the plane's nose. Once the radar had identified an enemy aircraft, the P.1079's offensive fire would come from four 30 mm MK 108 cannons housed in the fuselage under the cockpit.

As a pursuit aircraft, the P.1079 would have needed only minimal protective armor. Moreover, Heinkel's design group believed that the Allies had no fighters capable of overtaking the P.1079 with its 621 mph (1,000 km/h) speed. Attack, when it occurred, would likely come only from behind and in very narrow angles. Consequently, two fixed, 20 mm MK 151 cannons at the rear of the aircraft were considered suf-

ficient protection. To aim the German "back-aim-binocular" the pilot would control the aircraft the same as aiming ahead. Enemy aircraft would be visible to the radar operator as a lighted dot in the Braun telescope screen, and the pilot would fly in such a way as to make the lighted dot move in the scope's cross-hairs. The "back-aim-binocular" could be raised out of the fuselage electrically, and lowered in order to minimize drag when not needed.

For a while the Günter design team gave some thought to turning the P.1079 into an all-wing design and the team went so far as to develop a wind tunnel model without any vertical fin an attached hinged rudder. Wind tunnel tests were disappointing. The wind tunnel model showed definite loss of directional control. The model was modified by the addition of small, downward-bending side rudders on the wings themselves, similar to those found on their P.1078C. Some directional control was re-established, but Siegfried Günter found that although the downward-bending side rudders were sufficient for flight, the test model still showed small amounts of flutter, a kind of "duck walk" as it flew. Tests indicated that the downward-bending rudders would not give the aircraft enough stability in order to aim cannon fire at enemy aircraft - the same problem the Horten brothers' experienced with their Ho 9V2 all-wing prototype fighter.

Next, Günter's design team placed a "butterfly" type-tail on the P.1079, but even this rudder control arrangement was an untested concept. By 1945 the V-controlling surface had not been tested on any airframe, although the design team felt that the V-control had a great future and no opportunity should be missed to examine it thoroughly. Heinkel AG wanted to build the prototype P.1079 with an interchange-

The Heinkel He P.1079A. Scale model by Dan Johnson.

able fuselage end, so that the aircraft could be tested with the normal controlling surface as well as the V-controlling design. However, Günter was leaning toward a third controlling surface design in the closing months of the war, a Multhopp T-tail arrangement similar to the one developed through Deutsches Forschungsinstitut für Segelflug (DFS) research and used on their DFS 346 supersonic research aircraft. Günter was impressed with wind tunnel tests on a scale model having this arrangement similar to Hans Multhopp's Fw P.183 with its highly sweptback T-tail control surface and which also appeared on the Soviet copy of the P.183, the MiG-15.

Specifications: Project 1079

Engine	2xHeinkel-Hirth HeS 011 turbojets having 2,866 lb (1,300 kg) of thrust each, plus 4x2,205 lb (1,000 kg) solid-fuel boosters for takeoff assistance.
Wingspan	39.4 ft (12 m)
Wing Area	NA
Length	44 ft (13.4 m)
Height	NA
Weight, Empty	10,803 lb (4,900 kg)
Weight, Takeoff	32,808 lb (10,000 kg)
Number in Crew	2
Speed, Cruise	559 mph (900 km/h)
Speed, Landing	NA
Speed, Top	621 mph (1,000 km/h)
Radius of Operation	1,771 mi (2,850 km)
Service Ceiling	32,810 ft (10 km)
Armament	4x30 mm MK 108 cannon with 300 rounds of ammunition each, plus 2x20 mm MG 151 cannon with 200 rounds of ammunition each
Rate of Climb	NA
Bomb Load	NA
Flight Duration	3 hrs

PROJECT 1080

Early in 1945 Heinkel AG received an order from the Air Ministry to design a fighter aircraft powered by twin ramjets; however it was never built. The ramjet was perhaps the least popular form of propulsion tried during the 1930s and 1940s. Despite its simplicity and promise of high speed and performance, ramjet development was slow, for several reasons. A major drawback was that ramjets needed to be boosted to their self-operating forward velocity of about 149 mph (240 km/h). Furthermore, they were notorious for excessive fuel consumption. Since no aircraft had ever flown solely on ramjet power, it is not clear why the

The Heinkel He P.1079A look-a-like, the British Supermarine 508.

After the ban against aircraft manufacturing in West Germany was lifted in 1955, Heinkel came out with its He 211. It was configured with a butterfly tail and annular air intake for its two rear mounted turbojets which was a similar configuration planned for the Heinkel He P.1110.2.

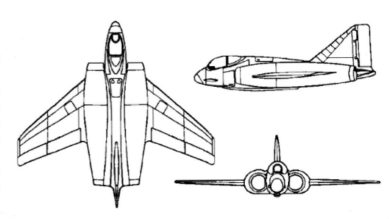

Full side view of the Messerschmitt ME P.1110.2. Scale model by Dan Johnson.

Air Ministry even bothered to issue specifications for a ramjet fighter-interceptor.

Ramjet experiments had been conducted in Germany since the late 1930s by Hellmuth Walter, Focke-Wulf, DFS, and the Luftfahrtforschungsanstalt (LFA), or Aircraft Research Establishment located at Ainring, but only Dr. Eugen Sänger at the LFA had actually brought ramjet development to the point of towing experimental designs behind Dornier aircraft. Even then, the ramjet was pretty much a laboratory curiosity, and no one really knew if it could ever become a reliable power plant for an all-round fighter aircraft.

There is very little information available about the P.1080, and Ernst Heinkel himself did not even mention this proposed aircraft in his memoirs. The original specifications called for two DFS ramjets of 3,440 lb (1,560 kg) thrust each. These DFS ramjets engines were about three feet (1 m) in diameter and mounted on either side of the P.1080's fuselage. The outer surfaces of the ramjet ducts were faired into the wing; it is assumed they were designed that way to maximize exposure to the airstream for cooling purposes. The ramjet-powered P.1080 design reflects an urgency to get the craft into the air against the Allied heavy bombers. For instance, it was to have used the same style of sweptback wings and elevator controls that were planned for Heinkel AG's P.1078 series. Design time was further reduced on this project through the use of a single orthodox tail fin and rudder.

Heinkel airframe designers placed the P.1080's pilot's cockpit well forward on the fuselage. The nose section contained the plane's radar unit and two 30 mm MK 108 cannons. Fuel was housed aft of the cockpit to the rear of the fuselage.

For takeoff, a jettison-undercarriage was to be used and boosting power was provided by four solid-fuel rockets each of 2,205 lb (1,000 kg) thrust. Landing was to be made on an extendable skid. The post-war Swiss N-20 prototype of the first Aiguillon (Sting) STOL jet fighter-bomber, with its four early SM01 engines bares a striking resemblance to the ramjet-powered P.1080.

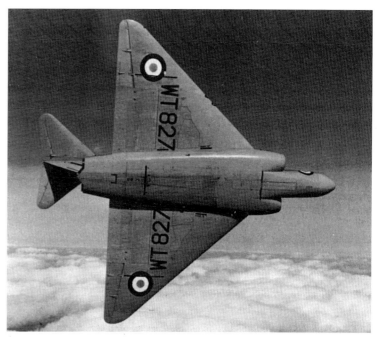

The British Gloster "Javelin" looking very much like the Heinkel He P.1080, shown here hard over in a turn.

Specifications: Project 1080

Engine	2xDFS ramjet engines having 3,440 lb (1,560 kg) thrust each, plus 4x2,205 lb (1,000 kg) solid-fuel boosters for takeoff assistance.
Wingspan	29.3 ft (8.9 m)
Wing Area	215.2 sq ft (20 sq m)
Length	26.8 ft (8.2 m)
Height	NA
Weight, Empty	NA
Weight, Takeoff	NA
Number of Crew	1
Speed, Cruise	NA
Speed, Landing	NA
Speed, Top	621 mph (1,000 km/h)
Radius of Operation	NA
Service Ceiling	NA
Armament	2x30 mm MK 108 cannon
Rate of Climb	NA
Bomb Load	NA
Flight Duration	NA

The Messerschmitt ME P.1110.2. Scale model by Reinhard Roeser.

10

Henschel Flugzeugwerke AG - Berlin, Germany

Secret Projects and Intended Purpose:
PROJECT 122- turbojet-powered tailless bomber
PROJECT 132- turbojet-powered dive-bomber
PROJECT 135- turbojet-powered tailless fighter

History:
By 1930 Henschel & Sohon of Kassel was one of the most prominent manufacturers of steam locomotives, trucks, buses, marine engines, and diesel electric railroad cars in the world. Looking for ways to diversify, in 1931 the family-owned and operated company under the direction of Oscar R. Henschel was seriously considering the manufacture and sale of commercial aircraft. To the Henschel family, aircraft production would be a natural extension of its involvement with transportation vehicles.

At the time of their decision to manufacture airplanes, the family had not decided whether to buy out, or merge with, an established aircraft manufacturer, or simply to start from scratch. In 1931 they discussed with Hugo Junkers the possibility of merging with his firm, which was on the verge of financial collapse; however, discussions broke off in February 1932 when the Reich came forward with some financial assistance for Junkers. Later Henschel would be criticized for not carrying out the merger, especially in view of the Reich's takeover of Junkers in 1934. Whether or not Henschel was responsible for breaking off negotiations with Junkers is not known, for details of the proposed agreement were never made public. Henschel was known for its aggressive business practices, and, whatever the reason for the break off with Junkers, the firm decided at that point to start its own aircraft manufacturing subsidiary, the Henschel Flugzeugwerke located in suburban Berlin.

Henschel's timing couldn't have been better: In mid-February 1932 the Air Ministry announced its Für die Luft (For the Air) program, which was in essence the Nazi Party's open invitation to major German industrial concerns to participate in the coming expansion of the Reich's air arm, the Luftwaffe.

On 30 March 1933, exactly 2 months after Hitler came to power, Henschel's new Flugzeugwerke had an experimental trainer ready for the Air Ministry's consideration. Known as the Henschel Hs 121, the trainer was a low-powered, single-seat, high-wing monoplane, a rather attractive aircraft for the firm's first venture in aircraft design. Oscar Henschel had assembled a group of talented individuals to help him launch his aircraft works. Heading up the Henschel group was Friedrich

Founder and guiding force behind wartime Germany's high-technologycompany, Henschel Flugzeug-werke was Oscar R. Henschel, photographed in 1938 at age 39.

The Henschel Hs 121 was the first aircraft to be designed and built by the new Henschel Flugzeugwerke under the "Für die Luft" (For the Air) program of the Luftwaffe. The Hs 121 turned out to have poor characteristics as a fighter trainer and was rarely used.

Hugo Junkers.

The Henschel Hs 126, which appeared in 1939, turned out to be the Luftwaffe's most effective short-range reconnaissance aircraft of World War II. The Hs 126, in addition to its reconnaissance duties, was also used for ground strafing and light bombing missions.

Henschel's successful anti-tank aircraft, the Hs 129, with its 75mm BK anti-tank cannon, was used extensively against Soviet tanks.

Designers gave the Henschel Hs 129 a narrow profile in an effort to present as small a target as possible to Soviet ground gunners seeking to defend their tanks from attact by Hs 129's.

US Army troops prepare a captured BK 75mm anti-tank cannon for shipment to the United States after the war. This cannon, used extensively on Henschel Hs 129, was one of the Luftwaffe's largest anti-tank weapons of World War II.

Nicholaus, formerly with the Hellmuth Hirth Versuchsbau. It was a heady time for the Henschel group, especially since Oscar Henschel had told his people that he would not stand for his company's being merely a manufacturer of the designs of established firms. He was interested in practical research, and he wanted his firm to be on the cutting edge of technology, to take ideas from research laboratories and universities and put them into practice. At the same time, the firm would also manufacture work-a-day aircraft for the Luftwaffe. Through Oscar Henschel's energy, resources, and drive to succeed in whatever his family undertook, the Henschel Flugzeugwerke prospered. In 1933, a year after the subsidiary's formation, construction began on an aerodrome and factory at Schönefeld near Berlin. By the end of 1936 Henschel was employing approximately 7,000 workers; most were engaged in the manufacture of their bi-wing Hs 123 dive bomber and Junkers' Ju 86 medium-range bomber, both of which were used extensively in the Spanish civil war.

Oscar Henschel, Friedrich Nicholaus, and chief engineer Erich Koch all believed that the future of Henschel Flugzeugwerke did not lie in manufacturing other manufacturer's aircraft under license. Henschel wanted to be known as Germany's high-tech weapons manufacturer and thus embarked on a course to develop new designs and concepts. The choice of Nicholaus as chief of design was a particularly good one for Oscar Henschel. Nicholaus had developed close ties with the Deutsche Versuchsanstalt für Luftfahrt (DVL) prior to joining Henschel. As a result, Henschel continued to maintain close ties, indeed, an active interest in DVL's work. This active participation, for example, included the design and development of high-altitude pressure cockpit cabins and in what would become a Henschel specialty, air-launched stand-off missiles. Helped along by DVL's Dr. Herbert Wagner, formerly of Junkers Jumo where he had been a leading pioneer in their 004 turbine engine

Secret Aircraft Designs of the Third Reich

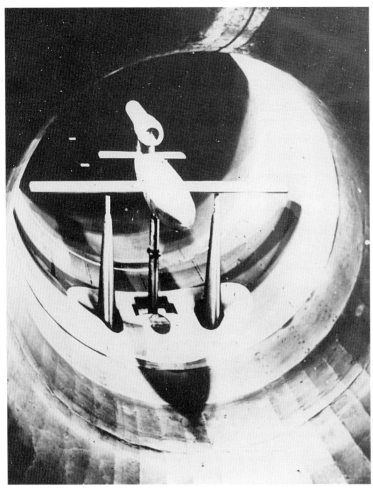

Henschel designers had full access to the giant wind tunnels at DVL-Adlershof. Shown here is a Fieseler Fi 103 in a DVL wind tunnel, the so-called "Buzz bomb," which terrorized Great Britain.

Wind tunnel at DVL-Adlershof in 1938. The huge nature of this wind tunnel can be gauged by the size of the man standing to the left of the window.

design development, Oscar Henschel gave Wagner all the resources needed in his missile work. Wagner's success in missile development was astonishing and later resulted in several advanced and comparatively successful weapons. One was the Henschel Hs 293 Egret, a rocket-propelled, remotely guided missile which, when launched by Dornier Do 217s, sank numerous transport vessels and warships. Another was the Henschel Hs 117 Schmetterling (Butterfly) ground-to-air missile.

Apart from Henschel's pioneering work in the field of ground-to-air and air-to-air missiles, the firm undertook considerable research on turbine engine powered aircraft. In the early 1940s Henschel had given a great deal of thought to producing its own turbine engine, but was talked out of it by the Air Ministry. In terms of aircraft projects, by war's end Henschel had a comparatively small portfolio. The reason was that Oscar Henschel saw the firm's future, as well as Germany's, in ground- and air-launched rocket's to bring down Allied long-range fighters and heavy bombers. Henschel Flugzeugwerke, however, had completed components for prototype turbojet-powered dive bomber prototype, their P.132. It was a novel design in which the pilot was placed in a prone position with a single dorsally mounted BMW 003 turbojet. It had not been fully assembled at the time the Red Army overran Henschel's suburban Berlin-Schönefeld advanced aircraft development

Professor Hubert Wagner.

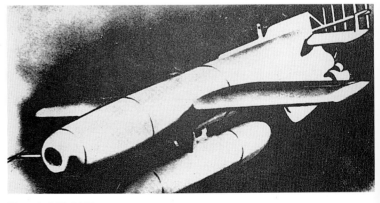

Henschel Hs 293D.

Henschel Hs 294B.

Henschel Hs 293. A glider bomb carried by the He 177's of KG 40 in anti-shipping operations. The 1,100 lb warhead was fitted in the nose of the missile underneath was the 1,300 lb thrust rocket engine which provided power. On the tailplanes were the posts which supported an aerial. This item received radio command signals from its launching aircraft, while a the rear of the missile were flares to assist tracking.

Henschel Hs 298 air-to-air missile. It carried a 106 lb warhead and a proximity fuse. It could be rail launched or could be carried aboard an Fw 190 fighter. Although tested thoroughly as prototypes, it did not go into production.

operations. Only two other turbojet-powered design projects are thought to have been on the drawing boards of Henschel at war's end. These included their P.122 which was a proposed twin turbojet tailless bomber expected to have a maximum speed of 615 mph (990 km/h) and their P.135 which was a proposed tailless fighter with a semi-delta wing configuration and powered by a single HeS 011 turbojet engine.

After the war Henschel lost both the Schönefeld and Johannesthal facilities to the Red Army. Both plants, which together covered 1.3 million square feet (125,415 sq m) and employed 14,000 workers, were completely dismantled by the Russians post-war and transported by rail to the USSR. Most of the Henschel people escaped the Red Army by going west to the city of Kassel. It was here, post-war, that Henschel engineers and employees regrouped and began producing non-aircraft transportation equipment and vehicles. In 1955 Henschel sought to re-enter the aviation business, drawing on the experience of the large number of former aircraft specialists on its staff, many of whom were experts in turbojet engine aircraft and missiles in the 1940s by concentrating on the manufacture of helicopters. However, the business climate

Focke-Wulf Fw 200 carrying two Henschel Hs 293 "Condor" air-to-ground anti transport vessel and warship missiles.

The Henschel Hs 295, another of the many members of the Henschel series of high-tech air-to-air anti-aircraft missiles.

A Heinkel He-177-A3 carrying two Henschel Hs 293 missiles.

On the drawing boards when the war ended was the very interesting Henschel Hs P.75. With its forewing or canard, its contra-rotating pusher propellers, and its wing tip rudders, this tailless fighter exhibited all the design characteristics US National Aeronautics and Space Administration officials were investigating for possible application to aircraft beyond the year 2000. Scale model by Dan Johnson.

Henschel Hs P.75. Scale model by Dan Johnson.

for Henschel's basic products such as diesel-electric railroad locomotives, trucks and buses, diesel engines, and tool machines had softened, and in 1957 the firm went through reorganization proceedings (bankruptcy) in order to settle the claims of its creditors. Under its reorganization and recapitalization, Henschel survived and continues to manufacture diesel electric locomotives and other heavy equipment. It never did get back into the manufacture of fixed wing aircraft again. Oscar Henschel died in Switzerland on 9 February 1982.

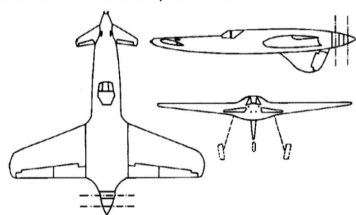

PROJECT 122

The Henschel P.122 was a proposed high-speed, night-fighter pursuit aircraft. For a time in late 1944 and 1945 the Air Ministry was interested in developing a twin-engine fighter capable of remaining aloft up to three hours at a time and reaching speeds up to 621 mph (1,000 km/h). The P.122 was Henschel's offering: Heinkel proposed their P.1079, and Dornier their P.256. The later also was entered in the competition held by the Chef der Technischen Luftrüstung, or Chief of Technical Air Equipment, in February 1945 for a combined night and bad weather

Curtiss XP-55, right side view on ground. November 1943.

Henschel Flugzeugwerke AG - Berlin, Germany

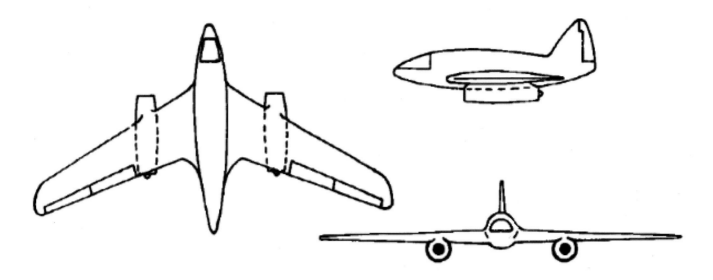

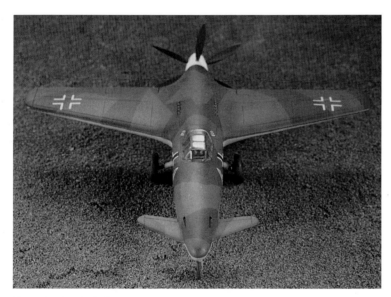

Henschel Hs P.75. Scale model by Dan Johnson

Curtiss XP-55. November 1943.

Germany's early military success in World War II was due in great part to the Junkers Ju 87 dive-bomber. But it proved to be no match for Allied high-performance fighters, and the RLM began looking for a turbojet-powered replacement.

Henschel Hs P.75. Scale model by Dan Johnson.

185

Blohm & Voss design team: Left to right: Otto Dahlke, Hans Pohlmann, Pre-design, Dr.Ing. Richard Vogt, Chief Engineer. Pohlmann, when he worked at Junkers early in his career designed the Ju 87 "Stuka." May 1937.

fighter. No action was taken on Henschel's design and no mockups or prototypes were constructed.

Henschel's P.122 was characterized by placement of its twin BMW 018 turbojets of 7,500 lb (3,402 kg) thrust each in nacelles under the wings. It also had an unusually large vertical stabilizer and rudder, reflecting the prevailing belief that, should one of the powerful, widely separated 018's turbines have to be shut down or quit, a large amount of rudder would be required to counteract the other turbine's thrust.

In spite of its cigar-shaped fuselage and its flush-mounted cockpit for its two-man crew, features that should have helped increase speed, Henschel's tailless fighter was not expected to reach the Air Ministry's goal of 621 mph (1,000 km/h). The best the engineers could do was an estimated speed of 615 mph (990 km/h); however, this was still considered acceptable by the Air Ministry because the aircraft could also be used as a high-speed bomber.

As a high-speed night pursuit aircraft, the P.122 would have needed only minimal protective armor for those chance encounters with enemy fighter escorts. As an offensive night fighter, it was to be fitted with six 30 mm MK 108 cannon, each having 120 rounds of ammunition. Four of the MK 108's were to be set in a fixed positioning the wing's leading edges. The other two were to be mounted in the fuselage, in a position to fire upward into the bellies of Allied heavy bombers.

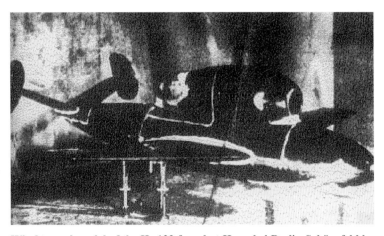

Wind tunnel model of the Hs 132 found at Henschel Berlin-Schönefeld by Soviet intelligence mid 1945.

This is about all the Soviets found at Henschel Flugzeugbau, Berlin-Schönefeld regarding the Hs 132. What they found was the Hs 132's complete fuselage out of wood and its BMW 003 engine.

Henschel's Hs 132 was offered as a replacement for the slow-moving and highly vulnerable Junkers Ju 87 dive-bomber in 1945. This photo is an airbrush impression by Gert Heumann. The Hs 132 was an incomplete prototype at war's end.

Henschel's adoption of the tailless design for its P.122, and later its P.135 was not an imitation of Alexander Lippisch, but was based on studies by their own engineers and designers and from their colleagues at DVL. Due to its projected size, the Hs P.122 would have been built of metal; only the wingtips would have been of wood. It is doubtful that the P.122 could have been built in early 1945, when the Air Ministry was considering the merits of this proposed aircraft against designs submitted by Heinkel and Dornier. With a loaded weight of around 5 ton, the P.122 would have required a great deal of steel and aluminum, both of which were in very, very short supply. Moreover, it would have been about 2 years before the 7,500 lb (3,402 kg) thrust BMW 018 turbojet engine the P.122 required would have been available. Four to six BMW 003s might have been used in the meantime but even then the extreme fuel shortage would have made this proposed design impossible to operate.

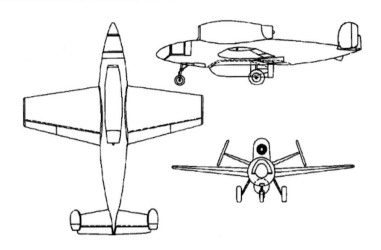

Specifications: Project 122

Engine	2xBMW 018 turbines having 7,500 lb (3,402 kg) of thrust each.
Wingspan	73.8 ft (22.5 m)
Wing Area	NA
Length	40.7 ft (12.4 m)
Height	NA
Weight, Empty	NA
Weight, Takeoff	33,290 lb (15,100 kg)
Crew	2
Speed, Cruise	581 mph (935 km/h)
Speed, Landing	NA
Speed, Top	615 mph (990 km/h) at 32,810 ft (10 km)
Rate of Climb	NA
Radius of Operations	1,243 mi (2,000 km)
Service Ceiling	55,777 ft (17 km)
Armament	4x30 mm MK 108 fixed forward cannon, and 2x30 mm MK 108 oblique firing cannon, all with 120 rounds of ammunition each
Bomb Load	NA
Flight Duration	3 hours

The cockpit of the Hs 132. Note its forward glass amour. Gert Heumann's air brushed Hs 132 impression which has been widely circulated in the aviation community does not appear to be correct given the very thick glass amour plate.

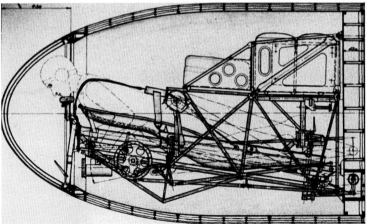

Engineering drawing of the pilot's couch for the prone flying position as developed by DFS.

A close-up view of the Hs 132's thick armored glass windscreen. It appears that the Hs 132 would have been flown with control system similar to "joy sticks."

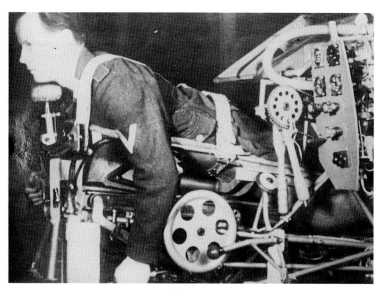

DFS research on prone pilot positions was extensive. Shown here is Rolf Mödel in position testing its fit, comfort and practicality.

Henschel Hs P.132 as seen from front and above.

PROJECT 132

The Hs P.132 was a single-seat ground attack aircraft and dive bomber powered by a single BMW 003 turbojet. A mid-wing with twin tail fins and rudders, it was expected to have a maximum speed of 435 mph (700 km/h) when carrying its 2,205 lb (1,000 kg) bomb load, and 485 mph (780 km/h) unloaded.

When the war ended in May 1945 and the Soviet's Red Army overran Henschel's Schönefeld facilities in suburban Berlin, all the component parts for the first prototype the P.132 V1, had been completed. However, even in May 1945 the P.132's wings had not yet been attached to its fuselage. Work on the prototypes appeared to be progressing slowly, especially the cockpit, and it is not known if and when the P.132 V1 prototype would make its first flight. Two additional P.132 prototypes, the V2 and V3, were in various stages of completion at Schönefeld. It is believed that Henschel had received an order for series construction of their P.132 by the Air Ministry only two months earlier, in March 1945. The Soviets dismantled Henschel's Flugzeugwerke and transported the three prototypes to the Soviet Union. It appears that the Soviet's never completed the P.132 V1 either. Photos obtained from the former Soviet state aviation archives show the P.132 V1 in a hangar with large square sections of its fuselage skin removed. This was a frequent practice of the Soviets when they wished to view and study design and engineering details. One sees photos of Me 163's and He 162's with squares of their fuselage-skin removed, too. Afterward, the dissected fuselage and wings were given over to be recycled or to the trash collector. The widely published air-brushed photo showing a fully assembled P.132 parked out front an unknown hangar is a false image.

Henschel's aircraft manufacturing division had been interested in dive bombers since its founding in 1932. One of the firm's first aircraft designs, the Hs 123, a bi-wing dive bomber, had replaced the Luftwaffe's older Heinkel He 50 and Arado Ar 65 bi-wing dive bombers. Henschel Hs 123s made up the famed Legion Condor, which the Luftwaffe created in October 1935 both to provide ground support for the Nationalist forces in the Spanish civil war, and at the same time to test theories on dive bombing under combat conditions. The Hs 123's reached the Spanish war zone in December 1936 and were extraordinarily successful, strafing and bombing with complete freedom directly over the battle zones throughout Spain. Although it was obsolete by 1940 due to its severe lack of power when pulling out of a dive. The Junkers Ju 87 Stuka, which replaced the Hs 123 in 1936, carried out all of the Luftwaffe's dive bombing assignments right up through 1945. Completely vulnerable at that point, the Ju 87 Stuka suffered considerable losses beginning with the Battle of Britain in 1940. However, during the winter of 1942-43 at the Russian front it was found that the Ju 87, when

Henschel Hs P.132. Scale model by Frank Henriquez.

Henschel Hs P.132 as seen from side and front. Scale model by Frank Henriquez.

armed with 37 mm cannon beneath the wings, was unsurpassed as an anti-tank aircraft. The cannon's shell, with its tungsten core, was able to penetrate armor plate 3 inches (8 cm) thick. At last an airborne "tank buster" was available to the Luftwaffe, and it was not long before the Russians came to recognize it as their deadliest enemy. Nevertheless, the Ju 87 Stuka lacked range and speed and operated at the limit of its capacity in the closing days of World War II. With Russian industrial production unhampered due to its location behind the Ural mountains, well beyond the range of Luftwaffe bombers, the Air Ministry in mid 1943 began looking for a new dive bomber, one with speed, range, and endurance, to replace the aging Ju 87s. Henschel believed its Hs P.132 was that aircraft.

In designing the P.132, Henschel engineers sought to make the aircraft as small a target for ground gunners as possible. An unusual feature of the aircraft was placement of its pilot in a prone position. This idea came about through Henschel's long association with the Deutsche Versuchsanstalt für Luftfahrt (DVL). As early as 1937 DVL researchers had been conducting tests in which pilots flew aircraft while lying in a prone position. One of DVL's earliest prone-pilot research aircraft was the Berlin B-9, twin-engine light plane on which the fuselage's frontal area had been reduced substantially due to the pilot's prone position. When Henschel began initial design studies on what would become the P.132 in late 1943, designers were able to draw on data provided by flight tests of the B-9, and they incorporated the prone position for the pilot in their turbojet-powered dive bomber. This feature added to the aircraft's sleekness, and it was thought that Russia and other Allied anti-aircraft gunners who were becoming successful at hitting the slow-moving Ju 87s would be unable to track the speedy P.132 (it was estimated that the P.132 would reach speeds in excess of 500 mph (805 km/h during a dive).

So that it could withstand the forces it would encounter while pulling out after a 500 mph dive, the P.132 was designed to a load factor of 12, compared with a load factor of 8 for a piston engine powered dive bomber. As required, Henschel engineers constructed the V1 prototype out of the simplest materials and the least possible amount of strategic materials. The fuselage had a metal framework with a steel covering, and the wings, as well as the tail fins and rudders, were constructed entirely of wood. The cockpit was extensively glazed and completely faired into the fuselage, and the single BMW 003 turbine was mounted above the fuselage in an arrangement similar to that used by Heinkel on their He 162. As on the He 162, the P.132's turbine was to exhaust out over the tail, and Henschel's designers placed inclined end-plate fins and rudders on the aircraft to keep the hot exhaust from burning off the aircraft's "tail feathers."

The Henschel Hs 135 featured a so-called "double cranked leading edge" similar to that employed years later by the SAAB J-35 "Draken" double-delta fighter - Swedish. The compound sweep of the wing decreased from 42 degrees inboard to 38 degrees at the center section with tips being reduced further 15 degrees to improve stall characteristics.

Three versions of the P.132 were proposed. The A model was to be powered by a BMW 003A turbojet having 1,764 lb (800 kg) of thrust, the B series by a 1,984 lb (900 kg) thrust Jumo 004B turbojet, and the C version by a 2,866 lb (1,300 kg) thrust HeS 011 turbojet. Henschel engineers believed their A model would reach a maximum speed of 485 mph (780 km/h) in clean condition (no externally mounted bombs). The more powerful C model would have been used to carry a 2,205 lb (1,000 kg) bomb and/or a group of air-to-ground armor-piercing rockets such as the SC 1000 "Hermann" or the PC 1000 RS rockets. All three models were to be armed with twin 20 mm MG 151 cannon and two 30 mm MK 108 cannon.

Specifications: Project 132

Engine	1xBMW 003A turbojet having 1,984 lb (800 kg) of thrust
Wingspan	23.6 ft (7.2 m)
Wing Area	159.4 sq ft (14.8 sq m)
Length	29.2 ft (8.9 m)
Height	NA
Weight, Empty	NA
Weight, Takeoff	7,496 lb (3,400 kg)
Crew	1
Speed, Cruise	NA
Speed, Landing	95 mph (153 km/h)
Speed, Top	485 mph (780 km/h) without externally mounted bombs at 19,686 ft (6.0 km)
Rate of Climb	NA
Radius of Operation	423 miles (680 km)
Service Ceiling	34,450 ft (10.5 km)
Armament	2x20 mm MG 151 cannon and 2x30 mm MK 108 cannon plus air-to-ground armor-piercing rockets
Bomb Load	1x2,205 lb (1,000 kg) bomb, or 2x1,102 lb (2x500 kg) SC or SD bombs
Flight Duration	1 hour

PROJECT 135

In February 1945 the Oberkommando der Luftwaffe (OKL), as part of its "emergency fighter program," requested proposals for a single-seat, single-turbojet-powered fighter. Five firms were invited to submit design proposals. Messerschmitt offered its P.1106, Junkers its P.128, Blohm and Voss its P.214, and Heinkel its P.1078C. The winner was Focke-Wulf, with Hans Multhopp's P.183, later designated the (Kurt) Tank Ta 183 by the Air Ministry. For a time Henschel had considered responding to the OKL's request with their P.135 tailless fighter. However, the design staff did not have enough time to complete design studies, and the P.135 was never proposed to the OKL.

The P.135's design called for a semi-delta wing planform and slightly upturned with tips. The aircraft was to have an internal farmwork built

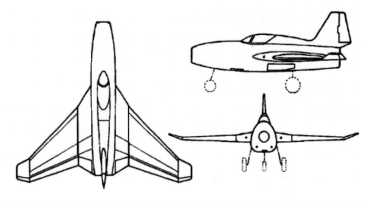

Heinkel He P.1708B.

SAAB J-35 "Draken."

Junkers Ju EF.128. Scale model by Dan Johnson.

Messerschmitt Me P.1106. Scale model by Dan Johnson.

Focke-Wulf Fw Ta P.183. Scale model by Günter Sengfelder.

of steel pipe and covered with plywood. A single-seater, it was to have a fully pressurized cockpit with a single-piece canopy and was to be located about midway back on the fuselage, flaring back into the base of its single vertical fin and rudder. The rudder was to extend aft of the central tail section. A single HeS 011 turbojet was to be mounted inside the fuselage, and a straight duct passing from its air intake in the nose to its exhaust in the tail.

Henschel's P.135 would have been a bit slower than the OKL's specifications called for, 611 mph (985 km/h) rather than 621 mph (1,000 km/h). Nevertheless, with its low wing loading of 41 lb sq ft (200 kg/sq m) the P.135 would, Henschel designers believed, have exceptional maneuverability. As specified by the OKL, the P.135 would have been armed with four 30 mm MK cannon. Two cannon would have been housed in the nose section, and one in the center section of each wing, close to the fuselage/wing root. A fully retractable tricycle landing gear was proposed, with the nose wheel retracting backward up into the belly of the fuselage and the two main wheel retracting flat up into the wing.

Although the firm was unable to meet OKL's deadline for submission of proposals, Henschel's design engineers continued to work on the P.135 tailless fighter design. With the fall of the Henschel Flugzeugwerke in April 1945, all design material was captured by Red Army intelligence and taken back to the USSR. It is not known if any of Henschel's turbine powered designs, including the P.135, were pursued by Russian aircraft designers. Probably not. Soviet aviation officials/designers generally rejected the tailless aircraft planform and did not pursue them. However, the Soviets apparently did some work based on the Junkers P.248, but not before adding a horizontal stabilizer and flying the aircraft designated as the MiG 1-270 (Zh) for a while before abandoning the bi-fuel HWK rocket drive aircraft. It is not known if the overall design of the P.135 never was duplicated in the USSR, or if its engineering and technical advances were investigated and applied to conventional Soviet turbine-powered aircraft.

A Heinkel He P.1079A. Scale model by Dan Johnson.

Dornier Do P.256.

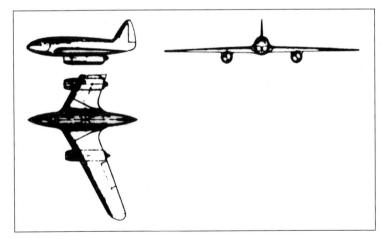
Henschel Hs P.122.

Specifications: Project 135

Engine	1xHeinkel-Hirth HeS 011 turbojet with 2,866 lb (1,300 kg) of thrust.
Wingspan	30.2 ft (9.2 m)
Wing Area	221 sq ft (20.5 sq m)
Length	25.6 ft (7.8 m)
Height	13.4 ft (4.1 m)
Weight, Empty	NA
Weight, Takeoff	9,040 lb (4,100 kg)
Crew	1
Speed, Cruise	NA
Speed, Landing	96.3 mph (15 km/h)
Speed, Top	611 mph (985 km/h) at 22,967 ft (7.0 km)
Rate of Climb	4,193 ft/min (1,278 m/min)
Radius of Operation	NA
Service Ceiling	45,934 ft (14.0 km)
Armament	4x30 mm MK 108 cannon
Bomb Load	NA
Flight Duration	NA

Color Gallery

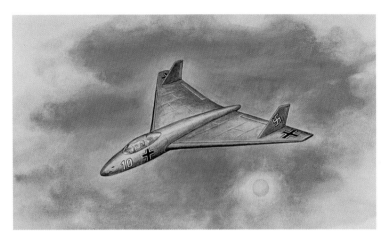

Arado P.NJ-1

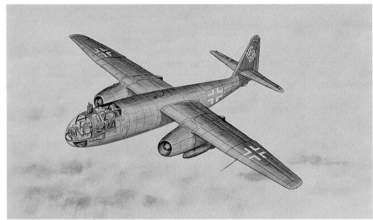

Arado 234B

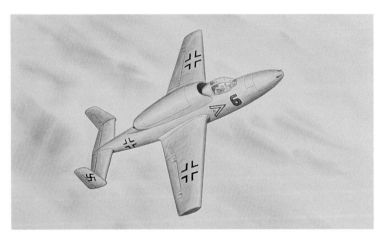

Arado P.580

Bachem Ba 349

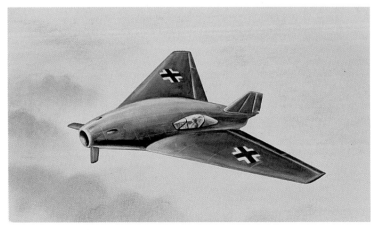

Blohm und Voss P.AE-607

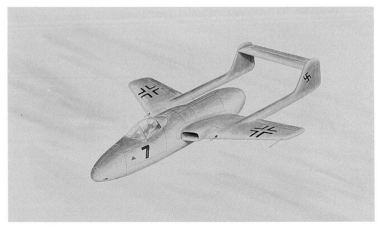

Focke-Wulf P.7

Color Gallery

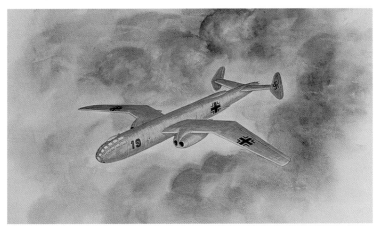
Blohm und Voss P-188

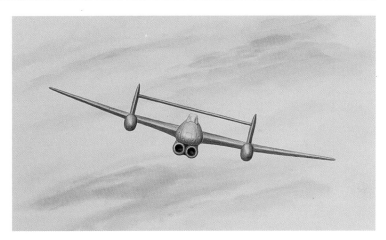
Blohm und Voss P-196

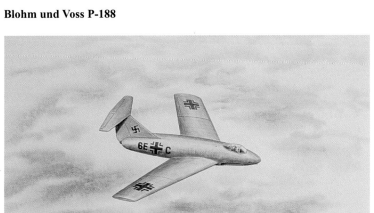
Blohm und Voss P-197

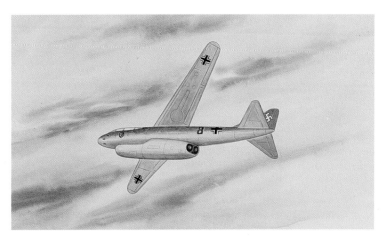
Blohm und Voss P-202

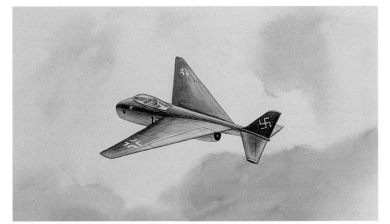
Blohm und Voss P-209

Blohm und Voss P-211

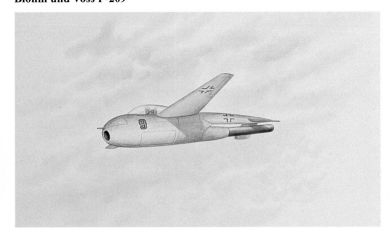
Blohm und Voss P-213

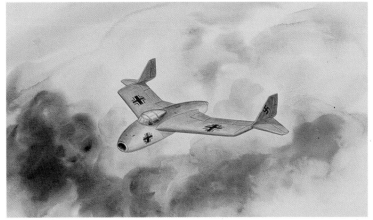
Blohm und Voss P-215

Secret Aircraft Designs of the Third Reich

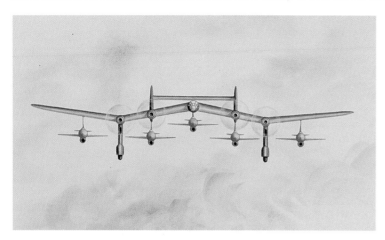
Focke-Wulf "Kamikaze" carrier

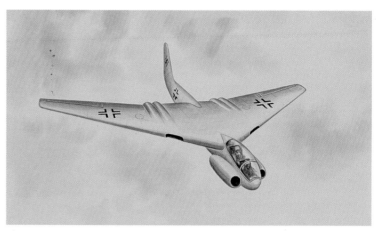
BMW Strahlbomber I

BMW Schnellbomber

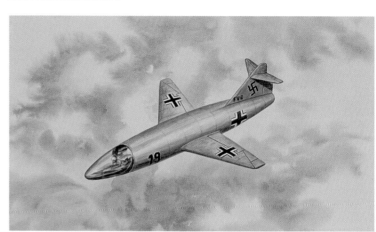
DFS P.346

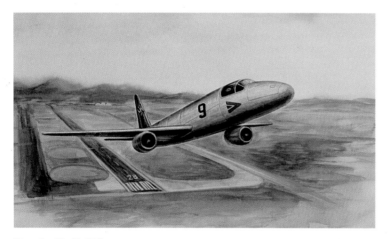
Dornier Do P-256

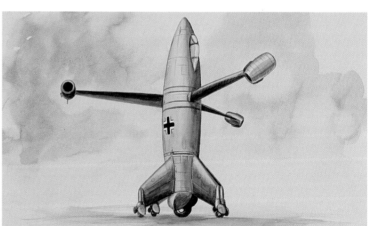
Focke Wulf Triebflügel

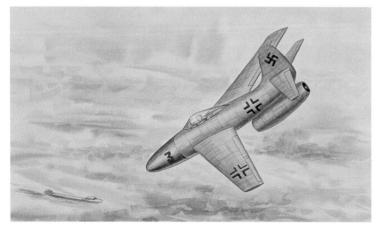
Focke Wulf "Kamikaze"

Focke Wulf TA P.183

Color Gallery

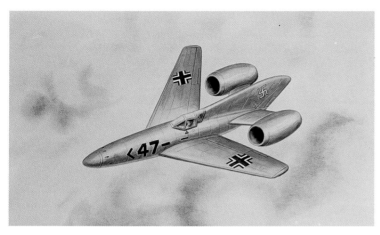
Focke Wulf TA P.283

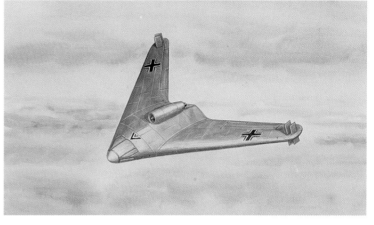
Gotha Go P-60B

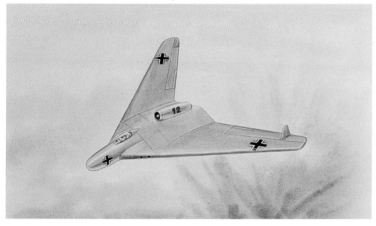
Gotha Go P-60C

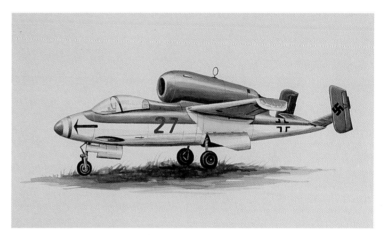
Heinkel He P-162

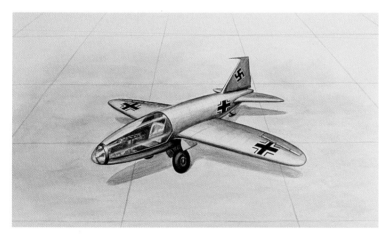
Heinkel He P-176

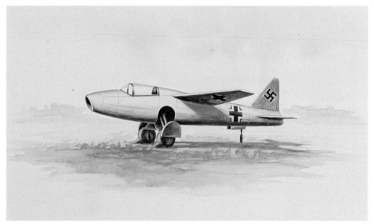
Heinkel He P-176

Heinkel He P-343

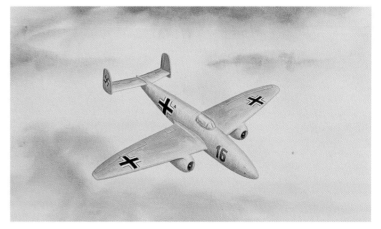
Heinkel He P-280

Heinkel He P-280

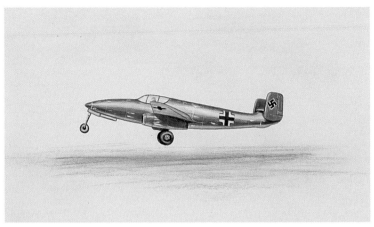
Heinkel He P-280

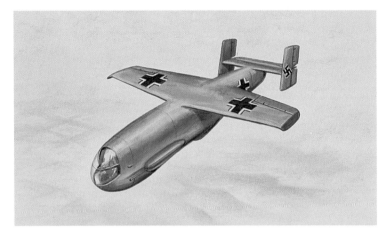
Heinkel He P-1077

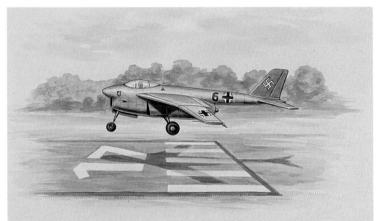
Heinkel He P-1078A

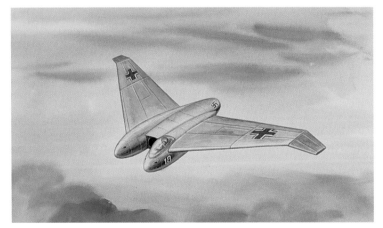
Heinkel He P-1078B

Heinkel He P-1078C

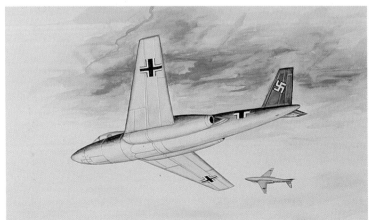
Heinkel He P-1079

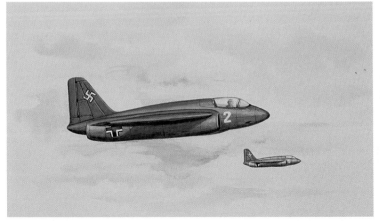
Heinkel He P-1080

Color Gallery

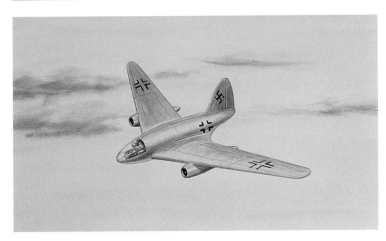
Henschel Hs P-122

Henschel Hs P-132

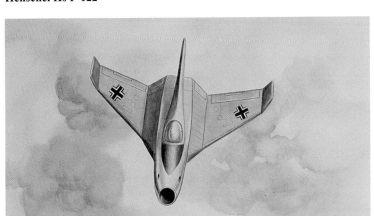
Henschel Hs P-135

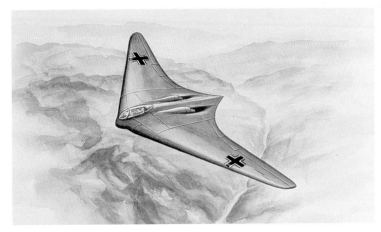
Horten Ho 9A

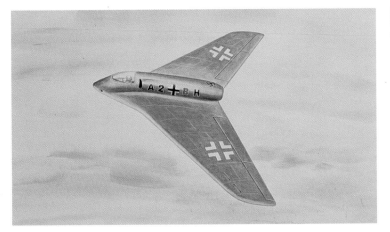
Horten Ho P-9B

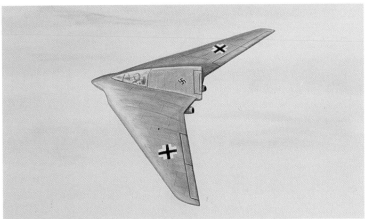
Horten Ho P-9C

Horten Ho P-10A

Horten Ho P.10B

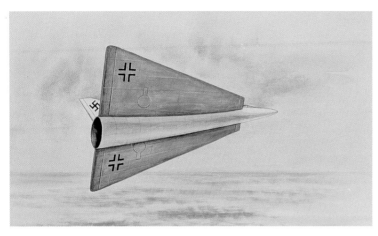
Horten Ho P.10C

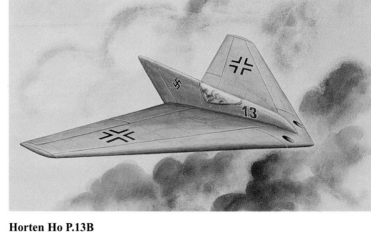
Horten Ho P.13B

Horten Ho P.18A

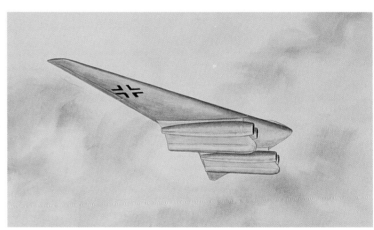
Horten Ho P-18B

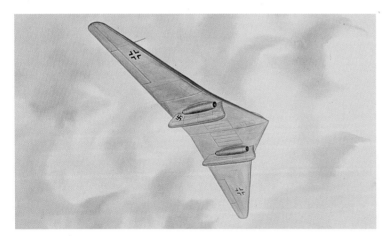
Horten Ho P-18C

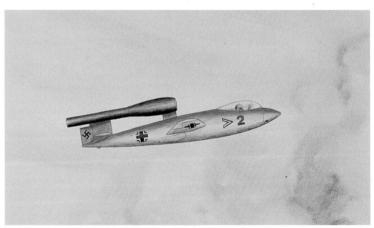
Junkers Ju EF 126

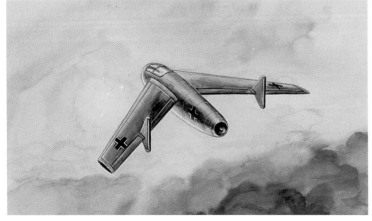
Junkers Ju EF 128

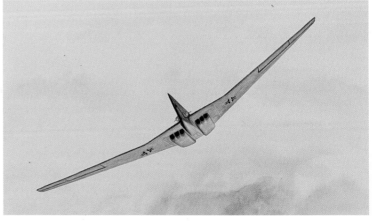
Junkers Ju EF 140

Color Gallery

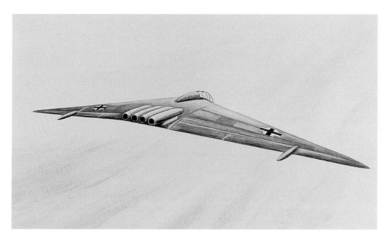
Junkers Ju EF 130

Junkers Ju P.248

Junkers Ju P-287-V1

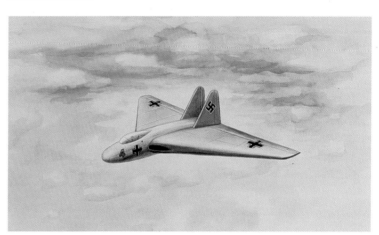
Lippisch Li P-11

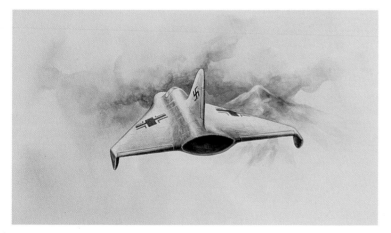
Lippisch Li P-12

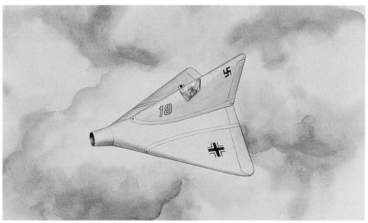
Lippisch Li P-13A

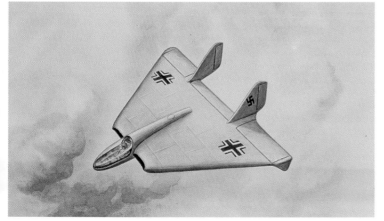
Lippisch Li P-13B

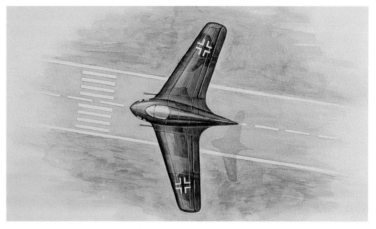
Lippisch Li P-15

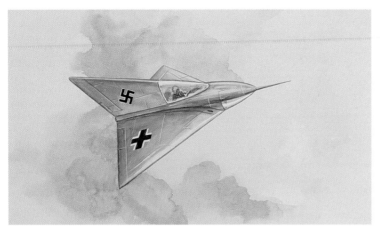
Lippisch DM-1

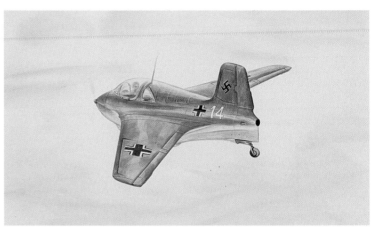
Messerschmitt Me 163B

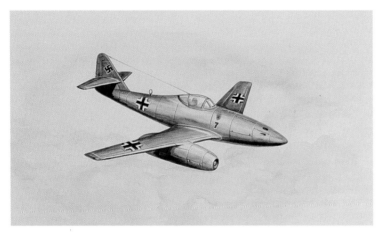
Messerschmitt Me P-262

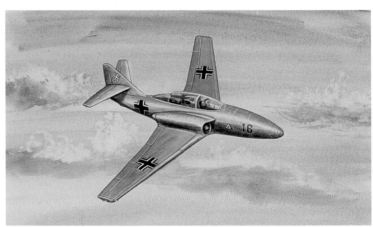
Messerschmitt Me P.262 HG2

Messerschmitt Me P-1101

Messerschmitt Me P-1106

Messerschmitt Me P-1106

Messerschmitt Me P.1107A

Color Gallery

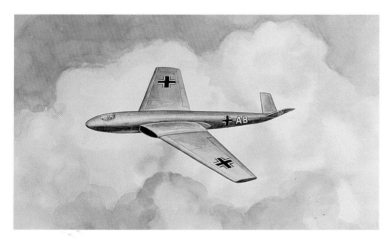
Messerschmitt Me P.1107B

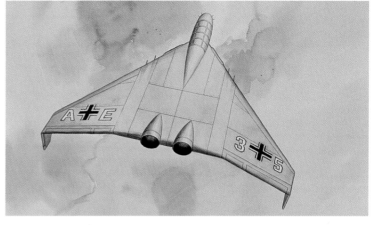
Messerschmitt Me P-1108

Messerschmitt Me P-1111

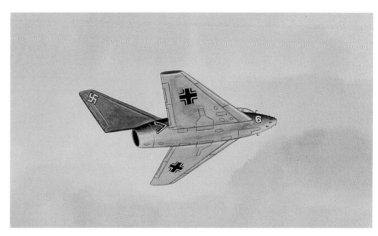
Messerschmitt Me P-1112

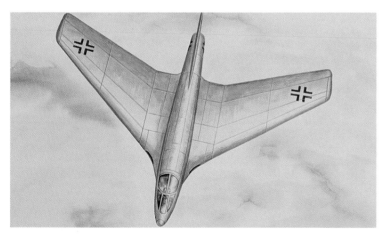
Messerschmitt Me P.1112

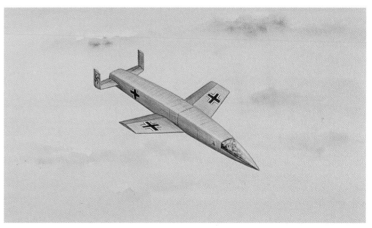
Sänger Orbital Bomber

Sombold So P.344

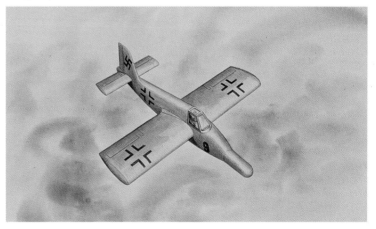
Zeppelin "Rammer"

Color Gallery

Color Gallery

Color Gallery

Bv P-212. Courtesy Mike Hernandez

Bv P-212. Courtesy Mike Hernandez

Bv P-212. Courtesy Mike Hernandez

Bv P-212-03. Courtesy Mike Hernandez; Photographed by Tom Trankle

Color Gallery

Bv P-212-03. Courtesy Mike Hernandez; Photographed by Tom Trankle

Bv P-212-03. Courtesy Mike Hernandez; Photographed by Tom Trankle

Bv P-212. Courtesy Mike Hernandez; Photographed by Tom Trankle

Bv P-215. Courtesy Mike Hernandez; Photographed by Tom Trankle

Color Gallery

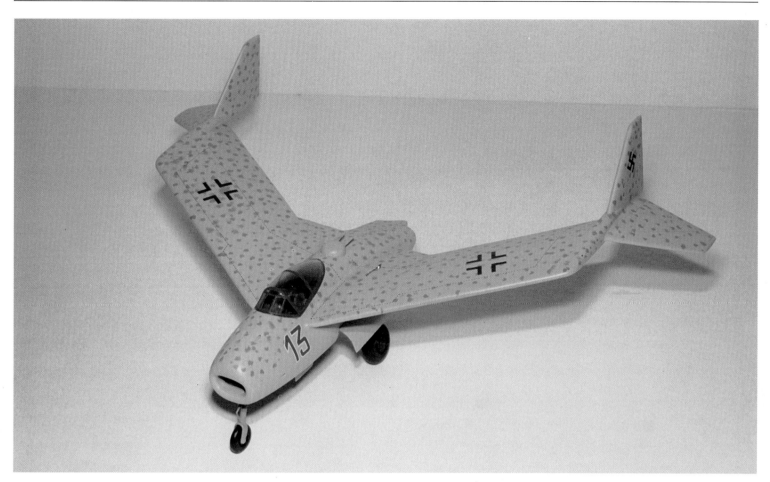

Bv P-215. Courtesy Mike Hernandez; Photographed by Tom Trankle

Bv P-215. Courtesy Mike Hernandez; Photographed by Tom Trankle

Bv P-215. Courtesy Mike Hernandez; Photographed by Tom Trankle

Bv P-215. Courtesy Mike Hernandez; Photographed by Tom Trankle

Color Gallery

Fw Triebflügel. Courtesy Mike Hernandez; Photographed by Tom Trankle

The Heinkel He P.1077 "Julia" leaving its vertical climb and leveling off to horizontal flight. Scale model by Jamie Davies.

Lippisch Glieter Bomber. Courtesy Mike Hernandez; Photographed by Tom Trankle

Lippisch Glieter Bomber. Courtesy Mike Hernandez; Photographed by Tom Trankle

Color Gallery

Lippisch Glieter Bomber. Courtesy Mike Hernandez; Photographed by Tom Trankle

Lippisch Glieter Bomber. Courtesy Mike Hernandez; Photographed by Tom Trankle

Lippisch P-13A. Courtesy Mike Hernandez; Photographed by Tom Trankle

Lippisch P-13A. Courtesy Mike Hernandez; Photographed by Tom Trankle

Color Gallery

Lippisch P-13A. Courtesy Mike Hernandez; Photographed by Tom Trankle

Lippisch P-13A. Courtesy Mike Hernandez; Photographed by Tom Trankle

Secret Aircraft Designs of the Third Reich

Fw TA P.283. Courtesy Mike Hernandez; Photographed by Tom Trankle

Fw TA P.183. Courtesy Mike Hernandez; Photographed by Tom Trankle

Color Gallery

Fw TA P.183. Courtesy Mike Hernandez; Photographed by Tom Trankle

Me P.1106. Courtesy Mike Hernandez; Photographed by Tom Trankle

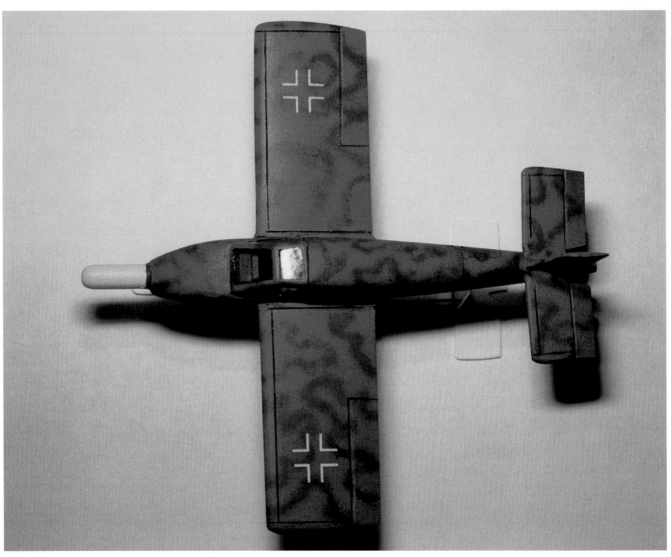

Zeppelin "Rammer." Courtesy Mike Hernandez; Photographed by Tom Trankle

Zeppelin "Rammer." Courtesy Mike Hernandez; Photographed by Tom Trankle

Color Gallery

Zeppelin "Rammer." Courtesy Mike Hernandez; Photographed by Tom Trankle

Zeppelin "Rammer." Courtesy Mike Hernandez; Photographed by Tom Trankle

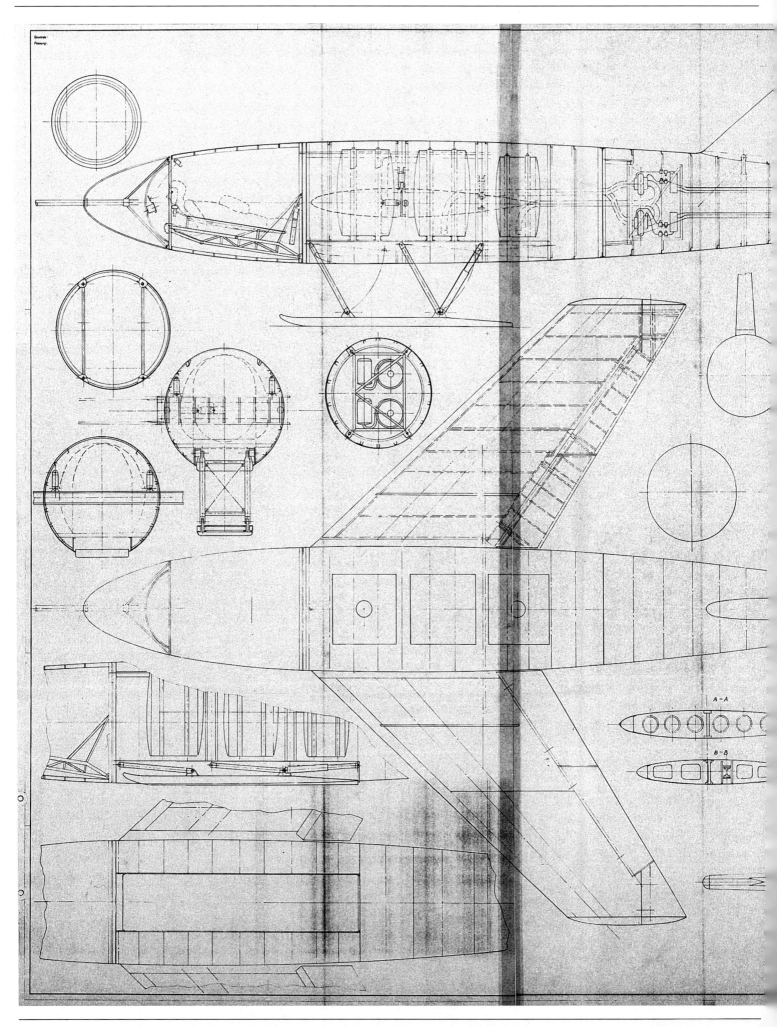

Color Gallery

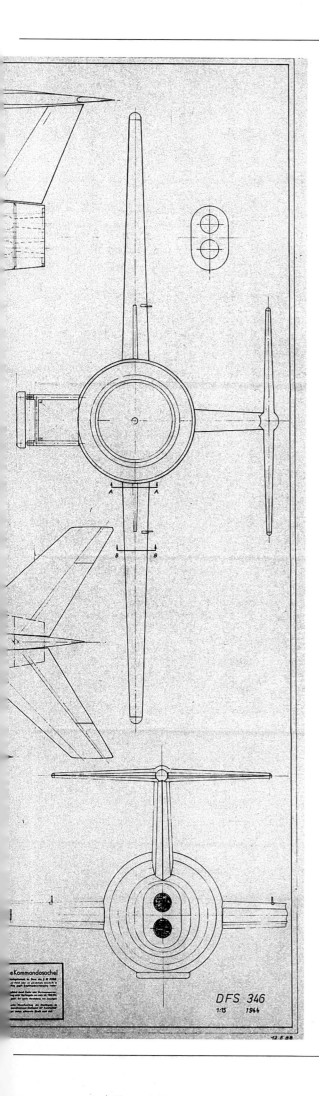

A set of structural drawings for the DFS-346 speed of sound research aircraft.

11

Horten Flugzeugbau GmbH - Bonn, Germany

Secret Projects and Intended Purpose:

Project 9A V2 - Prototype twin Jumo 004 all-wing fighter which flew approximately ten hours at Oranienburg December 1944 to February 1945. Crashed, killing its test pilot while attempting a single-engine landing

Project 229 V3 - First of series production of which forty were to be initially constructed. Located at the National Air and Space Museum, Washington, D.C. Unrestored.

Project 9B - Basically a modified 9A with a large vertical fin with an attached rudder to obtain greater directional control. Powered by a single Jumo 004. It had a side profile similar in appearance to the Lippisch DM-1 glider.

Project 10A - This design was submitted in response to the Volksjäger or "People's Fighter" program. Its turbojet engine was placed on the upper side of the aircraft. Some thought was given to have the turbojet rotate in the direction of travel.

Reimar Horten at the Bonn-Hangelar, 1934, standing in front of their Horten Ho 1.

Walter Horten, mid 1930s as a junior officer in the Luftwaffe.

Horten Flugzeugbau GmbH - Bonn, Germany

The Horten Ho 1, Spring 1934, being test flown by Walter Horten.

A Horten Ho 2, late 1930s. This particular Horten Ho 2 was flown in competition by the Luftwaffe.

Project 10B - The Horten's first delta planform fighter but it still lacked vertical rudder.

Project 10C - The first Horten aircraft with a vertical fin with an attached rudder. A single- turbojet powered fighter. Reimar Horten in the 1950s proposed a copy to the Argentines. Only a sailplane version was constructed and flown.

Project 13A - A non-powered research aircraft in the form of a sailplane with a 60 degree wing sweep-back. First flown in spring 1945 for a total of ten hours.

Project 13B - A single-seat fighter prototype with a 60 degree wing sweep-back with air intakes for its HeS 011 in the wing's leading edges.

Project 18A - The initial Horten proposal for the "Amerika Bomber" competitions. A six Jumo 004 turbojet-powered all-wing bomber which the Horten's guaranteed Hermann Göring the minimum 7,460 mi (11,000 km) range needed to deliver the German Atomic bomb on New York City.

Project 18B - Last war-time aircraft proposal. Project 18B was an improved version of their Project 18A and would have been powered by four HeS 011's with greater efficiency and range.

History:
This is the stuff of movies - three teenage brothers from Bonn, using their family dining room table as their first workbench, build and fly a series of all-wing sailplanes. Their achievements are amazing, but few people take them seriously. After all, famed advocate of tailless airplanes Alexander Lippisch, himself an accomplished sailpane-builder, has pretty much settled the matter: Aircraft can be tailless (that is, have no elevators), but they cannot be "all-wing," (i.e. they must have a vertical surface, or rudder, for directional stability). Undaunted, the brothers persevere, for they believe the nearest approach to the ideal flying machine is the all-wing design.

Financial backing is elusive, and by 1936, three years after they began their work, all three brothers are pilots in the Luftwaffe. They continue to experiment and fly their increasingly sophisticated all-wing sailplanes in their free time. In December 1944 the Horten brothers build and fly one of the most sophisticated aircraft built in the world. A twin turbojet powered all-wing fighter prototype. Known as the Horten Ho 9 V2 it flew numerous times at Oranienburg, a Luftwaffe airbase in suburban Berlin. The Ho 9 reached speeds of 500 mph at 70% throttle but was destroyed attempting a single-engine landing after one of its Jumo 004's stalled during flight. The Air Ministry had contracted with Gothaer Waggonfabrik to have them build as many as forty Ho 9 V2's. With serial production started in late 1944, the official designation of the all-wing fighter was the Horten Ho 229. Only one example survived World War II. It is at the National Air and Space Museum, Washington, DC in storage awaiting restoration. The Horten brothers in twelve short years went from a building and flying a crude all-wing wooden sailplane the Ho 1 to the turbojet-powered Ho 229.

The 1938 Lilienthal Award Ceremony for most innovative aircraft design. Winners: Walter and Reimar Horten. Walter was unable to attend because of Luftwaffe duties. Left to right: General of Flyers, Erhardt Milch, Italian General Jelice Porro, award winner, Reimar Horten, General Major Ernst Udet, and Professor Dr.Ing. Ernst Heinkel.

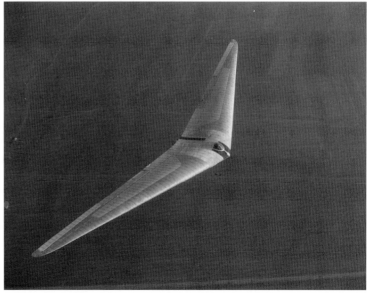

A Horten sailplane.

A Horten Ho 3G (motorized) for test flight, 1942 Göttingen.

JG26 officers and staff. Major Witt, commander of JG26, seated fourth from left and Walter Horten, technical officer, seated next to Major Witt. 1940.

The Horten Ho 5C in flight, 1944.

The "brothers" - Wolfram, Walter, and Reimar Horten - were the upper-middle-class children of Dr. and Frau Max Horten, he a professor of Oriental Culture at the University of Bonn. Wolfram, the eldest brother, was born 3 March 1912. In 1930 he went to sea with HAPAG (Hamburg-American Parcel/Packae Aktie Gesellschaft). He had thought of making the merchant marine his life work and for a while saved his wages in order to attend a helmsman school. Wolfram never did enroll. Instead, he became increasingly interested in sailplanes, and, catching his younger brothers' enthusiasm for the all-wing, came to believe that the all-wing concept, which had been vigorously applied to sailplanes in the 1920s but was out of favor by 1930, was worthwhile. Perhaps with a few modifications and experimentation, they might come up with the right combination and end up with some high-performance competition sailplanes. The brothers had no other goal but to build a more efficient sailplane and in the process win a few soaring contests at the Rhön Wasserkuppe each summer. The entire family provided funds needed to purchase supplies and materials. All the work was conducted in the Horten home in Bonn; frequently it occupied the entire family dining room for long periods of time.

In 1934 Wolfram entered the German Naval Academy. During his vacation time he continued to help Walter and Reimar with the construction of the Ho 1 and Ho 2 sailplanes. The Ho 1 was considered satisfactory and in fact won a prize for the most innovative aircraft at the Rhön Wasserkuppe Annual Sailplane Competitions of 1934. In all, it logged some seven hours of flying time. It was purposely set afire

The Horten Ho 4A. Göttingen, 1942.

The Horten Ho 5B.

after the Rhön competition when the three brothers could find no way to return it to Bonn. Shortly afterward the Horten's turned their attentions to building a much improved and larger version called the Ho 2.

The Ho 2 was built after 1934, also in the family house in Bonn. Like the Ho 1, the Ho 2 was of wooden construction, but it had greater wing span and aspect ratio (the ratio between the length of a wing and its width or chord) and a number of improvements suggested by experience with the Ho 1. Famed German test pilot Hanna Reitsch flew the Ho 2 in 1938. This good fortune was in great part due to the friendship Walter had with Ernst Udet, technical director of the Luftwaffe, Walter would later marry Udet's secretary. Reitsch reported that the Ho 2's handling characteristics were favorable, and noted that the Ho 2 was not vulnerable to spin or stall. This had to do with the twist, or wash out, designed into the wings. As air speed decreased the Ho 2's center section stalled first, while the wing tips continued flying. When this happened, the nose of the aircraft dropped and the entire wing became unstable again. Reitsch did note various deficiencies in the control system and found the pilot's accommodation very cramped.

Wolfram graduated from the Naval Academy in 1937 as a flight lieutenant. For a long time he flew a Dornier Do 18 Wal (whale), a long-range flying boat, weekly to the North Pole to conduct meteorological reconnaissance. With the outbreak of war in 1939, he began piloting a Heinkel He 59, a large twin-engined attack bi-plane flying boat. Patrolling the shipping lanes of the Baltic and North Seas he was credited

The Horten Ho 7. Scale model by Reinhard Roeser.

The Horten Ho 9 V1.

An impression of an early Horten Ho 9 design concept by Richard Keller. This early Horten Ho 9 was to have been powered by twin BMW 003s.

with sinking a Polish destroyer. After the fall of France in 1940, he moved up to an He 111 and laid sea mines in the North Sea. He died off the coast of France, in May 1940 when his He 111 loaded with sea mines exploded in a ball of fire during the laying of mines at sea. He was 28 years old.

Walter Horten was born 13 November 1913, and entered the Army Infantry (Goslarer Fighters) War Academy in April 1934. Later transferring to the Luftwaffe, Walter was trained as a pilot due to his extensive experience with sailplanes. Graduating in 1936 from pilot training, he became a lieutenant with the famed "Schlageter" fighter squadron as its technical officer. Known as the 3rd Group of JG26 and headed by Germany's future fighter commander Adolf Galland, "Schlageter" was the elite formation of the German Luftwaffe. JG26 pilots fought only on the Western Front and were known to the Allied fighter pilots as the "Yellow Nose Boys" because of the yellow-cowl on the Bf 109 and Fw 190 fighters.

After seeing considerable combat activity in the French campaign and later in the Battle of Britain, Walter was asked to join the Luftwaffe's Inspection of Fighters Command in Berlin. He was bright, extroverted, and a convincing talker. Seemingly always blessed with good fortune, he met a senior-ranking officer in the Luftwaffe's Quarter Master's Office by the name of Artur Eschenauer. According to Eschenauer, he and Wolfram had been close friends. Wolfram was constantly telling Eschenauer about his two younger brothers: Walter and Reimar and how he wished that he could do more to help them develop the all-wing planform for military use. It was Eschenauer who wanted to pickup where Wolfram no longer could, that is, help Walter and Reimar build a

Ex-Reichsmarshall Hermann Göring at the Nuremberg War Crimes Trial, 1945-46. Photo taken prior to his suicide on 15 October 1946.

The Horten Ho 9 V2-B being flight tested by Lt. Ziller, in 1945, Oranienburg.

A Horten Ho 229 V3. Scale model by Michael Emmerich. **A Horten Ho 229 V3. Scale model by Michael Emmerich.**

A Horten Ho 229 V3. Scale model by Michael Emmerich.

turbojet powered all-wing fighter-bomber. Through his power in the Quarter Master's Office, Eschenauer established a top-secret special detachment where the Horten's could go on building their all-wing aircraft. This detachment was called the Luftwaffe Sonder-Kommando #9 with headquarters at Göttingen. For the first time the Horten brothers had a serious sponsor, thus allowing them to continue development of their all-wing sailplanes and motor sailplanes. In addition to financial sponsorship, the Horten brothers had access to skilled labor and the use of the Peschke aircraft repair facilities at Minden for construction purposes. So cleverly did Eschenauer hide the Horten brothers all-wing activities, that no one ever learned of its existence. They began by making plans to have Reimar transferred into a secret commando detail within the Inspection of Fighters. Eschenauer next found work space for Reimar in little used military facilities. Most important, it was Eschenauer, who through his rank at the Quarter Master's Office, provided material, supplies, and workers to allow Reimar to develop and build what was being called Walter's German Mosquito, the Horten Ho 9 prototype. It was Eschenauer who obtained the extremely hard to obtain Jumo 004B turbojet engines.

At the time of the surrender Walter Horten held the rank of major. He found, as did many others, that former Luftwaffe officers were unwelcome in the atmosphere of defeat that pervaded Germany. Undaunted, he plunged into civilian work. He was successful in this transition and stayed in the private sector until 1952 and was invited to join in the planning for the birth of the new German Luftwaffe in 1955. This invitation came from his old friend Artur Eschenauer. Walter rejoined the planning group and participating in its reactivation in 1955. He retired in 1975 having spent over thirty years in the Luftwaffe.

Horten Ho 9 V2 with its Junkers Jumo 004B's engine cowling removed. Scale model by Reinhard Roeser.

A Horten Ho 229 V3. Scale model by Michael Emmerich.

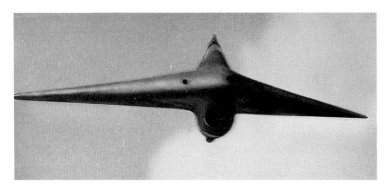
The proposed Horten Ho 9B. Airbrush by Gert W. Heumann.

The proposed Horten Ho 9B fighter. Scale model by Reinhard Roeser.

Reimar Horten was born 2 March 1915, and entered the Luftwaffe's Air-War academy in 1936 as a reserve officer. In 1938 he and Walter won the Lilienthal prize for aeronautical innovation, receiving a 5,000 mark award in addition to a trophy. The Hortens were released from their Luftwaffe duties to study aeronautics, with salary, at Berlin. While at the University the Hortens thought about their future and looked for facilities and support in order to continue their all-wing experimentation. In 1938 and 1939 they held discussions with Ernst Heinkel about an arrangement whereby Reimar would be in charge of a Heinkel design section for tailless aircraft. However, when Heinkel insisted that the Hortens would be employed in the design department, they broke off any further negotiations. Shortly afterward the Hortens received overtures from Willy Messerschmitt, but nothing came of them. Instead, Messerschmitt found that the Air Ministry had invited famed designer of tailless aircraft, Alexander Lippisch to join him, and they came together on January 1939, with Lippisch heading up a special secret section that led eventually to construction of the Me 163 HWK 509A bi-fuel liquid rocket-driven interceptor.

By September 1939, the Hortens had completed their university studies and both returned to the Luftwaffe. Reimar, like many other reserve pilots, was being called up for active duty. In the early 1940s the Luftwaffe became interested in transforming gliders into ammunition carriers, and Reimar set about converting a third sailplane design, the popular Ho 3, into wartime use. Five Ho 3's were converted into small arms ammunition carriers. In addition to his sailplane-building activities, Reimar served as glider pilot instructor, attaining the rank of captain.

In early 1942 the Horten brothers received what they considered a windfall of good luck.

Working pretty much in secret, it took Reimar Horten, his design assistants, engineers, and workmen, three full years to develop their all-wing turbojet-powered Ho 9. Later this prototype all-wing fighter would be known as the Ho 9A, which differed from the totally redesigned Ho 9B (the latter called for a large vertical rudder and a single turbojet hung in a nacelle beneath the semi-delta-shaped wing and looked surprisingly similar to Lippisch's DM-1 glider). The Ho 9A V2 prototype, was presented to Hermann Göring as a 3x1000 fast bomber. It first flew in December 1944 at the Oranienburg airfield northwest of Berlin. On the basis of its performance, Göring ordered 20 of these all-wings. They

The proposed Horten Ho 9B (single seat version). Scale model by Reinhard Roeser.

were to be manufactured by the Gothaer Waggonfabrik at its facilities in Friedrichroda. Gothaer had managed to nearly complete the V3, a nonflying prototype that was to test production techniques. Work started on the V4, which was to be the first flyable production model. Although Gothaer was to do the assembly, the all-wing turbojet powered fighter would still be designated as the Ho 229. However, U.S. Army Intelligence officers among the troops overrunning the Gothaer facilities in April 1945 assumed it was a Gothaer design, and referred to the series as the Go 229. It has been mislabeled since.

Gothaer engineers were never happy about the Ho 9A's aerodynamic stability, and were reluctant to continue production for the Ho 229 beyond the V6 prototype because they believed a better all-wing aircraft could be designed. Flight tests showed that the Ho 9A suffered from considerable "Dutch Roll" (a constant turning from side to side during level flight), making it practically unusable as a gun platform. Gothaer officials had just about convinced the Air Ministry to drop the Ho 229 and start over with their own new all-wing design called the Go P.60 series. Knowing the serious difficulties of the Ho 9A and its inability to function as a fighter, Walter Horten determined that the way to stabilize the aircraft would be to add a large vertical fin to the fuselage. Before a Ho 229 prototype could be modified with a vertical rudder, American Army troops captured the Friedrichroda district. Later, Walter and Reimar Horten were arrested and held in a prisoner-of-war camp until they were taken to London in the fall of 1945 for interrogation. Afterward, both were "draft contracted" to work in the Aerodynamic Testing Institute in Göttingen under British operation and supervision.

Released in January 1947, Reimar returned to college at the University of Göttingen, earning a Ph.D. in mathematics. After making fruitless inquiries about employment with Armstrong-Whitworth in England (he was told the firm didn't need any more talent) and the Northrop Corporation in the United States (his letter was opened, then resealed and returned unanswered), Reimar immigrated to Argentina in 1949 to work for President Juan Perón through the Instituto Aerotécnico (IAe) in Còrdoba. For the next 25 years he concentrated on design and construction of conventional-wing aircraft and worked on his all-wing sailplanes only in his free time. He did enter two unmistakable Horten types with the Argentinian sailplane team in the 1952 World Soaring Championship in Spain, but they did not perform well.

At war's end in May 1945, the Horten brothers had eight or more turbojet-powered aircraft on the drawing boards. The reason for the large number of projects stemmed from the rapid turnaround the brothers had made regarding the use of a rudder on their aircraft for directional stability. Between January and May 1945 the Hortens went through what could be called a radical redesigning of their turbojet-powered aircraft. They began including a vertical rudder in almost all of their design stud-

The proposed Horten Ho 9B.

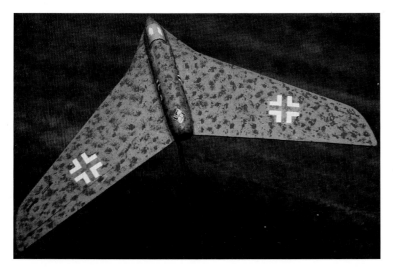

The proposed Horten Ho 10A "Peoples' Fighter" 1945. Scale model by Reinhard Roeser.

The proposed Lippisch Li P.13A.

The proposed Gotha Go P.60A.

The proposed Horten Ho 10A. Airbrush by Gert W. Heumann.

ies. In addition, they were abandoning the all-wing concept in favor of Lippisch's delta-wing planform; in their place came the delta-like tailless designs supporting huge vertical rudders. Lippisch had been correct.

Refinements kept coming in the Horten design studios. Project 10 represented two new designs. Planned for their P.10A was a single turbojet atop the wing surface directly behind the cockpit. Air intakes were built into the sides of the engine nacelle. This design represented the Horten's last attempt to perfect an all-wing military aircraft; it was abandoned in favor of a symmetrical, delta-wing design called the P.10B, which carried a large vertical rudder for directional stability. Several non-powered glider models were constructed, too. But in 1945 the Hortens feared that the large pure delta-winged aircraft would be underpowered with only its single Jumo, BMW, or HeS turbojet, and the design was shelved, to be revived when a more powerful turbojet engine like the BMW 018, with its 7,500 lb (3,400 kg) of thrust was available in 1946. With the postponement of their P.10 proposals, the Hortens turned to their Project 13 designs currently under consideration.

The Ho P.13A was a research sailplane with a 60 degree sweepback. It was tested as a glider in early 1945, but was damaged when it landed on a wire fence. On the basis of its performance, the Horten's added a large vertical rudder similar to the one on their earlier P.9B; however, the turbojet was fully enclosed in the aircraft's center wing section. Air intakes for the HeS 011 turbojet were slots built into the leading edges of the wings close to the center section. For additional interceptor capability, arrangement was made for a rocket booster housed above the HeS 011 turbojet engine.

The Hortens' final wartime turbojet-powered design project was their fast "Amerika Bomber," the Project18. In December 1944, they had been asked by the Air Ministry if a Horten all-wing in the form of bomber might reached 11,000 km. They believed that it could even when powered by six turbojets. Their P.18A was designed along the lines of their Ho 9. A further refinement came to be called the Ho 18B. With four HeS 011 turbojet engines its maximum speed was expected to be 559 mph (900 km/h); expected range was 7,500 mi (11 km), and expected surface ceiling was 52,493 ft (16.0 km). The Ho 18B known as

The proposed Horten Ho 10A. Scale model by Reinhard Roeser.

the "Amerika Bomber" was to have carried Hitler's atomic bomb to the United States. This bomb was expected to have an overall weight of 8,818 lb (4,000 kg).

Although the world thinks of the Horten brothers as the developers, builders, and flyers of the first turbojet-powered flying wing, the brothers attach very little significance to their Ho 9A. Then, as now, they considered themselves fundamentally sailplane builders and flyers. They did think that the Ho 9A represented an advanced stage of fighter design, a type they feel even now will be applied to future aircraft. But perfecting the all-wing design, not designing military aircraft, was their goal. Each advance came about through plain hard work, they asserted. During their interrogation by U.S. Army Intelligence officers in the summer of 1945 in London, the Hortens did not claim to have evolved any secret principles in designing all-wing flying machines. They believed their techniques constituted rather straightforward engineering. However, the development of their aerodynamic ideas had been obtained through flight experience. No wind tunnel tests were made on complete models because they did not have access to wind tunnels. Their Ho 9A (glider version) had been instrumented by DVL for lateral stability tests to demonstrate its suitability as a gun platform. Apart from this, the majority of the Hortens' flight research seems to have relied on test pilots' reports. For this reason more than any other serious students of aerodynamics such as Lippisch tended to ignore the work of the Hortens. Although the Hortens may have started their careers as aircraft designers in very practical ways without assistance from academic theory, they

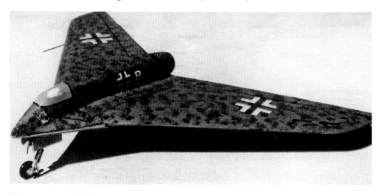

The proposed Horten Ho 10A. Scale model by Reinhard Roeser.

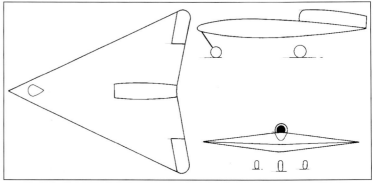

Three-view drawing of the proposed Horten Ho 10B.

The proposed Horten Ho 10C, with a delta wing planform.

The Horten Ho 13A. A pure research sailplane with a 60 degree wing sweep. 1944-45.

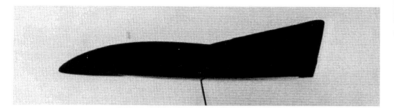

Side view of the Horten Ho 10C.

tended to rely mainly on what they found satisfactory. Critics in Germany and post war in the United States on the other hand found the Horten brothers' approach crude. Most people in aviation design tended to trust theory versus experience and felt that practice without theory was ultimately limiting.

Contrary to the suggestions of some writers that the Hortens were seeking to "show up" Lippisch and others who had, in effect, given them little chance of achieving stability with their all-wing aircraft, the Hortens never viewed themselves as being in an aerodynamic race with anyone. Reimar frequently commented that he preferred building sailplanes because he could do the complete design himself. He resented time spent in supervising staff on larger projects. Critics reviewing the Horten achievements have been generally impressed with the speed of their work and the utter irrelevance of much of it to the German war effort. The Hortens, too, generally acknowledge that much of their work on gliders was a dead loss to Germany and taught no useful lessons in terms of military or civilian designs. Indeed, much of the work on sailplanes between 1943 and 1945 was without the Air Ministry's consent,

and Reimar commented during interrogation with officials at Farnborough, England that the chief advantage of their dispersed workshops (the Hortens had eight) was that the Air Ministry could not find out what was going on, or how their money was being spent. Nor were the brothers in a race with Jack Northrop, America's own advocate of the flying wing. Instead, the brothers were soaring enthusiasts from their first interest in aviation and remained so throughout their lives.

From the Horten brothers' first introduction to gliders and soaring in the late 1920s, they had endeavored to improve the performance of sailplanes by a method other than the classical one, increasing the wing's aspect ratio, namely, by building all-wing sailplanes. The Hortens preferred the all-wing design over the conventional glider for a number of reasons - its lighter weight, slow flight characteristics, high performance, and low construction expense, to name a few.

The main goals of design to improve glider performance were to decrease weight and drag. A conventional aircraft, complete with fuselage and empennage, always would weight more than an all-wing airplane. The all-wing design would not only decrease weight, but also

The proposed Horten Argentine IAe37. Similar to the Horten 10C of 1944-45.

The proposed Horten Ho 13B.

The proposed Horten Ho 13B. Scale model by Hans-Peter Dabrowski.

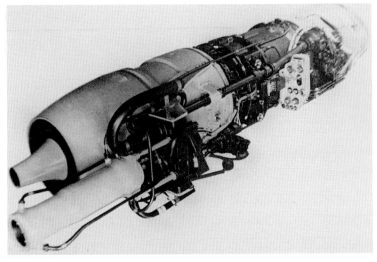

The BMW 003R composite turbine with its bi-fuel BMW 718 rocket engine. This was suggested for use on the Horten Ho 13B when immediate additional thrust was required, for example, rapid climb during night fighting.

decrease the parasite drag (i.e. the drag resulting from the shape or form of the aircraft).

Before the Hortens, engineers like them had spent countless hours at the drafting board, with the slide rule, at the wind tunnel, and in the testing laboratories, in an effort to design a plane with minimum drag. Early steps in this direction were the changes from bi-plane to monoplane designs, the elimination of external wing struts and wires, the incorporation of retractable landing gear, the enclosed fuselage cockpit, improved engine cowling, a full cantilever wing structure, and all-metal moncoque construction. Despite these advances, the average conventional airplane of the 1920s and 1930s was thought to have two to four times the drag of one created by a hypothetical craft whose functions were incorporated inside the wing alone.

Conventional aircraft produced more parasite drag and were inefficient. Only the wing contributes lift in return for the drag it creates; the fuselage, engine nacelles, tail surfaces, and other parts create drag, but rarely contribute to lift. Therefore, the theoretically ideal flying machine would be one that would be all-wing - a pure supporting surface in which every exposed portion contributes to lift. The parasite drag of all-wing aircraft is usually less than half that of its conventional counterpart, and the total drag is from 25% to 45% less, depending on the type, speed, and kind of power plant employed. This drag reduction is directly reflected in higher speeds or greater ranges or a combination of the two, because less drag means less power is required to overcome drag, and, therefore, smaller engines and less fuel are required for a given job. Or, with the same power and fuel, much higher speeds and greater ranges can be obtained. Although actual figures vary considerably depending on the types and purpose of the aircraft, in general the all-wing aircraft would have between 50 and 100 miles per our more

The proposed Horten Ho 13B. Scale model by Reinhard Roeser.

speed and about 25% greater range than a conventional airplane of the same gross weight and power.

Reimar Horten's tireless battle to increase lift, the sustaining power of flight and reduce drag, the force that resists a plane's movement through the air, had its roots in the theories of famed German aerodynamicist Dr. Ludwig Prandtl. Prandtl had developed several ideas of motion between an object and a moving fluid such as air. It was Reimar's belief that an all-wing aircraft would move through moving air far more efficiently than any other type of wing arrangement. Yet in the late 1920s and early 1930s, no all-wing aircraft, even the gliders of Lippisch and others - or even the Hortens' experimental craft - had gone beyond the experimental stage. It certainly was not due to a lack of interest. De-

The proposed Junkers Ju EF 140 "Amerika Bomber." Early 1945. Scale model by Hans-Peter Dabrowski.

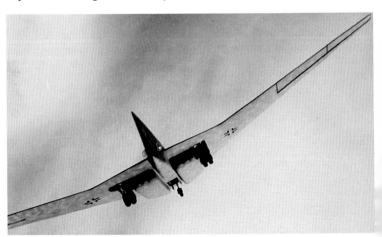

The proposed Junkers Ju EF 140 "Amerika Bomber." Early 1945. Scale model by Hans-Peter Dabrowski.

signers in several countries had been attracted to the challenge before the Hortens picked up on the idea.

It would be difficult to name the father of the all-wing aircraft, since its conception has been that of many men over a long period of time; several are strong contenders for that title. The designs of Britisher J.W. Dunne, during the period 1907-1919, though in the strictest sense tailless rather than flying wings, demonstrate that a self-stabilized wing system could be built and the tail eliminated. In doing so, Dunne established the fundamental principles of forward center-of-gravity, and of wing sweep and twist, that have been used on all tailless and flying wing aircraft up to the present.

Hugo Junkers of Germany is another strong contender as the father of the all-wing aircraft. In 1909 and 1910 he was investigating the significance of the relationship of lift to resistance or drag. And on 1 February 1910, he received a patent for an all-wing airplane, one that consisted of one wing that would house all components, engines, crew, passengers, fuel, and framework. Junkers' patent was also the origin of the "thick cantilever wing," which first was ridiculed as fantastic and, when introduced in practice, was rejected by the experts. Twenty-five years after the granting of the patent, the cantilever wing became a general feature of aircraft construction in all countries.

In America John (Jack) Northrop embraced the flying wing idea during the early 1920s. In 1928 he built a small aircraft of flying-wing configuration but having twin booms to support tail surfaces. By 1940 Northrop had built and flown an attractive and fairly successful powered all-wing design, the N-1M. He went on to build and fly some massive flying wings in the 1950s.

Although Junkers patented his all-wing designs in 1910, few people felt secure about an aircraft that lacked elevators or a rudder. During World War I the all-wing concept disappeared, dropping out of the design picture due to the rapid development of more powerful aero-engines. These engines could easily overcome the drag resulting from conventional configurations. Consequently, more powerful engines enabled the flight of aerodynamically less efficient airplanes. For this reason, pure flying wings like the Hortens' and Northrop's are relatively new and rare. But their half-brothers - tailless aircraft - dot the pages of aviation history. Several designers such as the Horten brothers, not content with a half-solution achieved through the brute strength of high-powered aero-engines, sought to develop super economical and efficient all-wing aircraft. The modern flying wing, with its submerged cockpit and engines and with all flight controls within the wing planform, began with the gliders (both powered and non-powered) of the Horten brothers of Bonn, Germany. The prone position for the pilot was simultaneously pioneered and used on most of their gliders.

The Horten brothers remained convinced of the potential of the all-wing aircraft and saw it as the logical evolution of the flying machine

Reimar Horten's proposed Argentine twin turbojet powered fighter, based on his World War II proposed delta wing fighter, the Horten 10C.

pioneered by the Wright brothers in 1903. The Hortens remind us that the history of aeronautical science has been largely an 100-year struggle to reduce the airplane's parasite drag. With continued advancement in computer-assisted flight controls automatic gyrostabilizer, eliminating the tail surface, and even the fuselage, is only a matter of time. What we will be left with is the wing itself.

PROJECT 9A V2

For a long time neither the Air Ministry nor the Luftwaffe was aware that the Horten brothers were secretly building a turbojet-powered, all-wing, fighter aircraft. The Hortens received material support from the Luftwaffe's Quarter Masters Office for the construction of specific aircraft, such as their Ho 9. Also a large portion of the manpower and equipment went into their all-wing sailplane research. The Ho 9A was one such private research project at the urging of Walter Horten, entirely unauthorized, indeed, unknown by the Air Ministry.

The Hortens' skill at keeping virtually all their projects secret was amazing. The Ho 9A was conceived in 1942, yet it was not until 1944 that the Air Ministry became aware of its existence - and only then because Hermann Göring asked if anyone company in Germany could give him an honest 3x1000 fast bomber, that is, 1000 km/h speed, 1000 km range, and carrying a 1000 kg bomb. It is at this time that the Hortens submitted a proposal to Göring stating that they had a project (Ho 9A), which, if sufficient funds and manpower were allocated to its construction, could be flying by 1945. Actually the Hortens had already completed a full-size glider version of the Ho 9A in order to test its flight characteristics prior to powered flight. Göring was immediately impressed. Whether or not he realized that the mere existence of such an airplane, built at Luftwaffe facilities with supplies provided by the Quarter Master's Office would have been reason enough to court-marshall the two brothers and their supporters throughout the Luftwaffe is not known. What is known is that the exotic aircraft captured his imagination. Suddenly the Ho 9A received the Reichsmarshall's personal backing, and he instructed the Air Ministry to provide the Hortens with what

The Horten Argentine IAe38.

The Horten Argentine IAe37.P glider.

ever they needed to get the all-wing aircraft into the air as soon as possible.

The Horten Ho 9A had a sharply swept-back wing, and its cockpit cabin was located aft of the wing's trailing edge, so that it formed a sharp rearward-facing point. It could be fitted with one or two seats, depending on the aircraft's intended use as a fighter or bomber, and an emergency ejection seat device was built into the cabin. Two Jumo 004B turbojets, each having 1,984 lb (900 kg) of thrust, were mounted close inboard and passed through the main wing spar, exhausting on the upper surface of the wing about 30% from the wing's trailing edge. To protect the wings from the exhaust, the wing surface was covered with metal plates aft of the turbojet; in addition, cold air was taken from the lower surface of the wing by a forward-facing duct and introduced between the turbojet's exhaust pipe and the wing surface. The installation angle was such that during high-speed flight the turbojets were parallel to the direction of the flight. All fuel tanks were in the wings.

Lateral and longitudinal control was achieved by single-stage elevon control flaps, a system also used on the Ho 7. Directional control was achieved by drag rudders, set up in two sections. At high speed a slight movement of the rudder bar by the pilot opened the small (outboard) section and would give sufficient control. At low speeds, when coarser control was necessary, large movements of the rudder bar by the pilot also opened the second spoiler, which started moving when the small one was fully open. By pressing both feet at once, the pilot could move both sets of spoilers simultaneously: it was expected this would be a good method of steadying the aircraft on a target when aiming guns.

Landing flaps consisted of plain trailing edge flaps (in four sections) on the wings; a 3% chord lower surface spoiler running right across the center section functioned as a glide path control during landing. The outer pair of plain flaps lowered 27 degrees, and the inner pair 30 degrees. The center section spoiler could be used as a high-speed brake. Dive recovery flaps were considered unnecessary. Since all the principal items of equipment were mounted in the wing's center section, the design had to be very compact - and this caused problems: When the two BMW 003A turbojets the Hortens initially intended to use proved to have larger diameters than the anticipated, they could not be installed through the spars of the wing. Therefore the aircraft was completed as a glider with a non-retractable nosewheel and test flown at Oranienburg in March of 1944. Because of their greater availability, two Junkers 004B turbojets, which had even larger diameters, were to be fitted to the redesigned Ho 9A V2. Beyond the V2, the plan was for Gothaer Waggonfabrik to complete seven prototypes and produce as many as 40 of the aircraft for military use which would be designated as the Ho 229. The war ended before this was accomplished, but it is doubtful that Gotha would have not have produced the 40 aircraft called for by their contract with the Air Ministry although Gotha designers were seeking to replace the Horten design with an all-wing version of their own called the Go P.60. In addition to Gotha's lack of faith in the Horten design, officials at the Air Ministry were becoming increasingly convinced that the designs of Alexander Lippisch, with their large vertical fin and rudder combination would be a more stable gun platform.

By war's end seven prototypes of the Ho 229 were under construction, including:

Ho 9A V1 - First flown in March 1944 as a non-powered glider, with Horten test pilot Heinz Scheidhauer at the controls. The V1 was also flown by Walter Horten and by the Hortens' second test pilot, Erwin Ziller. All in all, Scheidhauer did most of the flying (about 30 hours) at Oranienberg, while Horten and Ziller flew about 10 hours each.

Ho 9A V2 - First flown in December 1944 powered by two Jumo 004B turbojets. The aircraft had completed only 2 hours of flight time before it crashed while attempting to land with one dead engine, on 26 February 1945, destroying the aircraft and killing Ziller, the only person ever to have flown the turbojet version. In test flights, flown against an Me 262 and in the presence of Hermann Göring, the V2 achieved speeds of 404-435 mph (650 to 700 km/h) on about 2/3 throttle opening. These were the only flight test figures ever obtained, and this one demonstration was the basis for the Hortens' claim that the Ho 9A V2 was faster and more maneuverable with ability to climb steeper and faster, than the Me 262. During its fatal flight, the Ho 9A V2's port engine failed while down-wind on approach, crashing and killing its test pilot Lt Ziller.

Ho 229 V3 - Similar to the V2 and nearly completed when the war ended. The V3 was shipped to Wright Patterson Air Force Base in Dayton, Ohio, after the war as part of "Operation Seahorse" (whereby interesting but not necessarily flyable aircraft, supplies, and materials were taken to the United States for evaluation). Air force officials had intended to complete the aircraft, but when they discovered that it had been built not to fly but for purposes of studying the problems associated with mass production they abandoned the idea. Afterward the Ho 229 V3 was left outdoors, destined for the scrap heap. It was rescued in the 1970's by Paul E. Garber, an Air and Space Museum representative who was overseeing the moving of collectible aircraft from a Chicago, Illinois, site once planned by the U.S. Air Force as the official Air Force Museum to Washington, D.C. As aircraft to be saved were being loaded onto trucks, someone asked Garber if he wanted "that old German flying wing sitting over by the fence." The Ho 229 is in storage at the National Air and Space Museum's warehouse in Silver Spring, MD.

Ho 229 V4 - First production model for the series, close to completion when the U.S. Army overran Gotha facilities at Friedrichroda (armament, armor plate, and self-sealing fuel tanks were already in place). The V4's landing gear retraction system was completely changed from earlier versions. It appears that he aircraft was to have been a 2-seater model because the Air Ministry wanted all subsequent Ho 229's to become night fighters.

Ho 229 V5 - Similar to the V4. The wing's center section had been completed.

Ho 229 V6 - Same as the V4 and V5. Some assembly work had been started on the wing's center section.

Ho 229 V7 - Same s the V4, V5, and V6, but no construction work had been started.

Ho 229 V8 - Same as the V4, V5, V6, and V7, but no construction work had been started. The V8 was to be the final prototype completed prior to the delivery of the 20 aircraft of this type ordered by the Air Ministry.

Overall, the Ho 9A was a controversial aircraft when it was built in 1944 - and it remains so today due to designers' concerns about its overall stability and control. Reimar Horten had no fears about it as a fighter while Walter Horten the fighter pilot during the Battle of Britain, with seven kills, felt the Ho 9 would continue to have directional and stability problems in the absence of vertical tail/rudder. The Deutsche Versuchsanstalt für Luftfahrt (DVL) or German Gliding Research Institute had instrumented the Ho 9A V1 for drag and directional stability measurements. No information on drag was obtained because of trouble with the instrument installation. However, direction oscillation tests were completed successfully. The essence of the results was that lateral oscillation was of an abnormally long period - about 8 seconds at 115 mph (250 km/h) and damped out in about 5 cycles. At low speeds the oscillation was of the "Dutch roll" type (a swinging from right to left): at high speeds very little of the problem occurred. Many heated debates about desirable directional stability characteristics took place at DVL, Reimar Horten arguing that the pilot could damp out any directional swing with the aircraft's drag rudders and keep it perfectly steady for shooting. It was found that by using the drag rudders simultaneously when aiming, the aircraft could be kept somewhat steady, with high damping of any oscillation. Lateral control was found to be quite good with very little adverse yaw.

Longitudinal control and stability on the Ho 9A V1 was more like that on a conventional aircraft, it was reported, than any of the preceding Horten types had been, and the aircraft did not exhibit the longitudinal "wiggle" usually produced by flying through a gust. Flight tests showed that the aircraft was good at the stall, the aircraft sinking on an even keel. Test pilot Heinz Scheidhauer reported that directional stability was very good, as good as a normal aircraft. Scheidhauer had flown the Me 163 as a glider and was obviously very impressed with it; he was confident enough to do rolls and loops on his first flight. When asked how the Ho 9A V1 compared with the Me 163B, he was reluctant to answer, remarking that the two were not comparable because of difference in size. He did admit, however, that he preferred the Me 163, which was more maneuverable and a delight to fly. Scheidhauer called the Me 163 a "Spielzeug," or plaything, although he Ho 9A V2 was estimated to have a maximum speed of 590 mph (950 km/h).

Generally, the men who flight tested the aircraft considered the Horten's all-wing V1 and V2 to be design failures in terms of fighter aircraft due to their tendency toward Dutch roll. This one characteristic made it an unstable gun platform. It was with considerable reluctance that Reimar Horten decided to undertake a partial redesign of the Ho 9A in hopes of correcting the problem. The two notable modifications on the redesigned aircraft, known as the Ho 9B, were a delta-wing planform and, for the first time in a Horten design, a large vertical rudder. Reimar also continued to experiment with the basic Ho 9A design while working on the Ho 9B. Called the Ho 9A V3, Reimar removed the entire cockpit and its canopy and placed the pilot in a prone position at the very apex of the wing's center section. The entire frontal area of the aircraft was to be glazed over and looking similar to the Go P.60A. Other design changes made by Reimar included the removal of the pointed rear center section trailing edge and the installation of two vertical surfaces on the wing's trailing edge outboard the turbojet exhaust ducts. Walter Horten managed to construct a full-scale model of the Ho 9A V3 in late 1945, however, time ran out before the Air Ministry could evaluate the modified Ho 9A.

The Ho 229, retrieved from the scrap heap in Chicago is the only surviving example of the world's first turbojet-powered all-wing aircraft. Badly deteriorated from being left uncovered out of doors after its capture by U.S. Army troops in 1945, it is not known when it will be restored and then placed on display at the Smithsonian Institution's Air and Space Museum in Washington, DC.

Specifications: Project 9A V2

Engine	2xJunkers Jumo 004B turbojets each having 1,984 lb (900 kg) of thrust .
Wing Span	54.1 ft (16.4 m)
Wing Area	566 sq ft (52 sq m)
Length	24.6 ft (7.5 m)
Height	9.2 ft (2.8 m)
Weight, Empty	10,140 lb (4,600 kg)
Weight, Takeoff	18,739 lb (8,500 kg)
Crew	1
Speed, Cruise	429 mph 69 km/h) at 32,810 ft (10 km) at 2/3 throttle
Speed, Landing	NA
Speed, Top	590 mph (950 km/h) at sea level or 607 mph (997 km/h) at 39,372 ft (12 km)
Radius of Operation	1,180 mi (1,899 km) on internal fuel or 1,970 mi (3,170 km) with external drop tanks
Service Ceiling	52,496 ft (16 km)
Armament	4x30 mm MK 108 cannon
Rate of Climb	4,331 ft/min (1,320 m/min)
Bomb Load	2xSC 2,205 lb (1,000 kg) bombs
Flight Durations	3 hours

PROJECT 9B

After the Ho 9A V1 glider had been test flown in March 1944, the Horten brothers suspected that its "Dutch Roll" tendencies would keep it from being adopted as a fighter aircraft. Walter Horten was particularly interested in employing the Ho 229 in the defense of the Reich, but its side-to-side oscillations would reduce the aircraft's role to that of a bomber. There was even the possibility that the Ho 229 would be abandoned in favor of the Gotha Go P.60. In any event, the Hortens realized in March 1945 that the quickest way to correct the oscillation problem was to redesign the Ho 9 completely. Walter was more willing to make a radical change than was his brother Reimar, who believed that with time the problem could be worked out. It was Walter's idea to borrow several ideas from Alexander Lippisch particularly is use of a large vertical fin with a hinged rudder, and he even went so far as to build a mock-up for the Air Ministry to inspect.

The Horten designed IAe48. Wind tunnel test model, Córdoba, Argentina.

The proposed Ho 9B (as it came to be called) retained the wings from the original Ho 9 (now known as the Ho 9A), but on the "B" model they were to have a greater degree of backsweep than the 32 degrees employed on the "A." The Hortens proposed that the Ho 9B be a two seater, with the crew housed in a cabin built into the base of the large vertical fin. The crew were to sit in tandem behind a transparent (glazed) canopy on the leading edge of the vertical fin, an arrangement intended to ensure good visibility. (In this respect the Ho 9B shared the same side appearance as Lippisch's DM-1 glider as well as his proposed ramjet-powered P.13A.) During his interrogation by the British at Farnborough, England, after the war, Reimar Horten stated that he was unaware of the Lippisch design projects such as his P.13A with its large vertical fin and hinged rudder. The Horten's indicated that their selection of a vertical fin was based on the need to dampen the oscillations experienced in the Ho 9A and in no way were they copying Lippisch.

The Ho 9B was to have been powered by a single Jumo 004B turbojet having 1,887 lb (856 kg) of thrust. For ease of maintenance and/or replacement in the field, the turbojet was to hang directly beneath the wing. The Horten's hoped that the more powerful HeS 011 turbojet with its 2,866 lb (1,300 kg) of thrust, would be available by the time the aircraft was ready for flight (expected to be in 1946). It was predicted that with the HeS 011 turbojet, the Ho 9B could reach a maximum speed of 634 mph (1,020 km/h) and could carry a 4,409 lb (2,000 kg) bomb in addition to its basic armament of four 30 mm MK 108 cannon.

Wing structure on the Ho 9B was to follow typical Horten construction practices, which was to surround a main spar and one auxiliary spar of wooden construction with plywood covering. Wingtips were to be metal. Only the center section of the Ho 9B was to be built up from welded steel tube. The under-carriage would be completely retractable and of tricycle type, the front wheel folding backward and the main wheels outward into the wing.

Although the Air Ministry officials had an opportunity to review the Ho 9B mockup, it appears that they were not impressed, and no financial assistance for the construction of a prototype was offered to the Hortens.

Specifications: Project 9B

Engine	1xJunkers Jumo 004B turbojet having 1,996 lb (856 kg) or 1xHeinkel-Hirth HeS 011 turbojet having 2,866 lb (1,300 kg) of thrust.
Wing Span	54.5 ft (16.6 m)
Wing Area	566 sq ft (52 sq m)
Length	30.2 ft (9.2 m)
Height	9.5 ft (2.9 m)
Weight, Empty	NA
Weight, Takeoff	NA
Crew	1
Speed, Cruise	NA
Speed, Landing	NA
Speed, Top	634 mph (1,020 km/h) using the HeS 011 turbo jet
Radius of Operation	1,000 mi (1,609 km)
Service Ceiling	52,496 ft (16 m)
Armament	4x30 mm MK 108 cannon
Rate of Climb	NA
Bomb Load	1x4,409 lb (2,000 kg) bomb or combinations up to 2,000 kg)
Flight Duration	3 hours

PROJECT 10

The three proposed variations of the Ho 10 represented the beginning of the end for the Horten brothers and their attempts to adapt the all-wing concept to a military flying machine. By late 1944, the Air Ministry's enthusiasm for the Horten's all-wing and other unconventional aircraft in general was beginning to dwindle. Second-generation turbojets were being developed, and this meant that the day was not far away when aircraft would exceed the speed of sound. Aerodynamic research throughout the German aircraft industry was indicating that the best shape for aircraft operations beyond the speed of sound was the Lippisch delta-wing planform. Lippisch appeared to be right. The Air Ministry picked up on his dictum that an aircraft could be tailless and could have a wing formed int he shape of a delta or some modified form, but it had to have a vertical rudder for directional stability. Within the aircraft industry and the Air Ministry there was no preferred type of tail. It did not have to be mounted at the rear of the aircraft, where Lippisch usually placed his. Vertical surfaces could be mounted on the wingtips, as on Richard Vogt's Bv P.215, or inboard atop the wings, as on tailless aircraft designs by Junkers, Arado, Henschel, and Gothaer. Tailless, swept-back or delta planforms with wingtip rudders, their upward of downward bending, were acceptable. All that mattered was that the aircraft have some form of vertical surface, so that it could perform well and safely beyond the speed of sound.

PROJECT 10A

Initially, the Ho 10A was proposed to fill a request from the Chief of Technical Air Armament (Chef TLR) of the Air Ministry in late 1944

The Horten designed IAe48. Wind tunnel test model, Córdoba, Argentina.

for a light weight, single-engine fighter suitable for rapid mass production - the so-called Volksjäger or "People's Fighter." The Horten's were not even asked to submit a bid on their proposed Ho 10A design. The 10A was similar in some respects to the Horten's turbojet-powered flying wing, the Ho 9A, which was nearing flight status (it would fly in January 1945), but the planform was changed somewhat. One major change was mounting the single BMW, Jumo, or HeS turbojet atop the wing directly behind the cockpit. In customary Horten fashion, the pilot's cabin was placed well forward in the nose of the aircraft, but the pilot sat upright, a change form the typical Horten practice of requiring the pilot to fly the aircraft in the prone position. Air intakes for the turbojet were located on each side of the engine nacelle behind the cockpit and led directly into the engine, making a long duct unnecessary. No vertical control surface appeared on the Ho 10A, but the wings were given a slightly crescent shape unlike those on the Ho 9A. On the Ho 10A the wing leading edges were swept back some 60 degrees at the nose: about two meters down the leading edge the sweep changed to 32 degrees out to the wingtips. The entire trailing edge of the wing was occupied by two control surfaces. The outer surfaces gave lateral and longitudinal control, and the inner surfaces acted as landing flaps. For directional control, one large and one small brake flap were provided above and below each wingtip. The Ho 10A was expected to have a ceiling of 49,212 ft (15 km), a top speed of 684 mph (1,100 km/h) and a range of 1,243 mi (2,00 km).

To give the all-wing aircraft better directional control, the Horten's had thought about positioning the turbojet in such a way that the pilot could turn it in the direction he wanted to fly. Data from the first flights of the Ho 9A indicted that an all-wing flying machine lacking a vertical rudder would not provide an effective gun platform. With this, the Hortens pretty much abandoned further efforts to convert their all-wing aircraft into a military version, and they turned their attention to a planform more like that of the Lippisch delta.

PROJECT 10B

After abandoning the all-wing as a potential military fighter, the Horten's selected a modified form of the delta wing plan long advocated by Lippisch and very similar to his DM-1 glider and his powered version, the Li P.13A. The Horten's delta-wing Ho 10B was designed with a 70 degree sweep back and with only a 1.52 aspect ratio. (The Horten Ho 7, a research sailplane built in 1943, had an aspect ratio of 1:32.) The Ho 10B's wing section was designed to be symmetrical. Carrying on what had become a Horten tradition, the pilot was placed in a prone position to minimize aerodynamic drag. In one model the single turbojet was to be buried inside the delta's center section. In another design the turbojet was placed atop the wing in a fashion similar to the Ho 10A. In either case, the Ho 10B would have been powered by a second-generation HeS 011 turbojet having 3,307 lb (1,500 kg) of thrust. Estimated performance was 684 mph (1,100 km/h) in level flight. Air intake on the model with the buried turbojet would have been achieved by a flush scoop in the upper surface of the wing and aft of the cockpit canopy. Armament would have included two 30 mm MK 108 cannon and two 13 mm MG 131 cannon housed in the leading edges of the wing near the cockpit.

Although the Ho 10B design was a step in the direction then favored by the Air Ministry, the proposed aircraft still lacked a vertical rudder to assure its directional stability. The design was abandoned in favor of a similar model, but one that had a large vertical rudder and called for the pilot to sit upright in the cockpit.

PROJECT 10C

The Horten Project 10C, was a proposed single-seat delta-wing supersonic fighter with a vertical tail, represented the Horten brothers' ultimate submission to the prevailing attitude about the relationship between rudders and stability at high speeds. It was not easy for them to admit that the all-wing concept had shortcomings they were unable to correct (at least for the time being).

With their P.10C design, the Horten brothers offered the Air Ministry an inexpensive aircraft with superior speed (a BMW 003 or a Jumo 004 would be used). They had hoped to enter the design in the Air Ministry's competition for a Volksjäger or "People's Fighter," but it was not ready in time. Blohm and Voss initially won the competition with its Bv P.211; however Heinkel already had a mockup of its He 162, and the Air Ministry reconsidered and awarded a contract to Ernst Heinkel AG.

The Hortens believed that their P.10C was particularly suited to the Volksjäger design specifications. It would be purely a "wear and tear" aircraft with the ability to takeoff and land on a grass strip runway. A retractable tricycle landing gear was proposed; low-pressure tires would allow the aircraft to handle irregular ground (grass) surfaces safely and without damage. Even if the P.10C had been ready to compete in the Volksjäger competition, it is doubtful that the aircraft would have fared very well because its expected weight (6,000 lb or 2,722 kg) exceeded the maximum specified by the Air Ministry (4,409 lb or 2,000 kg). To its credit, however, the Ho P.10C would have presented a very small target to USAAF B-17 bomber gunners and could have been produced inexpensively by semi-skilled laborers.

The Hortens first proposed design with a vertical rudder would have had an internal frame of steel pipe similar to the Ho 9A and would have been covered with plywood. Taking into account the widespread bombing attacks on aircraft manufacturing facilities throughout Germany, the Horten brothers designed the Ho P.10C in such a way that the aircraft could have been assembled in one piece at an underground manufacturing facility and then transported on a regular truck "sideways" to its operational base the same way they had frequently transported their non-powered sailplanes to a launch site and back again.

The Ho P.10C was never built. However, Reimar Horten during his long employment with the Argentinian Instituto Aerotécnico (IAe) in Còrdoba did build a single-seat delta-wing supersonic fighter, the IAe 37, which was patterned after their P.10C. The IAe 37 was flight tested as a glider in 1954, with both prone and normal pilot positions.

Specifications: Project 10C

Engine	1xBMW 003 turbojet having 1,764 lb (800 kg) or Junkers Jumo 004 turbojet of 1,887 lb (856 kg) thrust
Wing Span	29.5 ft (9 m)
Wing Area	361 sq ft (33.8 sq m)
Length	NA
Height	NA
Weight, Empty	NA
Weight, Takeoff	6,000 lb (2,722 kg)
Crew	1
Speed, Cruise	485 mph (780 km/h) at 65 degree throttle
Speed, Landing	NA
Speed, Top	746 mph (1,200 km/h)
Radius of Operation	NA
Service Ceiling	49,212 ft (15 km)
Armament	2x20 mm MG 151 or 2x30 mm MK 108 cannon
Rate of Climb	NA
Bomb Load	NA
Flight Duration	NA

PROJECT 13B

In late 1944 the Air Ministry began thinking seriously about developing a supersonic fighter having speeds up to 1,119 mph (1,800 km/h). Although no design competition was held, the Air Ministry asked the German aircraft industry to start submitting some design ideas for a supersonic fighter. At this time there were no turbojet engines with enough thrust to propel an aircraft at the speed Air Ministry officials had in mind, but they believed that if a BMW, Jumo, or HeS turbojet were combined with a Walter HWK 509A bi-fuel liquid rocket engine, supersonic speeds could be obtained rather easily.

Control and directional stability frequently were problems at high speeds, and no one knew what would happen when an aircraft flew at the speed of sound and beyond. Prevailing opinion at the Air Ministry was that an aircraft with sharply swept-back wings would be needed and that an overall design planform similar to Lippisch's delta-winged models would perform the best a supersonic speeds. In contrast, Reimar Horten believed that his all-wing (Ho 9A) configuration would be a suitable planform for supersonic speeds, though the Ho 9A's wings would have to be more streamlined, perhaps up to 60 degrees of back sweep, to avoid control problems at such high speeds.

To see how an aircraft with a 60 degree back sweep on the wings would perform, the Horten's proceeded in their typical but controversial manner, building a full-scale prototype and flying it instead of trying out their ideas in wind tunnels. According to Reimar Horten, the idea was to build a non-powered glider, known as the Ho 13A, consisting of the wings from an Ho 3 fitted to a new center section so as to give a leading edge sweep of 60 degrees. All flight control surfaces in the wings were doubled to help compensate for the anticipated decrease of control the Ho P.13A might experience due to its wings being swept back at such an extreme angle. Although the Hortens had considerable faith in the correctness of their ideas, they would build the glider as a quick experiment concurrently with the construction of a fighter prototype which came to be called the P.13B.

Early in 1945 the non-powered Ho P.13A glider was completed. The pilot was placed in an underwing nacelle to give him a reasonably good view and to keep the upper surfaces of the wing as smooth as possible. All flight controls were inverted and hung from the roof of the nacelle. The undercarriage consisted of two main wheel built into the nacelle and a fixed nose skid attached to a stationary welded-steel tube.

The Ho P.13A was flown a total of 10 hours by Horten test pilots in early 1945. Flight testing came to an end when a test pilot landed the craft into a barbed wire fence. Although the glider did not appear to have been seriously damaged, it did not survive the war. American Army troops who discovered it at Göppingen in May 1945 found it had suffered additional damage, and after a thorough inspection the entire aircraft was burned.

The Horten's were happy with the flight test results of their P.13A, although its 10 hours of flight time did demonstrate that some sort of vertical surface was needed for stability. In laying out their design plans for the powered version, which came to be called the P.13B, the Hortens gave the aircraft a wing sweep of 60 degrees. Instead of placing the pilot in a nacelle beneath the wing's center section as they had done with their P.13A, the Horten's put the cockpit in the leading edge of a very large vertical fin, the end of which contained a rudder. Although Reimar Horten told the British interrogation team in 1945 that he personally was unaware of Alexander Lippisch's work, the Ho P.13B with its large vertical fin and rudder looks very much like Lippisch's DM-1 glider and his powered version, the Lippisch Li P.13B.

The Ho P.13B differed slightly from the earlier P.9B, although both had wings with 60 degrees of back sweep and tall vertical fins. In their P.13B, the Horten's had, in effect, refined the P.9B by burying the turbojet unit inside the wing itself. Air intakes were in the wing's leading edges, one on each side. To propel their P.13B up to 1,119 mph (1,800 km/h), the Horten's were planning to install the new BMW 003R combined HWK rocket and BMW turbojet engine. This new turbojet consisted of a conventional BMW 003, C, or D model having 1,760 to 2,420 lb (798 to 1,098 kg) of thrust, on which was mounted a HWK 509 bi-fuel liquid rocket booster to deliver a thrust of 2,750 lb (1,247 kg). Two of these modified units were tested in an Me 262 in March 1945, and according to BMW engineers the performance was sensational.

The BMW 003R combined turbojet and rocket engine would have been turned off and on at the discretion of the pilot. It was not intended for use during takeoff, but only when a very rapid climb was desired or for quick bursts of acceleration during level flight. Sufficient turbojet fuel was to have been carried to give an endurance of 20 minutes at sea level or 60 minutes at 30,185 ft (9.2 km).

The Ho P.13B had a tricycle landing gear, with the nose wheel retracting back and the two main wheels retracting forward and making a 90 degree turn to lie flat up inside the wing. Overall, the P.13B was one of the Horten's most radical departures from their all-wing sailplanes. In addition to the 60 degree back sweep of its laminar profile wings, the trailing edges had a 30 degree sweep while the vertical fin/rudder swept back over the aircraft at about an 18 degree incline. The Horten's did not have time to build a prototype of their P.13B for Air Ministry officials to consider in early 1945 because time was running out for the introduction of any new aircraft. Moreover, the Air Ministry felt that the brothers had no experience with conventional aircraft and looked more favorably on he designs of Alexander Lippisch, especially if the aircraft was intended to perform at supersonic speeds.

Specifications: Project 13B

Engine	1xBMW 003A, C, or D model turbojet have 1,769 to 2,420 lb (798 to 1,098 kg) of thrust plus 1xWalter HWK 509A bi-fuel liquid booster rocket engine having 2,750 lb (1,247 kg) of thrust.
Wingspan	39.4 ft (12 m)
Wing Area	(431 sq ft (40 sq m)
Length	39.4 ft (12 m)
Height	13.5 ft (4.1 m)
Weight, Empty	NA
Weight, Takeoff	17,637 lb (8,000 kg)
Crew	1
Speed, Cruise	NA
Speed, Landing	NA
Speed, Top	1,119 mph (1,800 km/h)
Radius of Operation	NA
Service Ceiling	36,089 ft (11 km)
Armament	NA
Rate of Climb	NA
Bomb Load	NA
Flight Duration	20 minutes at sea level 60 minutes at 30,185 ft (9.2 km)

PROJECT 18B

The final wartime aircraft proposed to the Air Ministry by the Horten brothers was their P.18B, a four-turbojet-engine, all-wing "Amerika Bomber." By late 1944 Göring pretty much believed that Germany had an atomic bomb to drop on America. Although he may have thought that Germany did in fact have a nuclear device, in reality it would not have provided an atomic explosion. In 1939 Otto Hahn and his associates had achieved the first stage of nuclear fission, and French and American physicists were seeking to duplicate his experiment just be-

Reimar Horten in retirement, Argentina, 1986.

fore the world plunged into World War II. It was on 2 August 1939, that Albert Einstein wrote to President Roosevelt stating his fears that the Nazis might develop a nuclear warhead. Einstein wrote: "A single bomb of this type, carried by boat and exploded in a port, might very well destroy the whole port, together with some of the surrounding territory." In April 1942, President Roosevelt decided to try to develop and manufacture an atomic bomb.

Prior to the Wehrmacht's great victories of 1940, French physicists working under Frederic Joliot-Curie had been conducting fission experiments with heavy water (deuterium oxide) that was manufactured in quantity in Europe only at the Norsk Hydro Hydrogen Electrolysis plant at Vemonk, Norway. The Germans, after occupying Norway, had ordered the facility to produce 1.5 tons of heavy water a year; in 1942, they raised the requirement to 5 ton annually, with the entire output being shipped to the Kaiser Wilhelm Institute at Göttingen University, Germany. Vemonk was repeatedly attacked by Allied bombers, British commandos, and saboteurs. It was never destroyed because it was well protected on all sides by mountain peaks and was dug in beside a granite overhang so that bombing was completely ineffectual. However, due to repeated attacks its operations did come to a standstill in 1943. Subsequently the Germans decided to move the entire store of heavy water to Göttingen. The transport ship carrying the heavy water was sunk on 21 February 1944, by a time bomb planted by a Norwegian saboteur.

With the loss of the heavy water, Germany's hopes for developing an atomic bomb were lost; nevertheless, German physicists, led by Werner Heisenburg (a man second only to Einstein in ability), never come close to the American efforts.

Nevertheless, the Air Ministry had been talking seriously in late 1944 with several aircraft manufacturing firms such as Arado, Messerschmitt, Junkers, and Focke-Wulf about developing a "Fernbomber," or a long distance bomber. After two design competitions, German aviation firms could not give Göring the 6,835 mi (11 km) range he said was minimum. It was at this time that Oberst Siegfried Knemeyer, the man who was filling Ernst Udet's old position, contacted the Horten brothers. Could they, he asked, build an all-wing bomber capable of providing a non-stop flight of 11km range to America and back? It was possible, they told him. It was then that Knemeyer requested that the Horten's submit a detailed design proposal for such a craft as soon as possible. The Horten's believed that the most suitable design for the "Amerika Bomber" would be an enlarged version of their Ho 9A all-wing fighter then under going flight testing a Oranienburg. Known unofficially as their P.18A, this all-wing bomber would have been powered by six turbojets, either Jumo 004s or BMW 003s. The turbojets would have been buried deep inside the wing in an arrangement similar to that of their Ho 9A, with the exhaust exiting out over the wing's trailing edge. The crew was to have been housed in a raised "greenhouse" type of cockpit canopy near the apex of the wing's center section. Directly behind the cockpit was to have been a manned gun turret containing two 30 mm MK 108 cannon with the ability to rotate 360 degrees. Directly beneath this top turret was a similar one on the all-wing's underside.

Overall, the P.18A's wing span would have been 131 to 138 ft (40 to 42 m), and its loaded weight about 4 tons. Maximum speed was expected to be about 559 mph (900 km/h), with a projected range of 7,457 miles far in excess of what Göring was demanding from the German aviation industry. Like other Horten aircraft, directional control would have been achieved through devices built into the wing's trailing edge. The P.18A would have had wooden wings and a center section/fuselage constructed of steel tube and covered with metal plate.

The Horten's were told to report to DVL at Berlin-Adlershof to discuss their P.18A with some of their aviation colleagues. At a conference held in February 1945 were representatives from Germany's major aviation manufacturers such as Arado, Junkers, Messerschmitt, Focke-Wulf and Heinkel. It was at this conference that the Ho P.18A was selected for immediate consruction. The Hortens were to work with Junkers and Messerschmitt to build the Ho P.18A. They were all to meet several days later at Junkers- Dessau. When Reimar and Walter arrived, they soon found that both Junkers and Messerschmitt people wanted to place a huge vertical fin with an attached hinged rudder on the Ho 18A.

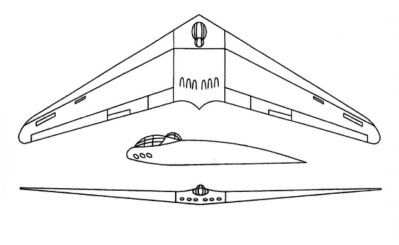
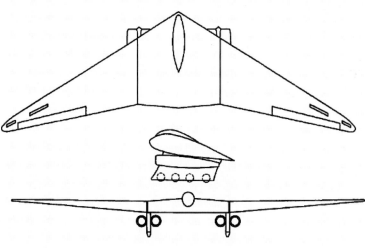

Walter Horten in retirement, Germany, 1986.

aircraft design with a 30 degree sweptback wing. Its three-man crew would have sat upright in a bubble-type cockpit near the apex of the wing. A major design change from previous Horten turbojet-powered aircraft was the placement of the turbojets. Reimar calculated that only four HeS 011 turbojets of 2,650 lb (1,200 kg) thrust each would provide sufficient propulsion. Since together they would weigh only about 5-tons, compared with 6-tons for six Jumo 004As, he could forget about the landing skid arrangement to be used on the P.18A and instead use two fixed main landing gear assemblies with an HeS 011 attached to each side of the two assemblies. Each of the two fixed landing gear would be made up of four single tires arranged in tandem. During flight the tires would be covered by "doors," which would make the landing gear streamlined to cut down on air resistance and drag. A nose wheel was not considered necessary. With a 13 ft (40 meter) wing span the P.18B would have been about the same size as the Instituto Aerotécnico's IAe 38 Reimar would later build in Argentina. Overall, the P.18B would have weighed about 35 tons. Empty weight was estimated at 16 tons, and it was expected the aircraft could carry 4 tons of bombs if needed and about 15 tons of fuel; crew weight would amount to virtually nothing. Reimar believed he could save about 1 ton of weight by not using a retractable landing gear.

With four HeS 0011 turbojets, the P.18B would have had a combined thrust of 10,600 lb (4,800 kg) and an anticipated maximum speed of 528 mph (850 km/h). With fuel for the four turbojets stored in the wing so that no outside auxiliary fuel tanks would be required, Reimar expected the P.18B to have a range of 6,835 mi (11,000 km), a service ceiling of 52,492 ft (16 km), and a round-trip endurance of 27 hours. Armament was considered unnecessary; nevertheless, Reimar proposed that two 30-mm MK 108 cannon be mounted at the apex of the wing's leading edge directly below the cockpit cabin.

The Horten brothers said to the committee that they were not interested in building a "committee bomber." It was the P.18A as it was accepted by Oberst Knemeyer or nothing. Reimar said that he was going to personally see Göring about the matter. This mattered not to the representatives from Junkers and Messerschmitt for they felt that the war was going to be over with in a few short months.

In February 1945, Reimar Horten spoke directly to Hermann Göring about his new proposed design for an "Amerika Bomber" with an 11km range which he called the P.18B. Göring instructed Reimer Horten to begin immediate construction. It wasn't much different in size from the P.18A, but unlike the P.18A it was to be powered by four HeS 011 turbojets. The proposed P.18B would have been a typical Horten all-wing

Specifications: Project 18B

Engine	4xHeinkel-Hirth HeS 011 turbojets with a combined thrust of 10,600 lb (4,800 kg).
Wing Span	130 ft (40 m)
Wing Area	NA
Length	NA
Height	NA
Weight, Empty	16 tons
Weight, Takeoff	35 tons
Crew	3
Speed, Cruise	NA
Speed, Landing	NA
Speed, Top	528 mph (850 km/h)
Radius of Operation	6,835 mi (11,000 km)
Service Ceiling	52,492 ft (16 km)
Armament	2x30 mmMK 108 cannon directly below the cockpit cabin
Rate of Climb	NA
Bomb Load	8,818 lb (4,000 kg)
Flight Duration	27 hours

12

Junkers Flugzeug und Motoren Werke AG - Dessau, Germany

Secret Projects and Intended Purpose:

Project 009 - An early concept for a turbojet powered target defense interceptor utilizing ten miniature Jumo turbojet engines.

Project 126 - Code named "Zlli," a project design for an infantry assault aircraft was powered by an Argus pulsejet engine having 1,101 lb (500 kg) of thrust. The pulsejet unit would have been mounted above the fuselage and supported by the tail rudder, and the cockpit would have been situated well ahead of the pulsejet engine near the aircraft's nose.

Project 128 - A single-seat, tailless interceptor-fighter was powered by one HeS 011 turbine. Junkers' designers wanted the P.128 to replace the Me 163B and its own improved version, the Ju 248.

Project 130 - Junkers' first entry into the Air Ministry's "Amerkia Bomber" competitions in late 1944. An all-wing planform with a 45 degree sweep to the wing, was to have powered by four HeS 011 turbines with a projected speed of 621 mph (1,000 km/h) and a range of 3,728 miles (6,000 km).

Project 140 - This was a modified version of the proposed six Jumo 004-powered all-wing Horten P.18A "Amerkia Bomber" of December 1944. The Air Ministry ordered Junkers and the Horten brothers to jointly

build several prototypes. The Hortens balked, angry over the addition of a tail to their all-wing by Junkers' designers. The Hortens went directly to Göring venting their objections. Göring excused Junkers and immediately directed the Hortens to build their second "Amerkia Bomber" project, their four HeS 011 powered all-wing Ho 18B.

Project 248 - Air Ministry's assignment to Junkers engineers the task of improving the Messerschmitt Me 163B Komet. Junkers' engineers were to redesign it to increase its flight duration, and to improve its fighting performance by adding a pressurized cockpit, a retractable undercarriage a cruising chamber HWK 509C bi-fuel liquid rocket drive, and greater fuel capacity - all the while seeking ways to reduce the aircraft's overall weight.

Project 287 - The swept-forward wing six turbojet-powered heavy bomber project.

History:

As the National Socialist German Worker's Party led by Adolf Hitler was coming to power in 1933, 74-year-old Hugo Junkers, the father of German aviation, was a democrat and a pacifist. For a long time Hitler had been openly promising the German people that once he seized power he would, in defiance of the Treaty of Versailles, found a powerful air force. After he was named Chancellor in October 1933, he told

The Father of German aviation, Hugo Junkers.

243

Adolf Hitler deplaning from a Junkers Ju 52 in the mid 1930s. He told Hugo Junkers that if he did not build war planes for the Reich, the Reich would simply nationalize the Junkers factories.

After signing over all of Hugo Junkers' aircraft and air motor factories to the Reich in 1935, Hitler appointed Dr.Ing. Heinrich Koppenberg as president of the new State run Junkers' factories.

Germany's top industrialists that it was necessary to establish Germany as an air power within one year. Hitler required 4,000 aircraft, among them heavy bombers that could "fly right around Great Britain under combat conditions," and he was looking to Junkers to fill his need for bombers. Dozens of new aircraft manufacturing firms sprang up overnight, and the Air Ministry warmly welcomed each one with contracts for trainers, fighters, dive bombers, and light bombers. Hugo Junkers did not respond.

Junkers flatly opposed any action contrary to the spirit of Versailles, and he would not allow his aircraft and aero-engines to participate in any way in the rearming of Germany. When the Reich told him that it also wanted a naval bomber for attacking Britain's shipping lanes, Junkers again refused. The Reich told him that if he persisted with his treason, it would simply nationalize his factories. German state officials believed the Reich had a strong claim on his factories anyway, for it alone had kept the company afloat during the worldwide economic depression that began in 1929. Still Hugo Junkers refused.

Professor Junkers owned and controlled his entire aircraft manufacturing and engine works and rather than turn to banks, he had always financed his firm's growth from internally generated profits. But in 1932 Junkers did encounter financial difficulties. Although company assets exceeded liabilities by $2.5 million, banks hesitated to give him short-term loans. The problem never came to a head because the Reich came to his assistance with operating subsidies in the form of cash grants and inflated orders for passenger aircraft for the government-owned commercial airline, Lufthansa.

In 1934 Erhard Milch, Hermann Göring's State Secretary for Air, delivered an ultimatum to Junkers: Start producing military aircraft (as Dornier, Heinkel, and Messerschmitt were already doing for the Luftwaffe) or face forfeiture of your factories. Hugo Junkers, now 75 years old and weakened by age, was placed under house arrest by the Gestapo (Geheim Stadts Polizei or Secret State Police). After six hours of steady interrogation by the secret police, the aged man gave in and agreed to sign over 51% of his company (including the Junkers Forchungsinstitut, Junkers AG, and the Junkers Flugzeug-und-Motoren Werke to the Third Reich. Two weeks later the Gestapo reappeared and continued its interrogation, issuing new demands and another ultimatum: Transfer all your aircraft and aero-engine patents to the Reich, and sell your remaining shares to us as well. Junkers refused - but later gave in under more interrogation, and a threat of criminal proceedings over some minor aircraft business he had conducted with the Soviet Union years earlier. On Nazi Party instructions, the ailing professor was then banished to Bavaria. He never saw his Dessau factories again, for he died in exile on 3 February 1935, at the age of 76, and putting a natural end to one of the most distasteful affairs in the annals of German aviation.

A young, determined looking Erhard Milch.

Junkers Flugzeug und Motoren Werke AG - Dessau, Germany

The Junkers' factories took on a whole new personality after they had been nationalized. Throughout the assembly halls pictures of Hitler peered down on the workers.

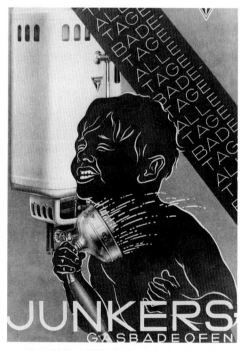

Among the Junkers' produced consumer products group were the highly popular hot water shower units. This ad for a Junkers' shower shows a young boy happily taking a shower on "Badetage" or bath-day with a Junkers' hot water system.

General Ernst Udet, right, and the general manager of the Junkers Flugzeugbau.

No man is more deserving to be called the father of German aviation than Hugo Junkers. His Dessau factories supplied more transport and commercial aircraft to airline companies throughout the world than any other firm. As common as Boeing aircraft are today, so it was with Junkers between 1920 and the mid 1940s. The firm had every intention of retaining its dominance of commercial aviation after the war. On the drawing boards in 1945 was a series of turbine-powered, all-wing passenger airplanes, including the four turbojet-engined EF 130, which was expected to have a top speed of 597 mph (960 km/h). Junkers' engineers believed that the 80-passenger EF 130 would have flown by April 1946 and would have been in commercial service by 1948. Had their predictions come to pass, their efforts would have far outpaced Boeing Aircraft's first commercial turbine-powered passenger aircraft. The famed 707, did not make its maiden flight until 5 June 1954, and did not enter regular airline service until 26 October 1958, when Pan Am introduced it on its New York to London run.

Hugo Junkers was born on 3 February 1859 at Rhegdt, Prussia. Educated at the Technical Institute of Berlin, he was internationally noted for his theories related to internal combustion engines long before he entered the aircraft industry. In 1888, as a young engineer with the German Continental Gas Company, Junkers began experimenting with internal combustion engines. Later he was given a workshop of his own as well as a number of technicians and helpers. By the time he left the company, he and his associates had developed a number of consumer items, including gas home heating furnaces and various instruments for measuring gases, their heat content, pressure, and flow. In 1895 the Dessau factory, where he had conducted pioneering work on gas appliances and measuring instruments, was registered as Junkers & Company. In 1897 he was called to the Chair of Thermal Studies at the Aachen

Junkers flugzeug-Dessau, 1940s.

Nazi banners adorn even the dining halls in the new Third Reich-run Junkers' factories.

Secret Aircraft Designs of the Third Reich

Hugo Junkers paid considerable attention to aerodynamic streamlining in order to achieve more efficient aircraft. He was one of the first aircraft manufacturers in the world to test his designs in wind tunnels. Two of Hugo Junkers' wind tunnels. Note the size of the wind tunnel in the foreground in relation to the man walking past the wind tunnel's mouth.

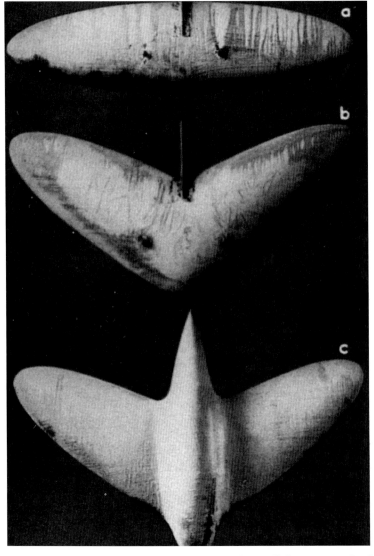

Early wing shapes Hugo Junkers found to be highly efficient as a result of wind tunnel testing about 1910.

Technical Academy, where he continued experimental work to improve the internal combustion engine.

At the same time, Orville and Wilbur Wright in the United States were beginning their study and experimentation with heavier-than-air machines. Little was known in Germany of the early flights of the Wright brothers. It was only when these pioneers came to Europe in 1908 and aroused the interests of French industrialists (then the leaders in construction of motor car engines) in the new science of aeronautics - and only when the people of Berlin, in the autumn of 1909, witnessed with amazement the Wrights' seven or eight mile "cross-country flight" from the Tempelhof Field to Johannisthal - that the "aeroplane" began to attract attention.

Generally, the men who took interest in flying were not so much the technicians and engineers as the intrepid sportsmen willing to trust their lives to weird-looking contraptions of wood and fabric and wire. But at the Aachen Technical Academy, two professors began conducting theoretical and practical experiments related to the heavier-than-air machine - Professors Reissner and Junkers. In April 1909 Reissner made tests (or rather, leaps in the air) near Aachen with his Ente (Duck), a monoplane having thin, zigzag-shaped wings. Professor Junkers, in characteristic fashion, worked at basic research to establish principles, as he had done in all his work, beginning at the German Continental Gas Company. Paying little attention to the work of others, Junkers began by making observations on energy and efficiency. He occupied himself chiefly with the significance of the relationship between lift and resistance, or drag. To his work he brought a special knowledge of, and interest in, metal-working, which others lacked. This combination led to an ingenious design that was granted on 1 February 1910 the All-Wing Patent DRP. No. 253, 788 - an airplane consisting of one wing which would house all components, engines, crew, passengers, fuel and framework. This was the origin of the "thick cantilever wing," which at first was ridiculed as absurd and later, when tested on prototypes, was criticized and rejected by the experts.

Although Junkers had devoted considerable attention to aircraft design, it was several years before the first Junkers airplane appeared, for he was concentrating on other work, especially between 1911 and 1913, on the problem of the large diesel engine. In the meantime, Junkers continued to ponder this basic idea - that any load and the power plant should be housed in the wing itself. This required, among other features, a large wing and lightweight construction, which Junkers believed could be achieved only with the use of metal, never with wood. In this way the idea of the all-metal airplane was conceived.

Junkers began preparation for aircraft construction in 1913, the diesel engine having progressed so well by then that it was possible to release labor and funds for the new project. He began with experiments in the Aachen wind tunnel on hundreds of models, not of airplanes but of elementary forms such as spheres, cylinders, and cones. He found that a wing could be shaped to combine a favorable lift-to-drag ratio with the practicability of housing all spars, struts, and framework inside the wing.

World War I, and the military interest in flying it created, speeded up airplane development. Junkers stopped all work on diesel engines, and in the autumn of 1915 unveiled his first experimental airplane, the J-1, an all-metal, unbraced monoplane made from sheet steel. At first the plane was rejected, for the military wanted bi-planes, not monoplanes, and even the tri-plane had arrived. Further, prejudice against Junkers' all-metal construction was considerable. No one believed the "metal horse" could fly. A famous World War 1 pilot once asked a Junkers test pilot, "Are you so tired of life that you are flying a machine without bracing wires?"

However, flight tests of the J-1 gave exceptional results. Although the aircraft weighed nearly a ton, equipped with a 120 horsepower en-

A wind tunnel model for a proposed design for a Junkers mono plane in the 1920s.

The Junkers G 38 was in essence a modified form of the Junkers all-wing concept which he had patented in 1909.

gine that Junkers had designed, it reached the unprecedented speed of 106 mph (171 km/h). On the basis of the J-1's performance, the Luftwaffe ordered several of the aircraft. But the military found that it needed aircraft capable of rapid high-climbing, and the weight of the J-1 made this impossible. The monoplane had proved the practicability of all-metal construction, but if Junkers was to obtain additional contracts, a lighter material must be found. Choosing a suitable lighter material presented problems, too, for each possibility had both advantages and drawbacks. Duralumin, for instance, could through existing technology, be joined only by riveting, not by welding, as was the common practice at the time. (Junkers disliked riveting because the rivet heads created too much air resistance.) On the other hand, Duralumin did have certain qualities of strength that permitted radical alternatives in design. Junkers finally decided on Duralumin, and work proceeded.

With the transition from heavy-metal to light-metal construction, quantity production of Junkers aircraft began. First to be built was the J-4, a monoplane in which engine, fuel tanks, and crew were sheltered in an armor casing one-sixth of an inch thick. This "Flying Tank" was enthusiastically received by line pilots in World War I despite the long distance it needed for takeoff. Once in the air, its minor shortcomings could be overlooked, for it flew well, it could climb, and it was fast. Even when it had encountered considerable enemy fire, the aircraft usually could land safely behind German lines, thanks to the metal construction. Junkers' success with all-metal aircraft now seemed assured, and he formed his own joint-stock company, calling his firm the Junkers Flugzeugwerke, AG. In the last 2 years of World War I Junkers delivered to the Luftwaffe 227 armored aircraft and 88 unarmored aircraft, for a total of 315 aircraft, all constructed from Duralumin.

With the Armistice of November 1918, activities in German aircraft factories largely ceased. Compulsory destruction of large numbers of war planes followed, and Junkers had to destroy parts of a projected two-engine bomber called the R-1, which was similar in design to the giant bombers manufactured by Gothaer Waggonfabrik (Gotha). Before long Junkers was making adjustments to accommodate the manufacture of civilian aircraft; and by 25 June 1919, his history-making F 13 monoplane was in the air. This single-engine aircraft was the true forerunner of civilian aircraft. Its appearance created a sensation. Similar in many respects to Junkers' wartime J 10 attack monoplane, the F 13 reflected a further practical application of Junkers' 1910 patent for wings of cantilever construction. The pioneer in commercial air transport, the F 13 was the ancestor of the metal transport aircraft that were to follow in the 1920s and 1930s. Its entire airframe, including control surfaces, was clad in corrugated Duralumin sheet. It soon was in service in almost every part of the world, a total of 350 aircraft being manufactured. The largest number of F 13s were used in Junkers' own transport company, Junkers Luftverkehr, which between 1921 and 1926 had upward of sixty F 13s in service. The F 13 was the first commercial aircraft to be fitted with seat belts. It carried a crew of two, plus four passengers, in a fully enclosed cabin at a speed of 87 mph (140 km/h).

Junkers F 13 made liberal use of corrugated duralumin on the fuselage sides.

Powered by four Junkers engines, the Ju G 38 had a wing which was 5 feet 7 inches thick and could house cargo and passengers.

Junkers initially had entered commercial aviation to market his proven products. Before World War I aviation had been regarded as the province of foolhardy adventurers and, during the war, as the patriotic activity of heroes. Junkers wondered how, from two such starting points, airplanes could become accepted as a means of transport for the masses. He believed such a transformation of ideas could be brought about only by extensive publicity. Regular air traffic seemed the most suitable means of popularizing aviation and of generally making it acceptable to the whole nation.

In perfecting and seeking a market for his aeronautical products, Junkers also was moved by another powerful motive. He believed that now that the war was over civil aviation could play a major role in fostering commercial and cultural relationships among distant peoples. To that end, Germany should use the skyways as a means of developing greater trade and commerce. Hugo Junkers was becoming a democrat and a pacifist. In 1925 he said in regards to the success of his commercial aviation activities and his growing pacifism,

"The geographical limitation to our creative horizon following upon the adverse issue of the war is removed. Once again we can think in terms of continents. No longer need our people lament over the destruction of the air fleets that own so many victories. They can rejoice at new and splendid peace-time flying achievements, and this strengthens their courage and confidence in the future."

By 1925 the work of promoting air traffic, originally regarded by Junkers as a way to promote cultural relationships, public acceptance, and economic trade, was taking a considerable amount of time. His airline operation involved 178 aircraft carrying 100,000 passengers and 650 tons of freight over 3 million miles each year. The complexity of the larger European operation - 99 routes extending over a distance of 35,500 miles - required a great deal of coordination between Junkers and other private and public commercial aviation firms throughout Germany, and all of Europe. In 1925 there were twelve airline companies in Europe, and the entire province of private enterprise was taking on a public character. Many believed that some unity of commercial aviation activities was called for. Junkers agreed that some coordination was needed, but he opposed any state-owned aircraft operations. When the Reich created the state-owned Lufthansa in 1926, Junkers withdrew from commercial operations in Germany (although he did retain some aircraft operations abroad).

Boyed by the quick acceptance of his F 13, Junkers next designed an all-wing large transport, the JG 1, which looked like his "Glider Patent" of 1910. The aircraft was expected to weigh about 9 ton, and its engine and fuel tanks were to be placed right in the wings. The wings and fuselage had been partially constructed when, in 1921, Germany was forbidden by the Treaty of Versailles to construct any kind of aircraft. This ban prevented completion for the JG 1, as well as the realization of bigger projects Junkers had in mind, including a 40-ton double flying boat, the Junkerissime, and a mighty land-plane of similar size. The ban was partially lifted in 1922; "war aircraft such as bombers and fighters" (including single-seat aircraft and other types that might be put to military use) remained forbidden, but peace-time aircraft such as transports and commercial planes were permitted. (Even then, only aircraft that could climb no higher than 13,000 ft (3.9 km), and carry more than 1,300 lb (590 kg), were allowed.) Back in business Junkers sold twenty-five F 13s in 1922 and sixty in 1923, and with the "conditions and definitions" of the Treaty of Versailles still in place in 1923-24, he introduced his first tri-motor civil airplane, the G 24.

The G 24 represented a continuation of Junkers' use of metal for aircraft, but his design was a first. The aircraft was a low-wing monoplane, and its outboard motors and fuel tanks were housed directly in the wing structure. The G 24 was followed in 1928 by the G 31 and then, in 1930 by the Ju 52/3m. The latter is rivaled perhaps only by the American made Douglas DC-3 for the honor of being the most famous transport aircraft of all time. Used by most of the world's airlines and air forces, the Junkers Ju 52 remained in service for nearly 40 years. Its design was the product of collaboration between Junkers and his design chief, Ernst Zindel. It was a low-wing cantilever monoplane with a fixed

Ernst Zindel. January 1897 to October 1978.

undercarriage. Intended primarily as a transport, the Ju 52's cargo hold had a capacity of 590 cubic ft (16.7 cubic m). When used as an airliner, it could carry fifteen to seventeen passengers at a cruising speed of 124 mph (200 km/h) for a maximum range of 796 miles (1,280 km).

Considered obsolete at the beginning of World War II, the Ju 52 soldiered on long past its time because it possessed outstanding flying characteristics. Its sturdiness and invulnerability in bad weather were almost legendary. Germany's loss of the war did not mean the end of the Ju 52. It continued to be manufactured as the AAC.1 Toucan by the Ateliers Aéronautiques de Colombes in France and was used by the French Aéronavale, the Armme de L'Air, Air France, and a number of other airlines. In 1949 and 1950 Ju 52s based in Hanoi were used in some number against the Viet Minh in French Indochina, and the aircraft was manufactured in Spain until 1954.

The success of the F 13 after its introduction in 1919 created an unforeseen demand for aero-engines. Initially the F 13 had been powered by a 160 horsepower Mercedes D 111a engine; later models used the 180 horsepower BMW IIIa. Existing BMW (Bayerische Motoren Werke) engine factories were not sufficient to meet Junkers' demand, so the Junkers Motoren Werke AG was established in 1923. It not only manufactured the well-known engines of stationary type, but also organized itself to meet the steadily increasing demand for the Junkers Flugzeugbau AG for light gasoline engines. The L-2, a standard water-cooled, six-cylinder, in-line engine with an output of 265 horsepower soon became prominent in aeronautical circles. Then came the L-5, which produced up to 425 horsepower, and the L-88, which was installed in the Junkers G 38 and gave an output of around 800 horsepower.

Despite the improvements in piston engines, many (manufacturers) were looking for another means of propelling aircraft, particularly turbine propulsion. Junkers' interest in turbines dated back to 1930 and his association with Professor Ludwig Prandtl, whom Junkers had hired as a consultant to conduct basic research on gas turbines. Ernst Heinkel, one of the most versatile of the German aeronautical designers, kept his firm interested in the subject, beginning that same year. In the mid 1930s engineers at BMW in München began preliminary research in turbine propulsion, and some beginning calculation on gas turbines began to take shape at Junkers' Dessau facilities. By 1936 Junkers' engineers, led by Herbert Wagner, head of aircraft development, had built a turbine engine but it could not run under its own power and in 1939 the project was abandoned. However, in 1938 the Air Ministry had taken an interest in turbine engines and had hired Helmut Schelp, a German who in 1936 had earned a masters degree in engineering from Stevens Institute in Hoboken, New Jersey, to head up a new department within the Air

The Junkers Ju 52 was designed by Ernst Zindel. Later he designed the Ju 90 four-engine bomber at the request of Walther Wever and then the Ju 88 twin engined fighter bomber of World War II fame.

Ernst Zindel shown here with a flight mechanic. His Junkers Ju 88 is perhaps Germany's most successful, durable, and accepting of various modifications and military roles as any German military aircraft then or since.

Ministry which was responsible for developing the turbine engine. Initially the idea was not well received in the Junkers Motoren Werke, for there had been considerable problems with the Wagner turbine engine. Schelp had persuaded Daimler-Benz at Stuttgart to enter the gas-turbine field, but its first design, by Professor Karl Leist, proved extremely complicated, and was abandoned after a short time.

Heinkel, on the other hand, working independently (from even the Air Ministry and Schelp), surprised everyone when on 24 August 1939, his HeS 3B turbine engine powered the world's first turbine flight. Later that fall the Air Ministry instructed Junkers to renew work on Wagner's engine, but the firm refused. Instead, a new Junkers group headed by Dr. Anselm Franz was formed. Financed by a grant from the Reich through the Air Ministry, Franz and his group were allowed to develop their own design. The engine they produced (with considerable assistance from BMW engineers) came to be known as the Jumo 004 (Jumo for the Junkers Motoren Werke). The design group (quite correctly) claimed that it had in 1940 the first successful axial-flow turbine engine in the world. The 004 was also the first turbine engine to be produced in volume, and the first to enter combat service.

Dr.Ing. Franz came to America after the war, having escaped the USSR's Red Army troops moving on Dessau by a wide margin of 2 weeks. Joining AVCO Corporation, Franz rose to become a vice president of AVCO Corporation's Lycoming Jet Engine Division. He retired in 1968, having designed several popular turbojet engines, including

249

Had this Junkers Ju 90, a four-engine bomber prototype designed by Ernst Zindel, been developed as planned, Germany would have been able to bomb Atlantic convoys as well as destroy Russian armament factories beyond the Ural Mountains. The Junkers Ju 90 project was terminated after the RLM's Chief of Staff General Wever died in a plane crash on 3 June 1936 and along with him the whole strategic bomber program.

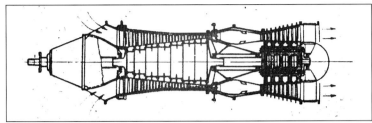

The Jumo 022 turboprop under development 1945.

Junkers Ju 248.

the T-55, the first jet engine with a high-bypass turbo fan, which came to be used on many passenger aircraft.

Development of the world's first axial-flow turbine had taken about 2 years. The design of the experimental 004A engine had been completed in the spring of 1940, and the engine had made its first start up on 11 October 1940. In December the engine was brought to full speed of 9,000 rpms, and in January 1941 a thrust of 946 lb (430 kg) was reached. At that point further development testing was delayed by failures in the compressor vanes. The stator was redesigned, and the thrust of 1,320 lb (600 kg) required by the Air Ministry contract was obtained in August 1941. In December of that year, a 10-hour run was accomplished and a thrust of 2,205 lb (1,000 kg) was demonstrated.

The first 004A approved for flight was tested on a Messerschmitt Me 110, and this historic flight took place on 15 March 1940. An Me 262 powered entirely by two 004A's flew the following 18 July. On the basis of extraordinarily promising results, the Air Ministry issued Junkers a contact for eighty engines of the 004A type to be used for further engine development and airframe testing. However, the 004A was not suitable for production, for it was too heavy, and it contained too high a content of strategic materials. The amounts of nickel, cobalt, and molybdenum that were required for the heat-resistant materials of this engine, for example, simply were not available under war conditions. With re-engineering and modifications, Junkers introduced in June 1943 with the 004B turbine, which was quite suitable for production, having a thrust of 2,006 lb (910 kg). An Me 262 powered by two 004Bs flew in October 1943, and some 5,000 Jumo 004Bs were produced before Germany's collapse in May 1945. Most of the 004Bs went into Me 262s; some went into Arado Ar 234s, and a few into experimental models of the big, four-turbine engine Junkers Ju 287. The Horten brothers obtained several 004B's for their pioneering Ho 9 V2 all-wing fighter prototype. Obtaining these scarce engines was possible for the Hortens thanks to a family friend in upper management of the Luftwaffe Quartermaster's Office. Later, in 1944-45, designs were completed and construction was begun on prototype units of the improved Jumo 004D and 004H versions, which had an eleven stage, axial-flow compressor (instead of the eight stage compressor in the 004B) and a two-stage turbine (instead of a single stage). Thrust from the 004D and 004H versions was more than double that of the production Jumo 004B, some 3,060 lb (1,796 kg).

While work culminating in the Jumo 004 was being done at the Junkers Motoren Werke, aircraft design was taking place in another Junkers division. Hans Wocke, Ernst Zindel, and Heinrich Hertel formed the nucleus of the aircraft design team at Junkers. Zindel had been with Junkers the longest, joining the firm in October 1920 and becoming chief of construction in 1922. He shared with Hugo Junkers the belief that the aircraft, if used properly, could become the chief means of transport to unite people all over the world. It was Zindel who designed the G 24, the first three-engine airliner, in 1923-24. The rapid growth of air passenger traffic renewed Junkers' interest in the all-wing aeroplane because he believed it could be flown more economically than a conventional aircraft. Junkers later built two very handsome all-wing prototypes called the J-1000, with a capacity for 100 passengers. It flew for 30 minutes during its first flight on 6 November 1929. One of the two prototypes, the Generalfieldmarshall von Hindenburg, was in service

Allied officials waste no time in dismantling a captured Jumo 004B in 1945, to investigate its inner workings.

for 10 years with Lufthansa. In 1930 Junkers and Zindel delivered another aircraft that was to become a worldwide success - the three engine Ju 52. The Ju 52 was to be their last great success, as a team, for in just a few short years Hugo Junkers was forced out of his own firm.

When Hugo Junkers was stripped of his factories in 1935, the Third Reich became sole owner, and Junkers' employees took all orders from officials in Berlin. Dr. Heinrich Koppenberg was appointed the managing director of both Junkers Flugzeugbau und Motoren Werke, AG. In his first address to the employees of the new Junkers in 1935, Koppenberg supported the Third Reich and its new direction:

"The Hitler government, which has worked wonders in providing employment for so many trades, has allotted to the long-suffering aircraft industry tasks that fill us with joy and pride. The past year has seen our tasks achieved, and has proved the spirit of loyalty and devotion to duty that animates every member of the staff - the draftsman at his board, the office man at his desk, the fitter at the bench, and the turner at the lathe. It is most gratifying to me to feel the purposeful spirit of progress again at work in our offices and shops, urging every worker forward."

Dr. Koppenberg remained managing director of Junkers until 1942, when Professor Otto Mader succeeded him. Mader, considered by some to be one of Germany's outstanding aeronautical engineers, died late in 1943. With Junkers working exclusively for the Air Ministry, all orders for aircraft production between 1935 and 1942 were placed by Ernst Udet. After Udet's suicide in 1942, all production and delivery instructions came from Generalfieldmarhsall Erhard Milch, Udet's successor. Junkers again got a new boss, Albert Speer in 1944, whom Hitler had appointed to take complete control of the German wartime economy and who filled that role until Germany's surrender in May 1945.

The arrival and departure of the several managing directors and outside bosses appeared not to have much impact on Junkers and its assigned wartime aircraft production, undoubtedly because the firm's assignments, until the end of 1944, were narrow and precise. It was to develop and produce only bombers, dive bombers, fighter-bombers, and other large aircraft which used gasoline or diesel engines. Junkers produced 30,390 airplanes between 1933 and March 1945, when its Dessau facilities were overrun by the Red Army. The following aircraft accounted for 80 percent of the firm's 12-year production:

Ju 52 transport	4,845
Ju 87 dive bomber	5,709
Ju 88 fighter	15,183
Ju 188 bomber	1,237

Prior to the end of 1943, Junkers' aircraft manufacturing and design groups were not allowed to use the Jumo 004 turbine being produced by Junkers Motoren Werke. Indeed, they had no need for them, since Junkers was a Grossflugzeug Firma (factory for large aircraft) and

The Soviet NK-12M, (first generation), perhaps the most powerful turboprop ever. Based on the Jumo 012/022. The Soviet 022-K was hand-built by a few ex-BMW turbine engineers in the USSR post-war beginning in 1946.

Side view of the Junkers Ju 248. Scale model by Günter Sengfelder.

was forbidden to design small-turbine propelled aircraft. Instead, the 004s were reserved by the Air Ministry for fighters, interceptors, and a few aerial reconnaissance aircraft. When Junkers' designers were able to show that turbines also would be advantageous for large bombers and long range reconnaissance aircraft, Junkers was ordered in March 1944, to complete the design for and then manufacture the turbine-powered Junkers Ju 287 heavy bomber.

The Ju 287 had begun as a new design on the basis of aerodynamic research work conducted by the Deutsches Versuchsanstalt für Luftfahrt (DVL). As early as 1935 German aerodynamicist Adolf Busemann had proposed backward-sweeping wings. He and others at DVL believed that using sweptback wings was the only practical way to reach speeds beyond the speed of sound. In 1940 Hans Wocke at Junkers, who was investigating wing designs for a proposed large, four-engine, piston-powered, heavy bomber for the Air Ministry, concluded that if turbines were used instead, the bomber would benefit by having the wings sweep forward instead of backward. Wocke's beliefs were borne out through wind tunnel tests, and in March 1943 the Air Ministry ordered from Junkers two prototypes of Wocke's unconventional design. Two prototypes were built, the Ju 287 V1 and V2.

The results of the Ju 287's flight tests in 1944 were considered outstanding, and mass production was started, but was stopped suddenly by the Air Ministry in the autumn of 1944 when the Reichsverteidigungs (Government defense) program was adopted. Work on other aircraft being produced by Junkers, including dive bombers and transports, also was halted, for the program called for concentration on fighters and interceptors. However, in February 1945 the Air Ministry reordered the Ju 287 into production with highest priority for what was being called Hitler's emergency program. It was only after Junkers' design projects for large aircraft were halted by the Reich's defense program that Junkers hd the opportunity to participate in fighter development. The firm did not have the wind tunnel experience to draw on when its designers set to work on high-speed fighters and interceptors; but with the cancellation of the Ju 287 turbine-powered heavy bomber, it did have a surplus of engineers with very little design work to do.

Designers at Junkers thought they could eventually overcome their lack of experience; for the near-term, they would play catch-up by learning all they could from the work of others, such as Alexander Lippisch, Richard Vogt, Ernst Heinkel, and Willy Messerschmitt. To get them started the Engineers from Junkers did not particularly like the Me 163 as a design, or its modification, Ju 248. They preferred to concentrate on their own fast-interceptor concept, the EF P.128 (Entwicklungs Flugzeug 128 or Development Aircraft 128). The EF P.128 would have been entirely different from either the Me 163 or the Ju 248. The HWK 509 bi-fuel liquid rocket drive would have been replaced by a HeS 011, and the single rudder would have been moved from the fuselage to the wings. Junkers engineers believed that these two changes would allow

Ground level front view of the Junkers Ju 248. Scale model by Günter Sengfelder.

higher speed with greater range; they also turned the plane into a virtual all-wing aircraft.

Other than the Ju 287, the EF P.126, and EF P.128, the Junkers organization produced very few creative or innovative turbine-powered aircraft designs. It was not noted as an innovative organization, and most of its projects involved obsolete pre-war designs with turbine units attached for increased performance. In addition to designs for these three aircraft, Junkers had only two other turbine-powered designs worthy of mention.

Junkers Flugzeug was truly a workhorse for the Air Ministry. When Junkers (Dessau) facilities were overrun by Red Army troops in April 1945, Junkers employed approximately 147,000 workers - 52,000 in the Flugzeugbau (aircraft group) and 91,000 in the Motoren Werke (engine group), and the remainder in administrative, research, and miscellaneous activities. About 46% of the employees (67,000) were slave laborers (that is, prisoners from concentration camps), prisoners of war, or foreign workers. Occasionally two shifts were maintained, and during the last 8 months of the war employees were compelled to work 70 to 72 hours a week. On rush jobs, or to relieve bottlenecks, a 30% second (overtime) shift was added.

The Allies bombed Junkers facilities heavily during the war. Between the time of the first Royal Air Force attacks of February 1940 and the last of the numerous attacks on the Junkers facilities. Thirty of the attacks occurred in 1944, a total of 3,465 ton of high explosive bombs, 646 ton of incendiaries, and 29 ton of fragmentation bombs being dropped that year. By 1943 virtually all personnel, technical and administrative, were being pressed into service to clear areas destroyed in bombing raids, and German Wehrmacht engineers were used when necessary. The policy was to continue work in undamaged portions of the facilities.

According to Junkers personnel, the bombing raids did not affect Junkers' overall production seriously, in large part because by 1944 Junkers was successfully carrying out the Air Ministry's order of the preceding year. It had dispersed its aircraft and aero-engine works and facilities by using existing plants, building new facilities in remote areas, and constructing underground factories in tunnels and caves. As the dispersed facilities were being established, Junkers took on double and triple suppliers to minimize the effects of bombing. Because many of the facilities were in remote locations, housing for workers and transportation of suppliers and parts became an acute problem. Junkers often had to erect houses for native, as well as foreign labor, and to initiate special transportation services to carry workers to the factories.

Recuperation from attacks apparently was speedy, reportedly ranging from only a few hours to a maximum of 2 weeks. According to Junkers' personnel, there was seldom a shortage of machine tools, and the wide dispersal of plants, the (sometimes triple) supply system, and stand-by facilities did much to offset the effect of air attacks. The biggest loss was caused by the widespread gasoline shortage. The shortage affected shipment of Junkers products, particularly after failure of the railway system. It also delayed testing of completed aircraft. Airplanes frequently sat at airfields for as many as three weeks awaiting gasoline for test flights. In the end, many were flown off before they were tested - a major reason why many new aircraft failed to reach combat status. Ernst Zindel and Heinrich Hertel were able to escape only a few days before the Russian occupation of the eastern sector of Germany on 1 July 1945. This is when the Red Army sealed the border and controlled travel in and out (including the travel of US Army troops, who were required to have Red Army escorts). Zindel and Hertel were helped in their escape by US Army Intelligence officers. Time ran out for Hans Wocke, however; he was trapped in the eastern sector and became one of the first of many former German aircraft designers, engineers, and scientists to be "hired" by the Russians and transported to the USSR in October 1946. Wocke, along with other former Junkers' personnel, was returned to East Germany by 1954.

As early as the spring of 1944 the Soviet Union had decided that its post-war military program would be directed toward the development of rockets and supersonic turbine-powered aircraft. These would be built on Russian soil - with materials seized from Germany. However, before moving the sophisticated precision tools and instruments used in the production of rocket drives and turbine engines, the Russians needed to learn how to use them. The best way was to watch the Germans. So the

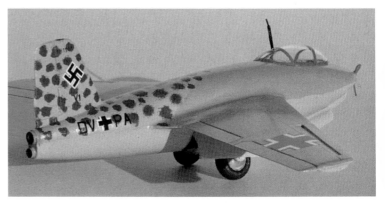

Rear side view of the Junkers Ju 248. Scale model by Steve Malikoff.

Ground level rear view of the Junkers Ju 248. Scale model by Günter Sengfelder.

Soviets reopened many of the German aircraft and aero-engine factories, even some where supersonic aircraft were being built, including Junkers Dessau with its Ju 287 highspeed, turbine-powered, prototype bomber. Hans Wocke was at the Dessau facility. Through the remainder of 1945 and continuing through October 1946, the Soviets watched as Wocke and others at Junkers continued their design work, their aerodynamics and flight mechanics, and their wind tunnel testing. They also observed and participated in advanced turbine design work at the Junkers Motoren Werke. While some individuals did manage to escape by underground routes through the forests into French and American sectors of Germany, Wocke did not; instead, he was transferred to Podberesje, USSR in October 1946.

Ernst Zindel, chief of construction at Junkers, found other work after the war and never did return to manufacturing aircraft. The designer of nearly all Hugo Junkers' aircraft after 1920, Zindel died on 10 October 1978, in Germany at the age of 81.

Dr.Ing. Heinrich Hertel, chief engineer at Junkers and responsible for building the prototype Ju 287 turbine-powered bomber, went to France after the war and worked for the French Aviation Ministry on proposed turbine-powered passenger (trans Atlantic) aircraft. Later he became heavily involved in vertical takeoff-and-landing aircraft with the French aero-engine firm of SNECMA. In 1955 he returned to Germany to teach airplane design and development at the Technical University in Berlin. There he also established a research institute for the study of air and space travel. Having begun the research institute in 1955 with three co-workers, Hertel retired in 1970 after seeing the institute grow to 53 workers. Born on 13 November 1901, he died in 1981, in Germany at the age of 79.

After Junkers Flugzeug und Motoren Werke was turned into state-owned, state-operated facilities, the Reich looked upon it as just one more company that could contribute to the war effort. Eventually much of Junkers' creativity, the spirit which the father of German aviation had steadfastly nurtured, for 20 years, disappeared because its new bosses did not value creativity and innovation. They merely wanted planes, planes, and more planes. After the war members of the Junkers team went their separate ways, but it is interesting to note that after the aircraft manufacturing ban in Germany was lifted in 1955, Wocke and Hertel plunged right back into designing and constructing large civilian passenger planes. Zindel, who was 58 years of age in 1955, felt he was too old to go back to full-time work as a designer; besides, he said, he had not done any design work for 10 years. But he never went very far away from designers and aircraft, for he was an active participant in the "Hugo Junkers Society." Although Junkers had died some years before, in 1935, most Germans remembered with admiration their country's aircraft pioneer, especially those workers who recalled the principles of pioneering he encouraged his colleagues to practice - in good times when everything was going as planned - as well as in dark and uncertain times. Professor Junkers once reminded his employees that:

"Working things out is an alternating process. From the idea emerges the line of action. Design and experiment follow. Then the idea is returned to. The line of action is modified and much experimental research done. Drawing-office work follows, and then practical application in the workshop. It does not do to contemplate dropping an invention or a design because it is not proving successful. When everything seems to be going wrong, then everything depends on the underlying spirit. It is on that, that success depends. Obstacles must not daunt; inner misgivings must be conquered; the goal must be doggedly pursued. That is what makes the real pioneer. It is not so much the technical as the spiritual achievements that are decisive in the long run."

The Junkers name has faded into oblivion, not to be seen on factories or aircraft since 1945. For a time after 1955 the Junkers organization in West Germany did try to return to aircraft manufacturing. But

Close up view of the cockpit of the Junkers Ju 248. Scale model by Günter Sengfelder.

Secret Aircraft Designs of the Third Reich

Russian troops captured a complete Junkers Ju 248 when they overran Dessau. Later the Russians added a horizontal stabilizer and test flew the rocket-propelled interceptor. They found that it did not fit into their overall aircraft program and the Ju 248 was abandoned.

The proposed Junkers EF 009 vertical take-off and landing interceptor. Scale model by Steve Malikoff.

things did not go as planned. Instead, in 1967 Junkers merged with Messerschmitt AG. Three years later, Messerschmitt merged with the Ludwig Bölkow organization, the Hamburger Flugzeugbau (formerly Blohm and Voss), and Siebel KG, forming the giant Messerschmitt-Bölkow-Blohm (MBB) as a world force in aeronautical engineering.

The merger with Messerschmitt meant that the Junkers Flugzeug und Motoren Werke AG ceased to exist. From the 1920s through the 1930s, more people flew in Junkers aircraft than in any other. Most of the world first experienced aviation via Junkers aircraft. It was truly the way Hugo Junkers wanted his aircraft to be used - helping people travel, in the hope that they would get to know and understand each other better.

PROJECT 009

The P.009 was known officially as a Hubjäger (lift fighter). Today it would be called a vertical takeoff aircraft, but it would have landed, not vertically, but as a sailplane, on skids built into the lower turbine engine nacelle and the tail. Landing speed would have been low, about 99 mph (160 km/h).

The Air Ministry's interest in the Hubjäger type of aircraft stemmed from a desire to find a replacement for the Me 163B "Komet" HWK bi-fuel, rocket drive interceptor. The Me 163 tended to suffer accidents during takeoff and landing and also from fire caused by malfunctioning of its HWK 509A bi-fuel liquid rocket drive. Eighty percent of all losses of Me 163s during World War II resulted from accidents. Junkers' engineers felt that a turbine-powered aircraft could be built having the same rapid takeoff capability as the Me 163 but presenting much less risk to the pilot and the aircraft. This idea led to the design of the P.009. The 009 was expected to have a greater rate of climb than the Me 163: 15,512 ft (15.7 km), in 3-1/2 minutes, vs 39,698 ft (12.1 km) for the P.009. It was also expected to have greater maximum speed: 621 mph (1,000 km/h) vs 596 mph (960 km/h).

The P.009's blazing rate of climb would have come from ten miniature turbine engines, to be built by Junkers Jumo. Although the entire internal wing area of the P.009 would have contained fuel tanks, it was estimated that there would be only sufficient fuel for 6 minutes of powered flight. The P.009 was never built because Junkers' idea for a miniature turbine engine was never pursued, perhaps, due to the rapid advancements in more powerful turbine engines at BMW and Junkers Jumo. The entire P.009 Hubjäger project was abandoned in late 1943.

To compensate partially for the enormous pressure on the pilot during liftoff, Junkers' designers planned to place the pilot in a reclining seating position. Armament for the P.009 was to be two 20 mm MG 151 cannon and, if desired, two 30 mm MK 108 cannon.

Specifications: Project 009

Engine	10xJunkers Jumo miniature turbines (type and thrust rating unknown)
Wingspan	13.1 ft (4 m)
Wing Area	NA
Length	16.4 ft (5 m)
Height	NA
Weight, Empty	NA
Weight, Takeoff	4,409 lb (2,000 kg)
Crew	1
Speed, Cruise	NA
Speed, Landing	99 mph (160 km/h)
Speed, Top	621 mph (1,000 km/h)
Rate of Climb	15,158 ft/min (4,620 m/min)
Radius of Operation	NA
Service Ceiling	51,512 ft (15.7 km)
Armament	2x20 mm MG 151 cannon and/or 2x30 mm MK 108 cannon
Bomb Load	None
Flight Duration	6 minutes under power

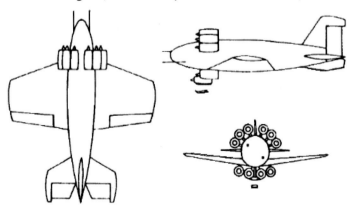

Rear side view of the proposed Junkers Ju P.009. Scale model by Steve Malikoff.

Junkers Flugzeug und Motoren Werke AG - Dessau, Germany

Side view of the proposed Junkers Ju EF 009. Scale model by Steve Malikoff.

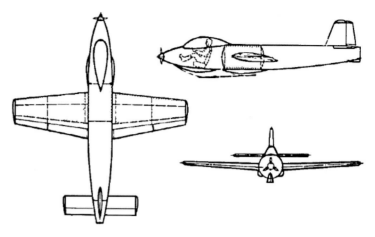

PROJECT 126

Development of the Ju P.126 was prompted by an Air Ministry's desire for an interceptor to replace the Heinkel He 162 turbojet-powered Volksjäger. The Air Ministry was looking for ways of minimizing the use of strategic materials, and, even more important, of producing an aircraft in a minimum amount of time. Officials estimated that using a pulsejet rather than a turbine and eliminating electronic gear could reduce construction time by as much as 450 man-hours over, for instance, the He 162.

So in November 1944, with the pulsejet still untried in an aircraft, the Air Ministry issued a requirement for the simplest possible fighter that could use an Angus-Rohr pulsejet, which produced 1,101 lb (500 kg) of thrust and was the propulsion unit for the Fieseler Fi 103 F-1 (V-1 Flying Bomb). Only three firms responded to the Air Ministry's so-called Miniaturjäger (Miniature Fighter) program, Blohm and Voss, Heinkel, and Junkers.

The anticipated performance of Blohm and Voss' submission, the P.213, was considered inadequate (about the same speed as Germany's faster piston-engined fighters then). Heinkel took a shortcut with its entry by proposing merely to fit its standard He 162 fuselage with the pulsejet (instead of its usual BMW 003E-1 turbine of 1,764 lb (800 kg) thrust). Heinkel's entry also was considered underpowered; although Heinkel officials later said it would be possible to place twin pulsejets on the He 162B, which would then give it double the thrust. In the end, Junkers was given the go-ahead to develop a prototype of its entry, the P.126.

Junkers managed to construct several prototypes before the US Army troops overran the Dessau facilities in April 1945. According to Junkers' engineers, at least one prototype was built and flown. Flight tests revealed that the aircraft would not have performed as anticipated, and the Air Ministry decided to continue with the project, not as an interceptor, but as a ground-attack aircraft.

Regardless of the purpose the P.126 would serve, the advisability of using the Argus-Rohr pulsejet for propulsion was questionable. The Fieseler Fi 103 F-1, actually a piloted V-1 bomb, had been involved in a number of accidents because the constant hammering of the pulsejet tended to loosen the glue of the plywood covering the wings. In fact, it is also generally believed that the main reason so many robot V-1's crashed before reaching their targets in Great Britain was the fact that the pulsejet virtually shook the tiny aircraft apart in flight.

Overall, Junkers' P.126 was similar in appearance to the Fi 103 F-1 piloted bomb. The Fi 103 F-1's cockpit was just behind the wing's trailing edge and directly below the air intake for the pulsejet unit, whereas the cockpit for the P.126 was positioned well forward on the nose and was enclosed in a bubble canopy. Otherwise, both aircraft looked very much alike. Both had a cigar-shaped fuselage, a rectangular fin and tailplane, wings of plywood construction with a single steel spar, and a simple Argus-Rohr pulsejet propulsion unit mounted above the fuselage, its bulged intake cowling starting just aft of the wing's trailing edge. On the P.126 the pulsejet's exhaust pipe passed along the top of, and beyond the tail fin and rudder. Unlike the Fi 103 F-1, which used a landing skid and was launched by catapult, the P.126 would have been mass produced with a conventional undercarriage. It would have been propelled up to the pulsejets operating speed of 149 mph (240 km/h) by solid-fuel rocket drives. The prototype P.126 had been tested with both wheels and skids, and the choice for the Junkers' engineers was to equip the aircraft with a tricycle landing gear in order to save pilots from possible spinal damage, even paralysis, due to hard landings.

There are no details of flight tests of the P.126 in April 1945; but it has been reported that the captured prototypes were test flown, unpowered and towed to flight altitude where, after several fatal accidents by German pilots, one a former top test pilot at Rechlin Test Establishment, the Soviets discontinued further experimentation.

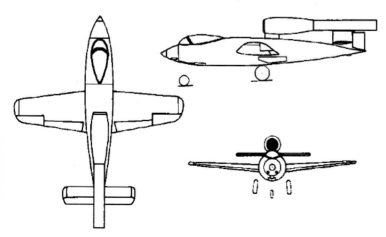

Overall view of the Junkers Ju 287 V1. Scale model by Reinhard Roeser.

Secret Aircraft Designs of the Third Reich

Hans Wocke's Junkers Ju 287 utilized the entire fuselage of a Heinkel He 177.

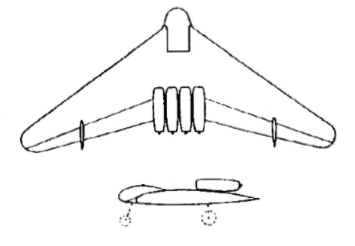

Specifications: Project 126

Engine	1xArgus-Rohr pulsejet of 1,102 lb (500 kg) thrust.
Wingspan	20.7 ft (6.3 m)
Wing Area	96 sq ft (8.9 sq m)
Length	26.3 ft (8 m)
Height	NA
Weight, Empty	NA
Weight, Takeoff	6,172 lb (2,800 kg)
Crew	1
Speed, Cruise	NA
Speed, Landing	NA
Speed, Top	484 mph (779 km/h) at sea level
Rate of Climb	26.3 ft/sec (8 m/sec)
Radius of Operation	217 miles (349 km) at 60% throttle, or 186 miles (299 km) at full throttle
Service Ceiling	NA
Armament	2x20 mm MG 151 canon
Bomb Load	2xanti-personnel bombs, or 12xPanzerbliz rocket missiles
Flight Duration	23 minutes at full throttle, or 45 minutes at 60% throttle

PROJECT 130

On the drawing boards in 1945 was a proposed replacement for the Ju 287 heavy bomber with its radically different swept forward wings. Designated the P.130, it was to be an all-wing, high-speed bomber equipped with four turbojet engines. Junkers' engineers perhaps had more practical experience in designing and constructing large, heavy aircraft than any other manufacturer in Germany. These people had been brought up by Hugo Junkers to believe that the all-wing aircraft would be the ultimate evolution of the airplane. Advanced as the Ju 287 was, Junkers regarded it as only a temporary design that tested certain principles. The next step was the all-wing, or flying wing.

It appeared to Junkers' engineers that the all-wing was the only design that would permit truly fast, high-altitude flight. When the Ju 287 was being developed in 1943, Junkers' engineers said that their knowledge and experience with high angles of back sweep and so on was not sufficient enough just then to proceed with the design and construction of an all-wing heavy bomber. Building the Ju 287 and giving it swept-forward wings was done in the interest of basic research for the all-wing. Junkers subsequently found that the advantage of swept forward wings did not outweigh the problems they created. In addition, tests with the Ju 287 V3 with its cluster of three turbine engines under each wing increased drag. Out of these tests came a design philosophy for the Junkers all-wing aircraft. Junkers' designers believed that they

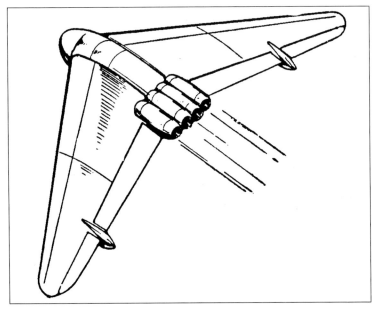

A pen and ink illustration of the proposed Junkers Ju EF130.

Horten Flugzeugbau graphic artist Richard Keller's illustration of their proposed six-turbojet engined Horten Ho 18A, "Amerika Bomber" of early 1945.

The Heinkel He 177.

had found a way to really minimize drag and compressibility problems and this could be achieved by making the turbines disappear into the wing as far as practical. About this time, the Air Ministry, too, was coming to believe that the USAAF was about to convert its heavy bomber fleet to the all-wing planform of Northrop. They instructed Junkers to construct a prototype P.130 all-wing heavy bomber.

Heinrich Hertel, Hans Wocke, and Ernst Zindel believed that they could have the P.130 in the air for a first flight by July 1946. In April 1945, when their Dessau facilities were overrun by US troops, design drawings had been completed and some high-speed wind tunnel model testing had been carried out. The P.130 was to be propelled by four advanced HeS 011 turbine engines of 3,307 lb (1,500 kg) thrust each. Performance of the P.130 was anticipated to 597 mph (960 km/h), with a range of 3,604 miles (5,800 km). Air Ministry officials viewed the P.130 as a long-range bomber to be used against the USSR and its war production facilities behind the Ural mountains as well as against England. Junkers viewed the P.130 as a potential commercial aircraft able to seat up to 80 passengers in relative comfort. Junkers' personnel felt that a turbine-powered commercial or transport airplane would be an instant success after the war and believed they could have the P.130 converted and ready for commercial service by 1948.

To help get the giant all-wing aircraft (wingspan of 79 ft or 24.1 m) off the ground, Junkers' engineers intended to use four JATO (Jet Assistance Take-Off) units, each having 3,300 lb (1,500 kg) of thrust, for approximately 45 seconds. These were to be dropped by parachute at approximately 328 ft (100 m) so that the casing could be reused. Addi-

tional tires also were to be used for takeoff. The two main landing wheels, basically designed to have only a single tire, were to be converted to a dual tire arrangement on the ground. When the aircraft became airborne the additional tires would drop off as the main wheels were retracting into the wings.

The Junkers Ju 287 V1 with its fixed and fully spatted wheel covers shown in April 1944.

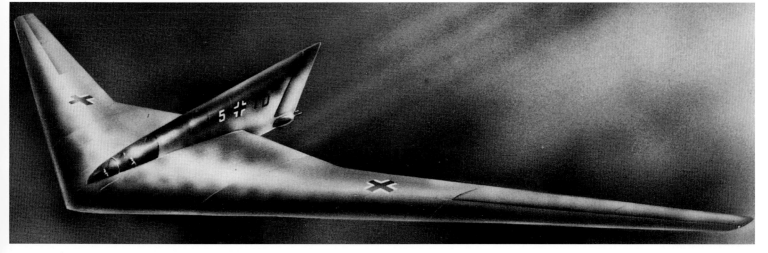

Junkers/Messerschmitt committee's redesign of the Horten Ho 18A, "Amerika Bomber." This committee modified the Horten Ho 18A with the addition of a huge dorsal fin with a hinged rudder attached. Early 1945. Airbrush by Gert W. Heumann.

Specifications: Project 130

Engine	4xHeinkel-Hirth HeS 011 turbojets, each having 3,307 lb (1,500 kg) of thrust, plus 4xJATO solid fuel assist rockets, each having 3,300 lb (1,500 kg) of thrust.
Wingspan	79 ft (24.1 m)
Wing Area	1,292 sq ft (120 sq m)
Length	NA
Height	NA
Weight, Empty	77,000 lb (35,000 kg)
Weight, Takeoff	89,000 lb (38,000 kg)
Crew	NA
Speed, Cruise	559 mph (900 km/h)
Speed, Landing	NA
Speed, Top	620 mph (998 km/h)
Rate of Climb	26.2 ft/sec (8 m/sec)
Radius of Operation	3,728 miles (6,000 km)
Service Ceiling	NA
Armament	NA
Bomb Load	4.5 ton
Flight Duration	NA

PROJECT 140

The final wartime aircraft proposed to the Air Ministry by Junkers was their P.140, a six-turbine, tailless "Amerika Bomber." Conceived in early 1945, this design came through the combined activities of Messerschmitt, Junkers, and the Horten brothers. In late 1944 the Air Ministry showed renewed interest in a long-range bomber capable of attacking targets in the United States of America with a load of conventional bombs, after Hitler's atomic bomb program had been destroyed in February 1944 with the loss of their atomic bomb-making heavy water (deuterium oxide) in Vemonk, Norway. Two design competitions were held but each time no single firm could promise Göring that their bomber design could provide the minimum 6,835 mi (11,000 km) range. After the failure of the second bomber competition Oberst Siegfried Knemeyer sought the help of his Luftwaffe sailplane-builders the Horten brothers in mid December 1944. About Christmas time, Walter Horten telephoned Knemeyer with the news that a Horten all-wing powered with six Jumo 004's could achieve 6,835 mi (11,000 km) non-stop bombing trip to America and back. This Horten "Amerika Bomber" was called the Ho P.18A. Göring gave the Horten's and their Ho P.18A immediate construction approval.

In order to get the Ho P.18A started as quickly as possible, the Air Ministry formed a committee composed of Voigt from Messerschmitt, Zindel, Wocke, and Hertel from Junkers, and Kosin from Arado. Junkers Flugzeug was selected by the Air Ministry to construct the Ho P.18A bomber in some caves in the Herz Mountains in eastern Germany. Reimar Horten had now become a reluctant participant in this highly unusual approach to designing aircraft in wartime Germany. As the "Amerika Bomber" committee met at the Air Ministry's cold window-less offices in Berlin, Reimar Horten was angry and uncooperative. Although the committee had selected the all-wing planform of the Horten brothers both Messerschmitt and Junkers participants wanted a large vertical fin with an attached rudder at the trailing edge for directional stability. Before Reimar Horten walked out, he argued that a vertical stabilizer on their huge all-wing would be a big mistake providing absolutely no benefits but merely adding additional weight and air resistance. However, Messerschmitt and Junkers' team members demanded the vertical tail. The committee also wanted to place the aircraft's six BMW 003's or Jumo 004's under the wing in two clusters of three turbines each. Reimar Horten strongly disagreed with this engine arrangement because of the great amount of air resistance it would create. Seeing that he was getting nowhere and that the modified P.18A, now being referred to as the Junkers P.140, would never make the 6,835 mi (11,000 km) range promised to Göring, Horten gave up fighting and walked out.

With Reimar Horten gone Messerschmitt and Junkers' members of the committee began calling their "Amerika Bomber" the P.140. The former Ho P.18A had really become a tailless bomber with a large, tall vertical fin with an attached hinged rudder. The cockpit, to hold a crew of six, was placed at the base of the proposed aircraft's vertical fin/rudder. The crew were to be seated in individual ejection seats. As a bomber, the P.140 would have carried its offensive load internally. The bomb bay was to be located between the two turbine-engine clusters. Anticipated maximum speed was 559 mph (900 km/h). To achieve this velocity, the committee chose six Jumo 004 or BMW 003 turbines having a combined thrust of 13,228 lb (6,00 kg). They were to be located in two clusters of three turbines each at the rear of the aircraft. Air intakes for the turbines formed the forward portion of each of the two engine

A rear side view of the Junkers Ju EF 140 "Amerika Bomber," 1945. Scale model by Hans-Peter Dabrowski.

As a flying test bed for the unusual swept forward wings, Hans Wocke hoped to prove that his design would demonstrate greater stability and performance than conventional styled aircraft.

Front overall view of the Junkers Ju EF 140 "Amerika Bomber," 1945. Scale model by Hans-Peter Dabrowski.

Junkers Flugzeug und Motoren Werke AG - Dessau, Germany

The Junkers Ju EF 140 "Amerika Bomber" project as seen from behind and below. Scale model by Hans-Peter Dabrowski.

clusters. A tricycle landing gear was to be provided, with the wheels retracting up into the wing's center section. The P.140 was to be armed with four remotely controlled barbettes, two located at the apex of the wing's leading edge and two at the rear of the vertical fin beneath the rudder. The barbettes were to fire 30 mm MK 108 cannon shells.

Junkers' design personnel were particularly pleased with their P.140's design. In addition to being a fast, heavy bomber, they also looked to the day when with a few modifications, it could be converted into an international passenger airplane with room for 80 or more passengers. No favorable action was taken on the P.140. After Reimar Horten walked out of the "Amerika Bomber" committee he and his brother Walter returned to Göttingen and redesigned the Ho18A. Changes included the substitution of the six turbines buried in the wing itself for four HeS 011 turbojets. The Horten's were now calling their "Amerika Bomber" design the Ho 18B. It would have a fixed streamlined landing gear with one HeS 011 attached to each side of the landing gear pylon. Reimar

Dr.Ing. Heinrich Hertel. Hertel was formerly of Heinkel then left and joined Junkers. Photograph taken post-war. At Junkers Hertel was becoming increasingly interested in tail-less and all-wing aircraft planforms.

Ground level front view of the Junkers Ju EF 140 "Amerika Bomber" project. Scale model by Hans-Peter Dabrowski.

Side view of the Junkers Ju EF 140 "Amerika Bomber" project. Scale model by Hans-Peter Dabrowski.

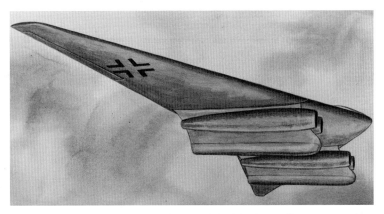

The "Amerika Bomber" which Göring picked for immediate construction, the Horten Ho 18B.

Horten went directly to Göring with his plans for the Ho P.18B. Göring gave the Hortens immediate orders to build it. The construction location was to be in the giant underground hangar near Köeingsbrun in western Germany.

Specifications: Project 140

Engine	6xJunkers Jumo 004 or BMW 003 turbojet engines with a combined thrust of 13,228 lb (6,000 kg).
Wingspan	138 ft (42 m)
Wing Area	NA
Length	62.3 ft (19 m)
Height	19 ft (5.8 m)
Weight, Empty	NA
Weight, Takeoff	NA
Crew	6
Speed, Cruise	528 mph (800 km/h)
Speed, Landing	78 mph (125 km/h)
Speed, Top	559 mph (900 km/h)
Rate of Climb	NA
Radius of Operation	6,835 miles (11,000 km)
Service Ceiling	52,492 ft (16 km)
Armament	4x30 mm MK 108 cannon, two mounted at the apex of the wing's leading edge and two mounted at the rear of the vertical fin/rudder
Bomb Load	4 ton
Flight Duration	15 hours

PROJECT 248

Although the HWK 509A bi-fuel, liquid rocket driven Messerschmitt Me 163 Komet was praised for its excellent flying characteristics, there was widespread agreement that it needed improvement in terms of flight duration and landing systems. In late 1943 the Air Ministry asked Messerschmitt to do some redesign work on the Komet in order to give

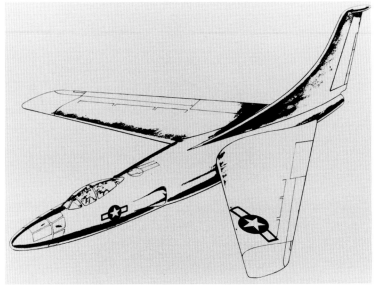

The proposed Convair XB-53 swept forward winged bomber of 1948.

it greater fuel capacity to go along with the new dual-chamber HWK 509C rocket drive. The Air Ministry also wanted Messerschmitt to replace the Me 163's landing skid system with a fully retractable tricycle undercarriage, both to eliminate the cause of a great many accidents and generally to save ground crew the time required to retrieve the aircraft after landing and return it to the hangar. However, not all the Me 163 pilots liked the idea of a tricycle undercarriage. To them, the landing skid meant that if for some reason they glided down beyond their assigned landing site, they could put the Komet down in a grass field or a pasture; they felt it would be a bit more complicated to land the powerless glider (the Me 163 carried only enough fuel for 8 minutes of powered flight beginning with its takeoff run) if it were fitted with wheels instead of the landing skid.

By early 1944 Messerschmitt engineers had not made much progress on redesigning the Me 163B. Believing that Messerschmitt's work was just too slow (at the time most of the design team was occupied with modifications to the Me 262 which the Air Ministry itself kept issuing), the Air Ministry handed over the Me 163B redesign job to Junkers. Under the guidance of Heinrich Hertel, by August 1944, the Junkers team at the Dessau facilities completed the work on the first prototype, which was now called the Junkers P.248.

Dr.Ing. Hertel was no stranger to rocket-driven aircraft. After completing his Ph.D. research on structural dynamics in 1930 and other postdoctoral work, he joined Heinkel AG in 1933. Eventually rising to be-

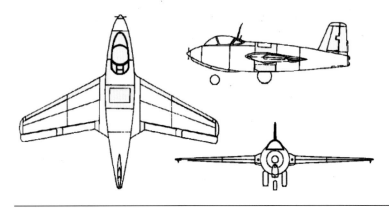

Left to right: Lehmann; Hans Wocke; seated, Professor Brunolf Baade; behind Baade, Professor Günther Bock; Kunzel; photographed in the 1950s at Baade's office in Podberesje, USSR.

Wind tunnel model of the Junkers Ju EF 128.

Head on view of the proposed Junkers Ju EF 128. Scale model by Dan Johnson.

come Heinkel's technical director for research, development, and construction, Hertel worked with Ernst Heinkel and Werhner von Braun on Heinkel's first liquid-rocket propelled aircraft, the He 176. Hertel had also worked on Heinkel's history-making He 178, the first aircraft powered by a turbine, which flew in August 1939.

Hertel left Heinkel AG in 1939, shortly after the flight of the He 178, to become director of aircraft development at Junkers. Although the pace was substantially slower at Junkers, and the work perhaps less stimulating, Hertel had been unhappy with the direction Heinkel was taking; he considered the concept of rocket-driven aircraft ridiculous, and the liquid rocket itself simply unsuitable for the many tasks and functions a fighter or interceptor must perform and the environment in which it must operate. Nothing changed his mind between the time he left Heinkel in 1939 and 1944 when the Air Ministry gave him and his group the job of redesigning the Me 163B. No sooner had they received the assignment than Hertel and his colleagues started a parallel project. In addition to redesigning the Me 163B, they would also design an entirely new tailless aircraft powered by a turbojet to take the place of the Me 163B and its liquid fuel HWK bi-fuel liquid rocket drive. This side project (the Me 163's replacement) was designated the P.128.

The Junkers team began its Me 163 assignment by starting on the fuselage. First they lengthened and reduced its diameter (the fuel tanks were transferred to the newly designed sweptback wings). Redesigned, the fuselage took on the appearance of a projectile. The team added a new bubble-type canopy for the cockpit, which was pressurized (for the first time) up to a flight level of 49,200 ft (15 km).

Hertel's group retained the Lippisch-designed vertical rudder arrangement. Gone, however, was the landing skid; in its place was a retractable tricycle landing gear. Because the main wheels were too narrow (they were the same width as the fuselage), Hertel retained the protective skids on the wing tips. The nose wheel retracted aft into a well beneath the cockpit, and the main wheels retracted vertically up into the fuselage sides. The twin 30 mm MK 108 cannon were retained from the original Me 163, but the number of rounds was increased to 150, and firing speed was increased to 75 rounds per minute. The substitution of the improved twin chamber HWK bi-fuel rocket drive was a major change on the Ju 248. During the mid 1940s, Hellmuth Walter's factory had conducted experiments with a two chamber rocket drive, called the HWK 509C. In this design increased performance was obtained by having a large chamber for takeoff and climb and small chamber for cruising and level flight; the 509C's large chamber would provide a step-controlled thrust up to 4,409 lb (2,000 kg), with the second chamber providing 660 lb (299 kg) of thrust. Flight duration also was improved, from about 8 minutes with the old, single-chamber drive to 15 minutes with the 509C. Part of this increase was the result of improved aerodynamic streamlining; also, the redesigned aircraft carried more fuel.

The Fieseler Fi 103 piloted suicide missile.

American soldiers inspecting a piloted Fieseler Fi 103. April 1945.

Secret Aircraft Designs of the Third Reich

Rear view of the proposed Junkers Ju EF 128. Scale model by Dan Johnson.

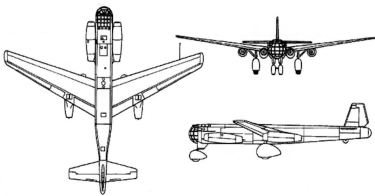

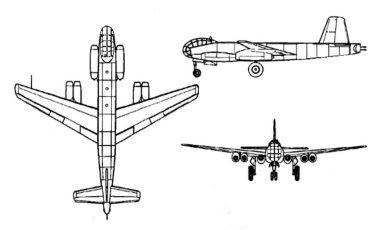

Junkers had completed handling trials (without the HWK 509C drive) by late 1944, and the new engine was ready for production by February 1945. But before powered flight tests could be made, the Red Army overran Junkers' Dessau facilities, and the P.248 prototype was captured virtually intact. Reportedly, the Russian design team of Artem I. Mikoyan and Mikhail I. Gurevich (hence the term MiG) discarded the Ju 248's tailless configuration, adding a horizontal stabilizer in the form of a Multhopp T-tail, redesigned the wing, and made other changes. Ultimately the former Ju 248 emerged in the USSR as the MiG-I-270 (Zh). The Soviets flight-tested their modified German HWK 509C rocket drive-powered interceptor in 1946 and quickly realized that it was unsuitable for their aircraft needs. The project was abandoned.

Specifications: Project 248

Engine	1xWalter HWK 509C bi-fuel, dual-chamber, liquid rocket drive with 4,409 lb (2,00 kg) thrust in the main or takeoff chamber and 660 lb (299 kg) thrust in the second or cruising chamber.
Wingspan	31.2 ft (9.5 m)
Wing Area	211 sq ft (19.6 sq m)
Length	25.1 ft (7.7 m)
Height	NA
Weight, Empty	4,640 lb (2,105 kg)
Weight, Takeoff	11,354 lb (5,150 kg)
Crew	1
Speed, Cruise	435 mph (700 km/h)
Speed, Landing	99 mph (160 km/h)
Speed, Top	590 mph (950 km/h)
Rate of Climb	3 minutes to 49,214 ft (15 km)
Radius of Operation	103 miles (164 km) at 36,091 ft (11 km)
Service Ceiling	49,214 ft (15 km)
Armament	2x30 mm MK 108 cannon with 150 rounds each
Bomb Load	None
Flight Duration	15 minutes (under power)

PROJECT 287

When the Red Army troops were moving through eastern Germany on their way to Berlin in early 1945, they had two main orders: push back the German Wehrmacht, and seize all industrial and scientific equipment that might be of use to the USSR. The latter was justified as "reparations." On 1 July 1945 when the Soviets received from the US and British occupational forces the eastern area of Germany which later came to be called East Germany, the Russians were particularly excited, for this area also included the city of Dessau and that was where Junkers had its major research, development, and manufacturing facilities. One thing the Russians hoped to find intact was the Ju 287 V1, Junkers' large multi-engined bomber prototype and powered by four Jumo 004B2 turbines of 1,980 lb (900 kg) thrust, which had flown at Brandis/Leipzig (also in the eastern zone) as early as 16 August 1944. Soviet military intelligence was fully aware of all the German Air Ministry's existing turbine and rocket-driven aircraft by April 1945. They were also aware of the Air Ministry's secret turbine-powered aircraft projects because when the Red Army began occupying sections of Berlin beginning on 22 April 1945 they finally had access to all the records and documents at the Air Ministry offices. An even better prize, if it could be found intact, would be the Ju 287 V2 prototype. It was rumored in the summer of 1945 to still exist at Brandis. The Ju 287 V2 was an improved version of the V1, with a new fuselage, retracting landing gear, and six BMW 003A1, 1,770 lb (800 kg) thrust turbines. On 1 July 1945, when the Russians received control of the eastern zone, they found the Ju 287 V2 still parked at Brandis; it had suffered superficial damage and although

The Junkers Ju 287 V3 (Ju EF 131). Scale model by Günter Sengfelder.

Side view of the Junkers Ju 287 V3. Scale model by Günter Sengfelder.

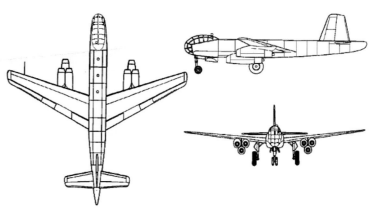

abandoned, Soviet aviation experts thought that it wouldn't take much effort to get it put back into flying condition besides it was such a small price to pay for a state-of-the-art piece of technology which they didn't have at all.

The Ju 287 aircraft had begun as an entirely new project in 1945. It was based on research conducted by the Deutsches Versuchsanstalt für Luftfahrt (DVL), or German Experimental Institute for Aviation, on sweptwing aircraft for high-subsonic flight. As early as 1935, in a paper he had presented at a conference in Rome, German aerodynamicist Adolf Busemann had proposed sweptback wings on high performance aircraft. He and others at DVL felt that the sweptback wing was the only practical way of reaching high Mach numbers.

During early 1940 Junkers aerodynamicists had been working on a bomber project headed by Hans Wocke. The aircraft being considered was to be powered by several turbine engines and to have wings sweptback 25 degrees. As the project proceeded, Wocke came to believe that the new bomber would greatly benefit by having the wings sweep forward instead of backward. Forward sweep would provide stability at low speeds as well as at higher ones. The wing tips, he reasoned, would tend to be the last section of the wing to stall and this would be good. This advantage was later borne out through wind-tunnel tests, and in March 1944 the Air Ministry ordered construction of a prototype of the bomber.

Since the swept forward wing was considered such a radical design, Junkers determined it prudent to test the wing on two hastily built, low-cost test aircraft while the new bomber was being developed. The low-cost test bed, which received the Air Ministry's designation Ju 287

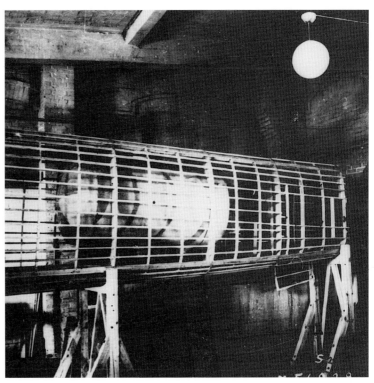
The aft wooden fuselage of the Junkers Ju 287 V3 without its plywood skin. Dessau, 1945.

Wooden mock-up of the Junkers Ju 287 V3's cockpit. Dessau, 1945.

The port wing of the Junkers Ju 287 V3 with three Jumo 004Bs attached in a group. Dessau, 1945.

A detailed mock-up of the Junkers Ju 287 V3 with actual components installed such as the rear gun turret.

The Junkers Ju 287 V3 and its three turbine cluster port wing. Scale model by Günter Sengfelder.

V1, flew at the Luftwaffe Air Base near Brandis/Leipzig on 16 August 1944. Parts of this prototype came from many sources - the fuselage from a Heinkel He 177, the nose wheels from an American B-24 Liberator, the main wheels from a Ju 352, and the tail section from Ju 388 parts.

The Ju 287 V1 was assembled at Junkers' Dessau facility, but was transferred to Brandis/Leipzig for testing because of the longer runways there. (About the same time, the Me 163 also at Brandis/Leipzig, was undergoing its first operational sorties against Allied bombers.) All in all, the Ju 287 V1 made about seventeen test flights. Considered extremely unpowered, it required assistance for takeoff from three HWK 501 solid-fuel booster rockets of 2,205 lb (1,00 kg) thrust each.

In its seventeen test flights this first swept forward-wing bomber, although plagued by repeated failure of its Jumo 004 turbine engines, demonstrated excellent flying characteristics. There were some negative effects associated with the revolutionary wing arrangement, however. When the aircraft was put into a tight turn, the trailing edge tip gained lift, resulting in a rolling effect. Some aero-elastic (wing flexing) tendencies were noted, but engineers thought this could be corrected through further structural stiffening of the wing. Overall, the swept forward concept, as demonstrated in the Ju 287 V1, appeared to be an aerodynamic advancement, and the Air Ministry ordered mass production to begin. It is not known if the Ju 287 V2 flew but, it was clearly intended to join the Ju 287 V1's flight test program at Brandis/Leipzig.

In 1944 the Air Ministry abandoned bombers in favor of all-out attention to interceptor fighter aircraft, and official backing for the Ju 287 was discontinued. However, in March 1945 the Air Ministry ordered the Ju 287 into production again. Junkers had not stopped work on the project, even when the Air Ministry had withdrawn. Work also had started on a third prototype, the Ju 287 V3, which was to carry an 8,820 lb (4,000 kg) bomb. Further development called for a version to be powered by two BMW 018A's advanced turbines of 7,497 lb (3,400 kg) thrust and to have an operational service ceiling of 54,120 ft (16.5 m)

In finding the Ju 287 V2, the Russians fell heir to an unorthodox example of technology. Soon after Soviet occupation of the Dessau facilities, rebuilding of the offices and workshops was initiated, and in a few weeks development of other Ju 287 prototypes was resumed. Manufacture and assembly of two prototypes the V3 and V4 was started. One was to be used for flight testing and the other was intended for static and load tests. Completed in Dessau in 1946, the Ju 287 V3 was engaged in a short test program carried out by Junkers' test pilot Hauptmann Jülge.

Ground level front view of the Junkers Ju 287 V3. Scale model by Günter Sengfelder.

Front side view of the Junkers Ju 287 V3. Scale model by Günter Sengfelder.

The Junkers Ju 287 V3 showing the cockpit. Scale model by Günter Sengfelder.

Soviet authorities then notified the Junkers team of its intention to carry out a completed flight test program in the Soviet Union. Flying the Ju 287 V3 to the USSR was considered but in the end the airframe was disassembled and transported by rail to Ramenskoje, near Moscow in September 1946. The entire Junkers design and engineering team, including chief designer Hans Wocke, chief engineer Brunolf Baade (who replaced Heinrich Hertel after the war), and the Jumo turbine specialist Ferdinand Brandner, followed in October (secrecy was such that the Junkers' personnel were not informed of their destination). In addition, workshops, drafting rooms, machinery, and all support facilities were dismantled and rebuilt as they had stood in Dessau, and all completed engines, spare parts, manufacturing jigs, drawings and documents were shipped to the USSR as well.

During the remainder of 1946 the Ju 287 V3 was flown more than two hours from Ramenskoje, mainly as a test vehicle for airframe and engines. Both the BMW 003 and the Jumo 004 turbine engines were evaluated as complete successes. Active testing of the Ju 287 V4 was used for extensive test under static loading and never flew.

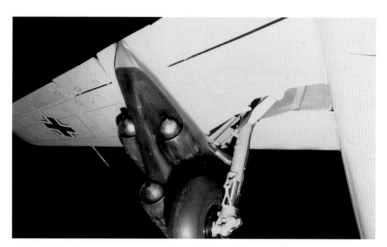

The Junkers Ju 287 V3 showing the exhaust cones of its three Jumo 004B's turbine cluster. Scale model by Günter Sengfelder.

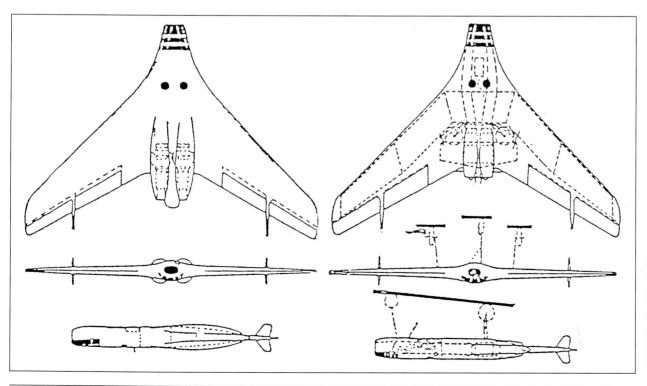

When Professor Hertel left Junkers at war's end, he went to France and worked for the French SNECMA where he produced several tailless designs known as the (left) SE-1801 and (right) SE-1802.

265

Secret Aircraft Designs of the Third Reich

Professor Brunolf Baade's post-war Ju 287 V3 design continuation in the USSR. Shown is the 287 V3 re-engined with two Lulko turbojets of 7,055 lbs (32,00 kg) thrust each. The Lulko turbines were the Soviet version of the British "Nene," and the aircraft was known as the Type 131. This aircraft still had its swept-forward wings and was extensively flight tested out of Ramenskoje, USSR 1946-47.

Professor Baade continued on with the Type 131 and redesigned the wings to make it into a conventional swept-back configuration and now known as the Type 140. Shown is an artist's impression of the Type 140 built by the Baade group in the USSR. 1947-48.

Junkers Flugzeug und Motoren Werke AG - Dessau, Germany

Professor Baade's Type 140 with its swept-back wings evolved into the Type 150. This long range bomber prototype was selected by the Soviets for series production.

The three leading personalities involved in development of the Ju 287 continued their aeronautical research in different ways. Heinrich Hertel, who had overall responsibility for development of the Ju 287 at Junkers, escaped to the West before the establishment of East Germany by the Soviets. He then spent several years in France with the aero-engine company of SNECMA. Later Dr. Hertel returned to Germany to become professor and holder of the chair for Aviation at the Technical University of Berlin.

Upon release by the Soviets in 1954, because they had "milked them dry," Brunolf Baade returned to East Germany, where he designed the Baade Type 152, a four-engined, turbine-powered airliner. The ill-fated aircraft had not progressed beyond the construction of three prototypes before all work was halted in 1958 by the East Germany government due to country-wide economic problems. In addition, the first prototype built crashed on its maiden flight being totally destroyed. Hans Wocke was also released by the Soviets in 1954, and was returned to East Germany, but later escaped to the West joining the Hamburg Flugzeugbau GmbH (HFB) in what was then West Germany. One of his creations was the HFB 320 "Hansa Düse," one of the few aircraft with swept forward wings flying today.

Specifications: Project 287

Engine	4xJunkers Jumo 004 turbines, each having 1,984 lb (900 kg) of thrust, plus 4xHWK 501 solid-fuel takeoff-assist rocket boosters, each having 2,646 lb (1,200 kg) of thrust.
Wingspan	65.11 ft (20.11 m)
Wing Area	628 sq ft (58.3 sq m)
Length	60.1 ft (18.3 m)
Height	17 ft (5.1 m)
Weight, Empty	27,557 lb (12,510 kg)
Weight, Takeoff	47,450 lb (21,520 kg)
Crew	3
Speed, Cruise	318 mph (512 km/h) at 22,695 (6.8 km)
Speed, Landing	118 mph (190 km/h)
Speed, Top	347 mph (559 km/h) at 19,686 ft (6 km)
Rate of Climb	1,500 ft/min (450 m/min) at sea level
Radius of Operation	932 miles (1,500 km)
Service Ceiling	35,433 ft (10.8 km)
Armament	None
Bomb Load	None
Flight Duration	NA

When Baade and several hundred other former ex-Junkers designers returned to the DDR in 1954, Baade was able to convince the Communist Government to design a high-tech airline passenger aircraft and sell it abroad. This was based on Baade's successful Type 150 in the USSR, and Baade's passenger version was now being called the Type 152. Several prototypes were built in the DDR but the entire aircraft program was canceled when the first prototype crashed due to pilot error.

13

Alexander Lippisch - Vienna, Austria

Secret Projects and Intended Purpose:
Project 11 - A single seat aircraft designed to be powered by twin turbojets and expected to achieve speeds of 645 mph (1,038 km/h). This was the fighter Lippisch wanted to build for the Luftwaffe instead of the Me 163 rocket plane.
Project 12 - A supersonic, ramjet-powered delta aircraft which designers believed would reach speeds up to 1,500 mph (2,414 km/h).
Project 13A - A project calling for two Lorin ramjets for propulsion. A mockup was completed but was destroyed in a bombing raid on Vienna, Austria.
Project 13B - An aircraft featuring a delta-style wing planform, a convention cockpit, and twin vertical rudders, to be powered by one coal-burning Kronach Lorin ramjet.
Project 15 - An attempt to turn the Me 163 into a turbojet-powered interceptor. To achieve the change, the Me 163's Walter HWK bi-fuel liquid rocket drive would have been replaced by a single Junkers Jumo 004 turbojet engine.

History:
As early as 1500 Leonardo de Vinci documented men's dream of duplicating the flight of birds, with all their grace and simplicity. In Germany between 1900 and 1920 there was considerable activity directed at perfecting an aircraft consisting almost entirely of a single wing and a tiny fuselage. Hugo Junkers, the father of German aviation, applied his design talents to the all-wing airplane, but his own tailless models, and those of his contemporaries, seldom led to truly successful flights. The problems were myriad, but most were related to the great stress put on the fragile wing during flight, which resulted in a torqueing, or twisting effect. As airflow over the wing increased, the wing began to deform badly, usually with the center section developing a negative lift and the wingtips starting to curve upward. As the twisting increased, the entire wing tended to become unstable, both fore and aft, and began looking like a shallow "U." Generally, the tailless aircraft then lost altitude and, gaining speed as it fell, went into an irreversible and fatal dive. Crashes were frequent, and skeptics argued that the concept of an all-wing aircraft was just naturally unworkable.

Helpers assist Orville Wright move his bi-wing aircraft into launch position at Berlin in 1909.

Claudius Dornier conferring with one of his aides while on a boat ride in 1930.

Alexander Lippisch, left and Gottlob Espenlaub, building their sailplane at the Wasserkuppe, winter 1921-22.

Pictured is von Opel's "Opel Rak.2" race car. Note the addition of what appears to be an air spoiler behind the front wheel, presumably for added control.

Among the few young aircraft designers in the early 1920s who remained attracted to the all-wing concept despite its poor reputation was Alexander Martin Lippisch. Born in München on 2 November 1894, Lippisch spent a good deal of his early childhood in Berlin, and later in Jena, in eastern Germany. He was first attracted to the field of aviation in 1909 when he and his father observed a flying exhibition by Orville Wright in Berlin. Wright, along with other European flying enthusiasts, was barnstorming through Europe that fall to stimulate public support for the fledgling aircraft industry. Lippisch recalled that he and his father witnessed nearly all of Wright's demonstrations, including his flights that broke the duration record (1 hour 45 minutes, at Potsdam) and the altitude record (700 ft or 225 m at Berlin). Lippisch was smitten and soon was constructing flying scale models, beginning with a model of Orville Wright's 1909 record breaker.

After completing his basic schooling in 1915, Lippisch was drafted into the Germany Army and served for a time on the Russian Front. He became ill with pneumonia and after his recovery was transferred to the aerial surveying group with the Army. When the war on the Russian Front ended in 1917 with the Bolshevik revolution, Lippisch transferred to Berlin. His superior officer was an engineer who had been employed

Alexander Lippisch standing near the front of Germany's first tailless powered airplane to fly. At the controls is Günther Groenhoff. This flight took place on 7 November 1929 at Berlin.

The Lippisch Storch-5 with Günther Groenhoff. 1929.

Günther Groenhoff hovering in one of Lippisch's tailless powered designs. 1929.

Wearing his ever-present leather flying helmet, 32-year-old Alexander Lippisch tried unsuccessfully to light a von Opel-sponsored, Lippisch-built rocket scale model airplane in 1928, while Fritz von Opel, in the white trench coat, walks away disappointed.

with Zeppelin Flugzeugwerke before the war. When the officer rejoined Zeppelin in 1918 after the war, Zeppelin's chief manager, Claudius Dornier, urged him to bring along an assistant. Lippisch was picked, not for his aerodynamic studies or experience, but for his strong engineering abilities. He was completely surprised by the offer and felt somewhat guilty about his lack of knowledge of aerodynamics except for that gained from constructing flying models, but he needed work and accepted the job nonetheless. He told the Zeppelin people he would be three weeks late reporting for the job. Since Lippisch was still in the Army, Zeppelin thought the time was needed to process his discharge papers, but Lippisch, knowing his discharge was only a matter of days, had other plans for the time. He spent most of the three weeks reading every textbook he could find on the new science of aerodynamics.

Thus beginning his professional work as an assistant aerodynamicist with the Zeppelin Werke (later Dornier) at the age of 24, Lippisch soon became interested in tailless aircraft. Gaining on-the-job experience, he came to believe that the key to success was abandoning the notion that man could, in effect, build a wing like that of a bird and fly as a bird does without a rudder. He considered the all-wing a valid concept and felt success would come with just a little effort to minimize the aircraft's natural tendency to twist and become increasingly unstable.

The Opel-sponsored Lippisch designed and built rocket scale model leaves its launch ramp. 1928.

A Lippisch-designed rocket powered aircraft appears nearly airborne.

Alexander Lippisch in his favorite headgear, late 1920s.

The Opel-sponsored, Lippisch-built rocket powered plane in flight. Late 1920s.

Lippisch Li Delta 2 - single seat with a 24 hp pusher-type engine.

The idea that an all-wing aircraft could fly without vertical surfaces for stabilization was a notion Lippisch fought throughout his professional career. He argued that flying creatures are successful because they exercise continuous control over their movements, that if a bird in its flight attitude were tested in a fixed position, it would be found that the all-wing configuration has no inherent stability. Any control movement (controls in the wings), he argued created a departure from the flight attitude, and this required another controlling movement (by a tail) to maintain the new position.

A bird can easily do that, Lippisch maintained, because it can react instantaneously by using its brain, and because its extremely flexible control system permits many more alterations in shape than any aircraft control system could. Thus, he concluded, the bird could not be used as a prototype in any way.

A man-made aircraft, Lippisch argued, must have a certain amount of inherent stability, provided by a three-dimensional control system. It must be designed to be stable in pitch, roll, and yaw, so that if it is disturbed from its original position, it tends to come back to that position, if the controls are in normal positions.

Because an all-wing aircraft had to be simple enough to produce and had to have great structural rigidity, the natural shape of the bird's wing had to be simplified to a more geometric design, with the curved lines replaced by straight lines, and the wings becoming tapered and sweptback.

These were pretty radical ideas at the time, and Lippisch remained throughout his life obsessed with resolving the all-wing's problems. Unable to test his ideas while working for Zeppelin, he left there in 1921 and, together with his friend Gottlob Espanlaub, formed his own small aircraft design firm. This allowed him some freedom to construct

This rare photograph shows the Lippisch design team for the Delta 1 at the Wasserkuppe in 1930. Left to right: Beverly Shenstone, British aerodynamicist; Günther Groenhoff, Lippisch test pilot; Fritz Kraemer, aeronautical engineer; Lippisch; Hans Jakobs, engineer; and Wiegmeyer, a Lippisch test pilot.

Lippisch's Project X (DFS 194).

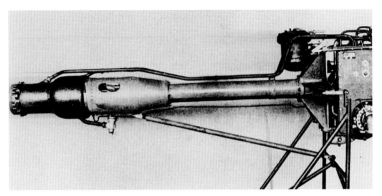

An early version of the HWK Hellmuth Walter bi-fueled rocket engine.

Powered by a Walter HWK 509A-1 bi-fuel liquid rocket motor as shown here, a Messerschmitt Me 163 could reach 12,100 km (39,698 ft) in 3.4 minutes. The location of the Walter rocket motor inside the Me 163 can be seen in the cut-away drawing showing the pilot's position and the fuel tanks containing the T-Stoff and C-Stoff.

test models, but it was not until 1925 that he was able to fly his first manned tailless aircraft, a glider. It was a design success and lived up to his expectations.

To counteract the torsion experienced by previous tailless designs, Lippisch gave the wing's inner section more stiffness as well as a normal cambered effect, and the outer portions a reversely cambered section. To form a firm aerodynamic center, Lippisch placed a single rudder at the tip of the wing. Perhaps his most creative innovation was the backswept of the wing itself. Overall, Lippisch's innovations reduced the wing's twisting and made the all-wing safer. Test pilots familiar with other all-wing aircraft found Lippisch's model to have the best flying characteristics and stability of any flown before. He had achieved a technological breakthrough, yet skeptics continued to scoff at the all-wing and most people remained unconvinced.

In 1925 Lippisch became director of the technical department for aerodynamic research of the Rhön-Rossiten Gesellschaft (RRG), or German Sailplane Institute. Most of RRG's activities involved glider flight, and designers there were having problems with stability for their training gliders. It was not long before Lippisch had turned out a primary glider trainer. Shortly thereafter, he developed the trainer into an advanced, high-performance glider that Günther Groenhoff piloted 375 km. Groenhoff eventually became Lippisch's teat pilot, but died a few years later while flying Lippisch's Delta III all-wing aircraft. Another Lippisch team test pilot for gliders was Heini Dittmar, who late in World War II became one of the first test pilots for Lippisch's rocket-powered Messerschmitt 163 Komet.

As Lippisch began achieving greater stability for RRG's glider fleet, his successes began to receive wide attention. Groenhoff and Dittmar, flying Lippisch's planes, won many national soaring contests. Others built successful gliders, too, but Lippisch was the innovator. About this time two teenage brothers, Walter and Reimar Horten, became involved with sailplanes and from time to time in the early 1930s sought Lippisch's

One of the Messerschmitt Me 163s test flown at Peenemünde in the early 1940s to gather flight test information.

A pilot dressed in a fire-retardant flight suit enters a Messerschmitt Me 163 for an operational flight.

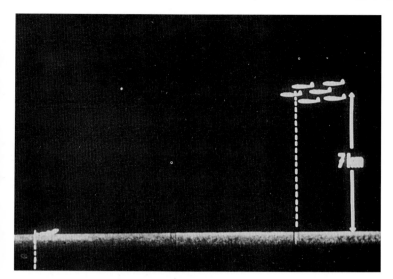

The Messerschmitt Me 163 ready for a bomber force to approach.

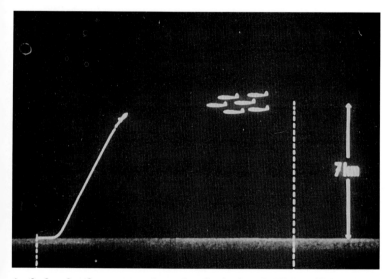

As the bomber force moves closer the pilot ignites the Walter HWK Bi-fuel rocket. In moments the Me 163 is climbing almost vertically to 22,966 ft (7,000 km), the same altitude as the bomber.

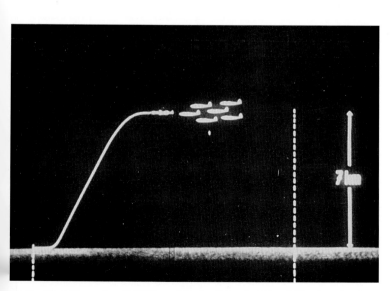

As the Me 163 reaches 7,000 km altitude, it levels off and flies head on into the bomber force.

advice. He could be difficult to work with, especially when his opinions about the use of a rudder on a tailless aircraft were disputed. In the late 1930s, Lippisch and the Horten brothers became philosophically separated as the brothers pursued a rudderless all-wing aircraft.

Having achieved success with gliders, Lippisch turned again to his driving personal ambition. He wanted to prove that an aircraft shaped like a "flying wing" could be practical. Now he had time at RRG to devote to the effort, and his research led, in 1931, to the first delta-wing aircraft produced anywhere in the world.

At the same time Lippisch was achieving his design breakthroughs, there was considerable interest in rocket propulsion. The leader in the field, both financially and in engineering, was the well-known automobile manufacturer Fritz von Opel. Von Opel pursued his interest mostly for fun, with very little practical application in mind, but he did think it would be entertaining to apply rocket power to aircraft and automobiles. His major work was carried out through the Verein für Raumschifahrt (VFR), or Society for Space Navigation, which funded very generously. In fact, rocket propulsion research occurred as it did in the early 1930s only because of financial support from von Opel and others who found the unexplored field a real crowd pleaser. Von Opel especially felt that it might be of some public relations value if rocket power could be applied successfully to a glider or to one of his automobiles.

The rocket research conducted by VFR chiefly involved solid-fuel rockets, and to a lesser extent liquid-fuel rockets. By 1928 VFR had perfected several high-thrust, fast-burning powder rockets for initial thrust, combining them with lower thrust, slower burning rockets to maintain velocity. Once the practical research had been completed, von Opel and some of his cohorts sought out Lippisch at RRG to discuss the possibility of his supplying one of his new, radically designed gliders to "test a radically new engine." The aircraft Lippisch and his colleagues at RRG developed for von Opel's new engine came to be called the Ente (Duck). When it first flew on 11 June 1928, it was Lippisch's first experience with rocket-propelled aircraft, and a seed had been planted for his future high-speed aircraft designs.

Between 1928 and 1930 a variety of rocket-propelled Lippisch gliders were designed, built, and flown. But in early 1930 a number of engine malfunctions and explosions caused the death of several pilots and ground crew, sending the publicity-conscious von Opel retreating as fast as he could from any further rocket research. The immediate result

Alexander Lippisch's secrecy over his "Project X" angered and frustrated Willy Messerschmitt, pictured here, and the two men quickly separated.

Junkers Ju 248. Scale model by Günther Sengfelder.

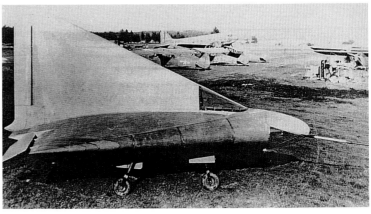
US aeronautical officials believed that Alexander Lippisch's humble DM-1, with its delta wing, was a design breakthrough. The DM-1 is shown here prior to being shipped to the United States after World War II.

was a drying up of virtually all research funds. With von Opel gone, Lippisch, too, abandoned his experimentation with rocket-propelled aircraft, and he returned to his work on aircraft with a delta wing.

Despite the ups and downs of public support and private interest in aircraft research, Lippisch was able to obtain needed funds at the right time. Herman Koehl, the famous trans-Atlantic pilot, had been impressed by Lippisch's early design work and began funding delta-wing activities. The first flyable result was the Delta I, an all-wing design, but a design featuring a wing having pronounced backswept. Powered by a 30-horsepower Bristal Cherub engine, the Delta I flew for the first time in May 1931. Later it was flown from the Rhön-Wasserkupp to Berlin, attaining a speed of 97 miles per hour. Although the Delta I won praise, the German Air Corps was not impressed enough to give Lippisch any financial support so that he could carry on his experimentation. Lippisch was bitter and felt the lack of support stemmed more from his lack of university degree and the fact that he was not part of the professional clique than from any shortcomings in the delta's design.

Although in test flights the Delta I had shown clear superiority over conventional aircraft of the time, opposition to the all-wing could not be overcome, partly because its performance figures could not be replicated in wind tunnel tests. The reasons for this were several: the limited size of existing wind tunnels, the practice of testing models in highly turbulent air streams, and the prejudices of aerodynamicists themselves. Lippisch claimed that when scale models used in wind tunnels became larger, approaching the full-size models, their performance changed considerably. His Delta I tailless aircraft, for instance, had shown unfavorable wind tunnel results because too low correction factors (Reynolds Numbers) were used. (The Horten brothers experienced the same wind tunnel problems when testing their flying wings.) More than 10 years would pass before the scientific world would see, as Lippisch had observed, that measurements obtained in wind tunnel testing as it was being done, in a highly turbulent air stream, were not comparable with actual flight measurements.

As the revolutionary delta wing showed increasing promise, Lippisch kept up a steady program of design improvements. Delta I was followed by Delta II and Delta III. By 1928 Lippisch was receiving orders for his Delta III and had entered into agreements to have Focke-Wulf build the tailless aircraft. This radical aircraft had two engines driving a pusher and tractor airscrew, two seats in tandem, a small foreplane (canard), and downward-sloping rudders.

In the autumn of 1933 RRG merged with the Deutsches Forschungsinstitut für Segelflug (DFS), or German Research Institute for Glider flight, at Darmstadt-Griescheim. With this move, Lippisch was able for the first time to devote all his energies to developing his

Shipped to the United States after World War II for testing, this Messerschmitt Me 163 rocket interceptor was later turned over to the Air Force Exhibit Unit, Wright-Patterson AFB, Ohio, to be put on public display in various cities throughout the United States.

Lippisch DM-1 shown at war's end at Priem am Chiensee.

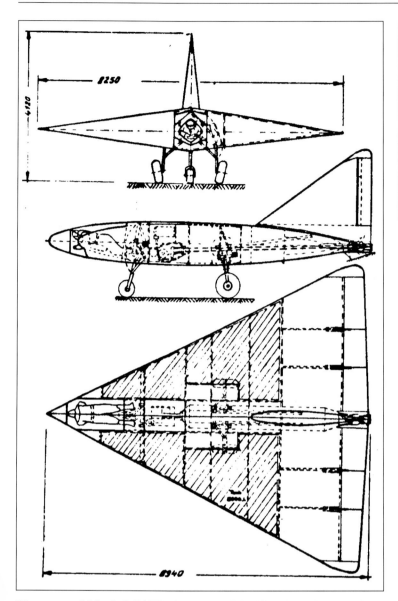

The proposed Lippisch Li DM-2 (never built) differed considerably from the DM-1. This design called for a prone pilot position and elimination of the huge dorsal fin shown on the DM-1. The DM-2 would have had no cockpit canopy for better streamlining. Directional control would have been achieved through a vertical fin with a hinged rudder.

Lippisch Li DM-1. Scale model by Steve Malikoff.

tailless designs. However, shortly after he joined DFS two fatal crashes of the Delta III within 2 weeks led the Air Ministry to ban any further testing of the unorthodox aircraft. This put the development of tailless aircraft of any design in doubt. Had it not been for Dr. Walter Georgii, DFS's director, who had great influence with the Air Ministry, it is possible that Lippisch's research would have been scrapped entirely. Lippisch believed that a prejudice against his unconventional designs, not the dual crashes, was the major reason for the ban.

Alexander Lippisch was viewed throughout the air industry as something of a loner, a crank, and the holder of silly notions, and people steadfastly believed that an aircraft without a tail must, of necessity, be unstable. Undaunted, he continued to perfect the delta concept. He recalled later Mark Twain's quip that "a crank is a man with a new idea - until it succeeds." With the debut of the Delta IVc in 1936, he was able to provide positive proof that the tailless configuration did not possess inherently dangerous handling characteristics.

Lippisch's Delta IVc was truly revolutionary. It featured a wing having a broad chord (width), a sharply tapered leading edge and a straight trailing edge occupied entirely by control flaps and elevons. Small canard (foreplane) had been provided, with wing tip rudders. The craft was propelled by a pair of 75-horsepower, Pobjoy "R," seven-cylinder, air-cooled engines, one mounted in the nose of the short fuselage, and a tractor and the second one in the rear of the fuselage as a pusher. In 1936 Lippisch obtained a full certificate of airworthiness for the Delta IVc. A few years later it became the craft powered by a top-secret Walter HWK bi-fuel rocket engine.

Hellmuth Walter, too, had been conducting his own research, and by 1939 he had perfected an engine having 1,300 lb (2,866 kg) of thrust and weighing only 220 lb (100 kg). With persistent lobbying he had attracted the attention of the Air Ministry's Technisches Amt (technical

Lippisch Li DM-1. Scale model by Steve Malikoff.

Lippisch Li DM-1. Scale model by Steve Malikoff.

The Lippisch Li DM-1 glider minus its huge vertical rudder.

office), which had created a special group to oversee his work on applying rocket engines to aircraft. With this generous support, he was able to establish the Walter-Raketentriebwerken in Kiel, and Walter went on to develop his rocket engines.

The decision to use Lippisch's tailless designs as a frame to house Walter's hydrogen peroxide rocket engines was a blow to Ernst Heinkel. For a long time Heinkel had lobbied the Air Ministry to finance his own rocket plane experiments on the basis of successful flights of the He 176 in September 1939. The decision to pass over Heinkel for Lippisch came about through the recommendations of Dr. E. Lorenz of the Air Ministry's Technisches Amt. Never very interested in linking rocket engines with manned aircraft, the Air Ministry felt that the He 176 was too conventional a configuration to take full advantage of this new unconventional form of aircraft propulsion.

But there was more to the Air Ministry's decision. Air industry experts throughout Germany felt Heinkel had made a fundamental error in calculating the amount of wing area needed on the He 176, that the wings were not large enough to keep the aircraft airborne and maintain satisfactory performance. Lippisch's experience with gliders told him what had gone wrong with the He 176, and he was never guilty of that mistake with his own rocket-propelled airframes. According to Lippisch, he was well aware that a constant thrust power plant such as a rocket would offer much less equivalent power for takeoff than would an orthodox piston engine. He had concluded that the high-speeds portion of the flight envelope demanded a very low lift coefficient in order to avoid early sonic shock. To Lippisch this indicated a pretty low wing loading, a point missed by the Heinkel AG team. After Heinkel's miscalculation became known throughout Germany, the Air Ministry's interest in Heinkel's rocket-propelled aircraft diminished considerably.

With the lift coefficients improving, technicians made a few more modifications to the Lippisch Li DM-1, adding a cockpit and a vertical fin-rudder and installing sharp leading edges.

Side view of the highly modified former Lippisch Li DM-1 glider in its final test form, which achieved a very satisfying lift coefficient of 1.32.

Consolidated Vultee (Convair) XF-92 "Cutlass."

The Air Ministry rightly concluded that an all-wing design might be best suited for rocket power. Alexander Lippisch, who nearly a decade earlier had been responsible for the design of rocket-powered gliders, undoubtedly had more experience with the tailless configuration than any other German designer. Therefore, Lippisch was approached by the Air Ministry with a proposal that an airframe based broadly on the design of the piston-powered Delta IVc should be developed for flight testing of Walter's HWK R1 (for Hellmuth Walter, Kiel, Rocket #1) hydrogen peroxide rocket motor. Speeds up to 280 to 310 mph (450 to 500 km/h) were envisioned.

Lippisch had not designed airframes for rocket engines since he had abandoned rocket-propelled gliders along with von Opel in 1930. He had, however, maintained his contacts with developments in rocketry. The Air Ministry's offer did not come as a great surprise, and after being briefed on the progress of Walter's work, Lippisch accepted the offer. Initially, the aircraft project was referred to simply as "Project X." To conceal the craft's development, the contract was drawn up by the Research Department, and work began in utmost secrecy in a specially built, closely guarded room at the DFS.

In view of the nature of the power plant to be used for Project X, an all-metal fuselage was considered essential. Since suitable facilities for its construction did not exist at DFS in Darmstadt, the work was subcontracted to Heinkel AG at Marienehe. DFS retained responsibility for construction of the wooden wings.

When the Delta IVc underwent high-speed wind tunnel tests, it was found that flight characteristics could be improved if the aircraft's delta wing were replaced with a swept wing with dihedral, and with a centrally mounted vertical stabilizer. In addition, tests showed that the end-plate vertical surfaces used by the Delta IVc tended to cause some flutter at the upper end of the performance scale. The changes suggested by these tests were so radical that Lippisch felt he had to build a full-size, powered model of similar configuration to serve as an aerodynamic test bed. This model, designated DFS 194, was to have a light, air-cooled engine mounted in the rear portion of the fuselage, driving an airscrew aft of the vertical stabilizer by means of a short extension shaft.

Throughout 1938 Lippisch faced constant problems with the management at DFS, and his work was impeded by strict security measures ordered by the Air Ministry. In addition, the subcontract work being

Taking the highly modified Lippisch Li DM-1 glider as their starting point, Convair engineers designed a pure delta tailless fighter, shown here in mockup, to fly at Mach 1.2. This aircraft called the XF-92, it first flew on 9 June 1948.

Consolidated Vultee (Convair) XF-92 "Cutlass."

With the flight testing a success of the XF-92, Convair came out with an operational version of the F-102 "Delta Dagger." It was the most advanced defense aircraft in the world when it first flew on 24 October 1953.

Convair consulted with Alexander Lippisch in the 1950s during the design development of its delta-wing F-102A and its B-58 "Hustler" supersonic delta-wing bomber.

done by Heinkel was proceeding unfavorably. In desperation, Lippisch elected to assume responsibility for building the entire aircraft. But the small DFS factory and workshop proved to be unsuitable in terms of the security measures demanded and the sophisticated shop machinery required to build the aircraft. Lippisch decided he had to leave DFS in order to get the job done, and he began looking for a large manufacturer that could ensure production of the aircraft he was designing within a reasonable period of time.

After some negotiations with Heinkel AG and Messerschmitt AG, Lippisch accepted Messerschmitt's offer to move his entire operation to Augsburg. On 1 January 1939, Lippisch, along with twelve of his coworkers, arrived at Augsburg and immediately formed Department L ("L" indicating Lippisch). Shortly afterward the DFS 194 airframe arrived, and it was decided to discard its piston engine and adapt the aircraft to accommodate the HWK R1 rocket engine. One of Messerschmitt's first acts after Lippisch and his team arrived was to redesignate the aircraft. Messerschmitt chose to assign the RLM "8" number, 163, to the type. (The number 163 had been used previously in the Messerschmitt factory, for the Bf 163, which had lost out in competition with the Fieseler Fi 156 Storch. Messerschmitt decided to reassign the Air Ministry "8" number because he thought the low number would conceal the advanced nature of the secret project. (The low number also had much to do with his ego in his efforts to show up Heinkel).

Shortly after Hitler invaded Poland in September 1939, experimental work slowed as the German war machine gathered momentum. In 1940 Hitler directed that all development work that could not reach the production stage within a year was to be halted; sine the war was to be short, he reasoned, any effort not connected with the immediate war effort was expendable. Thus, the Me 163 project was shelved, and drawings were put in storage.

Lippisch and his co-workers knew that the project would not be resumed if something were not done to keep it going. Moreover, Lippisch knew that many still doubted the validity of tailless aircraft; if the program ended, so, too, might the rapid development of the delta-wing aircraft he was so interested in. His solution was to convert the DFS 194 airframe into a test bed for Walter's 882 lb (400 kg) thrust R1-203 rocket

The British Avro 698 "Vulcan" turbojet bomber incorporated the delta-wing principle long advocated by Alexander Lippisch.

Four delta-winged Convair F-102A supersonic, all-weather interceptors influenced by the hard-work of Alexander Lippisch beginning in the late 1920s.

The wind tunnel test model of the proposed Lippisch Li P.11. The aircraft's twin vertical fin/rudders were to be located outboard of the center section at the point where the two Junkers Jumo 004 turbojets exhausted. Air intakes were to be located in the leading edge of the wing, one on each side of the nose section.

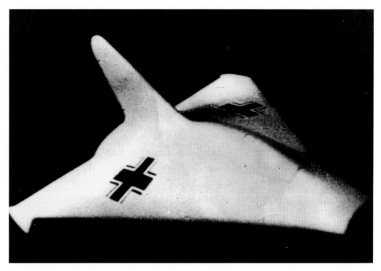

Front side view of the wind tunnel scale model of the proposed Lippisch Li P.12.

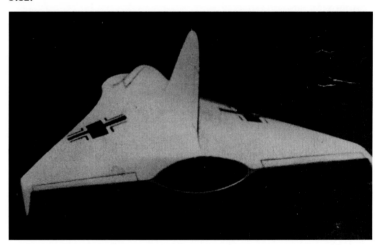

Aft view of the wind tunnel scale model of the proposed Lippisch Li P.12.

engine. Although the DFS 194 had not been designed for a rocket engine, the installation proved relatively simple because the 194's initial piston engine occupied about the same space as Walter's R1-203 rocket engine.

The DFS 194 was taken to Peenemünde-West, and by August 1940 Lippisch's test pilot, Heini Ditmar, had begun a flight test program. These successful tests created new interest among the Air Ministry officials and Lippisch was allowed to continue development. As work and testing resumed, Lippisch felt he stood on the verge of breaking the 621 mph (1,000 km/h) barrier, not to mention setting a new world speed record.

In May 1941 a wooden mockup of the Me 163 was delivered to the Walter-Raketentriebwerken's Kiel factory for trial installation of Walter's new improved rocket engine, the R2 rocket drive. In October of that year Walter's R2 rocket drive had been installed and Lippisch test pilot Dittmar was towed aloft. As he fired the R2 rocket drive, the Me 163's airspeed quickly pitched up, and soon the speed indicator was registering 623.9 mph (1,003 km/h), or Mach 0.84. At that speed the Me 163 began experiencing compressibility problems and the rocket engine stalled due to tremendous negative g's, but Dittmar regained control and brought the tailless aircraft safely back to the ground.

The immediate result of Dittmar's record-breaking flight and several subsequent flights was Ernst Udet's wish to develop a rocket-propelled interceptor. This proposal did not sit well with Lippisch, who claimed he did not favor the proposal because the potential shortcomings of such a warplane were many. Lippisch was interested in highspeed research aircraft; seeing his test bed turned into a gun platform stripped him of any further enthusiasm for the project. His deteriorating relationship with Messerschmitt also contributed to his loss of interest; furthermore he said, "I was more interested in designing another high-speed research aircraft, but this time more along the lines of a pure delta."

From the beginning a strong conflict of personalities was apparent between Lippisch and Messerschmitt. Certainly Lippisch was not a con-

A pen and ink drawing of the proposed Lippisch Li P.11.

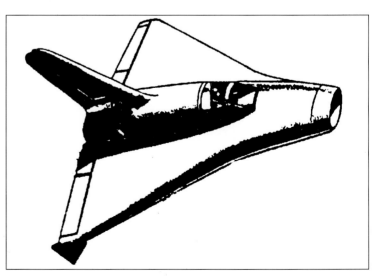

A pen and ink drawing of the proposed Lippisch Li P.12.

The proposed Lippisch Li P.20. Scale model by Steve Malikoff.

The proposed Lippisch Li P.20. Scale model by Steve Malikoff.

ventional person. Unlike Messerschmitt, he did not seek fame or wealth, and he was largely self-educated. He gained his Ph.D. only in 1943 from the University of Heidelberg, by writing, at the urging of the dean of the Department of Natural Science, a thesis on his delta-wing concept.

It did not take long for friction to develop at Augsburg between Lippisch and Messerschmitt partly because of the intense pressure put on Lippisch. The Luftwaffe, which was funding the Me 163 project, expected a great deal from Lippisch. Although it was a radical design, Lippisch never had any doubt about the rocket-propelled aircraft's ultimate success. However, taking the two concepts - the tailless airframe and the rocket-drive and making them compatible was the task at hand. It involved the application of load forces to a plane that lacked a traditional tail and use of newly conceived aerodynamic concepts.

From the beginning Lippisch and Messerschmitt labored under widely differing goals. Though both were interested in research, Lippisch was not concerned with the economics of the project, as was Messerschmitt, the businessman. Then there was the problem of just how much to tell Messerschmitt about the project and of the Me 163's progress. So secret were the communications between the Air Ministry, Hellmuth Walter, and the Lippisch team, that team members joked that any sealed correspondence they received should be labeled "to be burned before reading." Although the Lippisch team was a division within the Messerschmitt organization, members were reluctant to give anyone at Messerschmitt information about the Me 163. In effect, Lippisch and Messerschmitt were in competition, and this frequently led to conflicts between two massive egos. Usually Lippisch would say no to Messerschmitt's request for information, and Messerschmitt would counter by withholding the manpower, materials, and supplies Lippisch sorely needed to complete the project.

The proposed Lippisch Li P.13A. Scale model by Jamie Davies.

The primary source of the problem between the two men was that the Me 163 had not been fathered on Messerschmitt's drawing boards. Messerschmitt originally thought that he would be able to control the entire Me 163 project, but later discovered that just the opposite was true. With the arrival of the Project X team at Augsburg, Messerschmitt quickly found himself becoming more of an unwilling foster parent, a role he seldom, if ever, accepted. After a short time, Lippisch felt Messerschmitt began displaying complete disinterest in the Project X team; relations between the two designers deteriorated steadily until by early 1943, Lippisch considered the situation insupportable. On May 1 he left Messerschmitt to become director of the German government-controlled aviation research institute (the Luftfahrtforschungsanstalt) Wien in Vienna, Austria. Before he left, however, Lippisch sold all his patents to Messerschmitt, for 300,000 marks (at that time about $100,000). He invested part of the money in forest land and a glass works near Famlitz, Germany, a small community where his parents had retired, with the intention of producing and constructing pre-fabricated houses after the war. He had no intention of running the glassworks, but looked upon it as an investment. However, his entire investment was threatened for a time while the SS investigated the propriety of the Messerschmitt-Lippisch transaction. Nevertheless, Lippisch lost the investment after the war when the Russians seized the property and converted it into a prisoner-of-war camp.

While in Vienna, Lippisch continued his research on all-wing aircraft, attempting to improve controllability and simplify the structural design. His work led him into even more sophisticated high-speed concepts such as a triangular wing with a sweptback leading edge and a straight trailing edge. Such a wing today is called simply the "delta wing," a name Lippisch used on the first model of this type in 1930. The successful flight of the British-French Concorde prompted Lippisch to write Dr. Adolf Busemann, an old friend and colleague who developed theories and did calculations on supersonic flight while Lippisch worked on designs, that the supersonic commercial passenger plane was proof that their ideas in the 1930s and 1940s had been correct. Certainly the shape of the Concorde's wing though modified, is an offspring of the Lippisch delta.

Lippisch believed that the delta wing was the best design to improve the low-speed performance of a slender-bodied aircraft having a highly sweptback wing. Flight tests of several aircraft whose wings had greater degrees of sweep back had revealed a marked tendency for wing tip stalling at low speeds. The sweptback delta wing, Lippisch felt, could be flown without such trouble, even beyond its stalling speed. It was not difficult, Lippisch noted, to derive the configuration of a supersonic aircraft - a long slender body with a delta wing.

The proposed Lippisch Li P.13A. Scale model by Steve Malikoff.

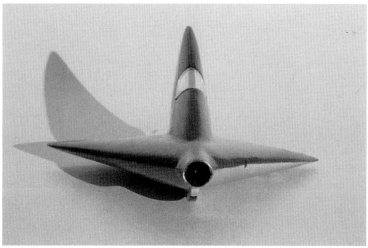

A head-on view of the proposed Lippisch Li P.13A. Scale model by Steve Malikoff.

Eventually the war began closing in on Lippisch, affecting his work in a major way. Although he sustained a minor eye injury in an air raid during a visit to Berlin, he was never seriously hurt in the bombing of Vienna during the last few weeks of the war. His greatest blow came when forty-five members of his staff were killed in an air attack on his Vienna design/work shops in April 1945. The raid also destroyed the partially completed twin-turbojet, delta-wing P.11 and the coal-fueled ramjet P.13.

With the Russian Army advancing rapidly into Austria during March 1945, Lippisch gave considerable thought to the question of whom he and his team should work for after the war. Knowing the Nazi dictatorship, they did not wish to live under another totalitarian government. On Easter Sunday in April 1945 at 2 a.m., Lippisch left Vienna, returning to Prien, Germany, with only a handful of belongings and research papers. Papers, Lippisch could not carry, he destroyed.

After his escape from Vienna and the Russians, Lippisch gave US Army Technical Intelligence (USATI) officers information on his P.11, P.13, P.20 and the rest of his Project series of high-speed turbojet-propelled designs he had been working on. In January 1946 Lippisch was invited to the United States under the government's "Operation Paperclip" and was assigned, along with fifty other German scientists, to Wright Field, Dayton, Ohio. Tons of captured German war documents and a variety of airplanes and related equipment accompanied the scientists.

About a year later Lippisch received news that pleased him immensely in the form of a National Advisory Committee for Aeronautics (NACA) report on its investigation of his DM-1 glider, which had been captured and returned to the United States. NACA tests of the glider, in its full-scale wind tunnel at Langley Field, Virginia, had proved that the delta wing was everything Lippisch had claimed it to be. Initial tests of the DM-1 had been disappointing, for they showed that its maximum lift coefficient was considerably lower than Lippisch had indicted. However, with some modifications to the aircraft, NACA was able to obtain a maximum lift coefficient of 1.32 and concluded that the modified configuration was highly desirable for supersonic flight. Subsequently, on request by NACA and with the assistance of Lippisch, Convair produced a series of delta-wing designs, including the F-102 "Delta Dagger" and the B-58 Hustler bomber.

In 1947 Lippisch transferred to the Naval Air Material Center in Philadelphia. He considered his 3 years with the Navy the least productive and generally the unhappiest of his career, and he regretted he had not chosen the Air Force's research offer instead. He had little to do, and spent his time on problems involving the landing of jets on aircraft carriers.

In 1950 Arthur Collins, president of Collins Radio, Cedar Rapids, Iowa, was developing integrated flight systems and autopilots for jet aircraft. He needed someone experienced in response controls for aircraft, and offered Lippisch a job. Later Lippisch became chief of Collins

Rear side view of the proposed Lippisch Li P.13A. Scale model by Steve Malikoff.

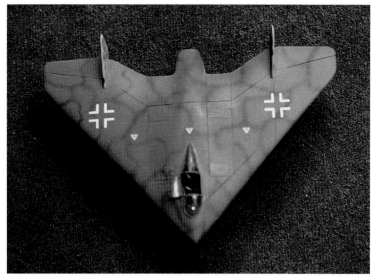

The proposed Arado Ar E.581.4. Scale model by Dan Johnson.

The proposed Horten Ho P.10B.

The proposed Horten Ho P.10C.

Aeronautical Research Laboratory where he spent most of his time at the Collins hangar at the Cedar Rapids airport. Here he conducted experiments on a sophisticated smoke tunnel for the study of airflow over various wing shapes. Although he had worked throughout his professional career to prove the aerodynamic validity of the all-wing aircraft, Lippisch now concentrated on a "no-wing" concept which he named the "Aerodyne." This design, he felt, was the ultimate form all aircraft would one day evolve. The aerodyne, as he envisioned it, would be a sleek, missile-like carrier with no wings, a vertical takeoff-and-landing machine able to achieve speeds exceeding 2,000 mph (3,219 km/h).

In 1964 Lippisch retired from Collins, but, having formed the Lippisch Research Corporation, he continued working on his aerodyne under the sponsorship to the Dornier Company. He believed his aerodyne would surpass the delta wing in aviation significance. The father of the delta wing, he recalled the skepticism and opposition that had greeted his delta-wing concept thirty years earlier. Personally, he gave more weight to persistence than to raw knowledge in problem-solving. He believed, like other inventor-pioneers such as Thomas Edison, that 95% of genius was persistence.

While working for Dornier, where his professional career had begun in 1918, Lippisch and his wife were able to return to Germany, near Lake Constance for two years. He also consulted for the Rhein-Flug, a West German firm that had contracted to continue development of another of his projects, the "Aerofoil." Lippisch believed water could be used like air to provide very fast, efficient transportation. He rejected the hovercraft, which creates its own lift by means of huge fans blowing in a downward direction, and preferred the "Aerofoil," which gets its lift through the aerodynamic shape of its hull. A test model called the X-112 "Flying Flounder" was tested in Germany in the mid 1970s. When he died in Cedar Rapids, Iowa, on 11 February 1976, at the age of 82, Lippisch was working on a technical book about aviation history and the delta wing. He had already completed his memoir and was particularly proud of the 13 part series, "The Secret of Flight," which he wrote and presented in 1965 for National Educational Television.

Though by no means a recluse, Lippisch preferred to work alone with a small staff, to perform key experiments himself, and to build small or full-scale models to test his concepts. However, Lippisch had his dark side in that he was extremely jealous of individuals/companies

The proposed Focke-Wulf Fw P.3x1000C.

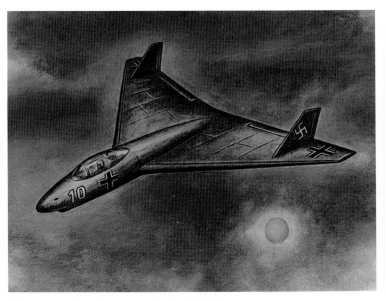

The proposed Arado Ar P.NJ-1 (night fighter).

The proposed Henschel Hs P.135.

The proposed Messerschmitt Me P.1108.

who were interested in tailless and all-wing designs too. He could tolerate no competition and used his influence frequently to keep would-be competitors to a minimum. Overly protective of his research and data, Lippisch tended not to publish his research, nor would he readily offer helpful advice to other all-wing enthusiasts. He remained a closed, uncommunicative man throughout most of his life; however, in his later years he became more of an educator. Although many of today's advanced aircraft such as the Concorde and the US Space Shuttle Columbia embody ideas which originated in Lippisch's mind and labs, his method of working alone and his reluctance to publish papers (until later in life) have largely obscured the extent of his contributions.

PROJECT 163

Only a few aircraft developed during World War II can genuinely be described as "sensational," exhibiting the combined qualities of unusual design, outstanding performance, noteworthy in their intended roles. Lippisch's liquid fuel rocket-drive interceptor Me 163 Komet undoubtedly was one of the sensational.

The Me 163 was a single-seat, tailless interceptor, driven by a Walter HWK 509A-1 bi-fuel rocket drive which could, at least in theory, provide 3,748 lb (1,700 kg) of thrust. Its wings, which were swept back 23.3 degrees on the quarter chord line, were made of wood (8 mm plywood skinning covered by fabric) and featured a fixed leading-edge slot some 7 ft (2.13 mm) in length which terminated about 1 ft. (30 cm) from the wing tips. Lateral and longitudinal control was provided by individually operated, fabric-covered elevons. Large longitudinal trimmer flaps were fitted inboard of the elevons and behind the landing flaps. The fuselage was an oval-section, stressed skin, light-alloy structure. A two-wheel takeoff trolley was attached to the rear portion of the landing-skid housing by two lugs which engaged mechanical catches.

The Me 163 Komet had a dismal accident record, largely because of its motor. The rocket drive ran on two types of fuel. One type used immense quantities of hydrogen peroxide, or T-Stoff, which was stabilized with 20% phosphate, or oxyguinoline; the second type used a solution of hydrazine hydrate in methanol, or C-Stoff, which acted as a catalyst. The slightest irregularity in the set ratio of C-Stoff to T-Stoff being fed to the motor, such as might result from a momentary interruption in the flow of one or the other, could produce an explosion that left little of the aircraft or its pilot. A bad bump on landing could produce equally disastrous results if fuel were remaining in the tanks. But there were hazards other than those stemming from the fuels. Trimmed tail-heavy, the Me 163 took off with the twin-wheel trolley attached to the housing of the skid, which was extended. The lugs connecting the trolley were automatically disengaged when the pilot retracted the skid. It was standard practice to jettison the trolley at an altitude of 20 to 30 feet (7 to 9 m); if the pilot miscalculated, the trolley could bounce up off the ground and hit the fuselage, rupturing the fuel tanks or even hooking onto the skid again. If the jettisoning gear malfunctioned, the pilot had to abandon the aircraft, as the chance of making a successful landing with the trolley in position was negligible.

Landing speed for the Me 163 was high, around 160 mph (257 km/h). Since touchdown was effected with power off, considerable pilot judgement was needed. If the pilot failed to touch down within a reasonable distance of the correct spot on the runway and skidded onto rough ground, chances were great that the aircraft would flip over and explode. Even if the remaining fuel did not explode, the pilot was still in considerable danger of having the "left-over" fuel seep into the cockpit ad eat through his protective asbestos overalls.

Alexander Lippisch's "Flying Flounder" undergoing flight tests at the Dornier Flugzeugbau-Germany in the mid-1970s.

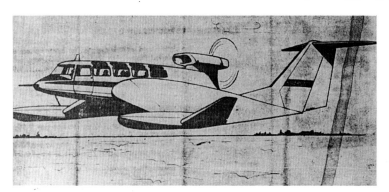

A pen and ink sketch demonstrating Alexander Lippisch's idea for a mass transit over water, an adaptation of his "Flying Flounder" idea.

Alexander Lippisch's no-wing aircraft design called the "Aerodyne." He believed that all aircraft ultimately would evolve into the shape of the "Aerodyne." This would have been of considerable interest to Professor August Parseval, one of Kurt Tank's professors of aeronautics. In this photo Lippisch (left) stands beside a model of his "Aerodyne" at Cedar Rapids, Iowa, in 1958.

According to German sources, 80% of all Komet losses resulted from accidents during takeoff or landing, 15% from either fire in the air or loss of control during a dive, and 5% from combat. Bailing out frequently was not an option, because at speeds exceeding 250 mph (400 km/h) the pilot's canopy could not be jettisoned safely.

What was it like to fly the Me 163? Once in free flight the Komet handled beautifully. Pilots cited its ability to climb, almost "rocket-like" at a speed of 435 mph (700 km/h) to 39,698 ft (12,100 m) in something short of 4 minutes. At this speed the Me 163 could get well above enemy bombers with no trouble. Because the craft did not have a pressurized cabin, 12.1 km was just about its operational ceiling. But even 30,000 ft (9.1 km) was enough altitude to set up a fast dive without power, straight through the intruding bomber stream, blasting away with its pair of 30 mm, 60-round MK 108 cannon. With a tactically usable speed of Mach 0.82, the Komet could not be touched by Allied escort fighters as it passed through B-17 bomber formations, banging away with its cannon. But while the speed at which the Komet closed in on its prey was very high, the maximum distance at which it stood a chance of hitting a target the size of the B-17 Flying Fortress was 1,952 ft (595 m). Since the Komet pilot had to break off the attack some 600 ft (183 m) from the bomber in order to avoid collision, he was left with something less than three seconds in which to operate his slow-firing MK 108 cannon and bring down the B-17.

After this pass came the critical moment for the Komet pilot: He had to relight the Walter rocket drive and zoom up to altitude for his next pass, climbing through the bomber formation as he went. The total time available at full throttle was about 4 minutes; since 2 minutes had to elapse between an engine cutoff and relight, the Komet pilot's cockpit stopwatch was vital for survival. Furthermore, the pilot had to take care that he had enough fuel in his tanks to elude pursuit by escort fighters after making his final firing pass.

The Me 163 probably was more lethal to its pilots than to its enemies, and, on balance, the tremendous research effort that carried it to service status may not have been justifiable. Such observations angered Lippisch, the craft's designer. In 1972, after a long silence, he stated that what had been written about the Me 163, the soundness of its basic design and the advisability of its development, was, in effect, garbage. The design that came to be called the Me 163, he noted, was never intended as an interceptor.

The original purpose of the Me 163's design, according to Lippisch, was to test high-performance aircraft in the transonic speed range of 621 mph (1,000 km/h). Lippisch and those who worked for him at DFS, and later at Messerschmitt, had no intention of developing a combat aircraft. "Personally, I was not much in favor of the rocket-propelled interceptor," he said, "because it was not too difficult to see the potential shortcomings of such a warplane." According to Lippisch, the idea of using the Me 163 as an interceptor was conceived when Ernst Udet witnessed the flight of the original Me 163 prototype at 624 mph (1,003 km/h) on 2 October 1941. Until that time the purpose of the Me 163 had been threefold: to gain data on the performance of aircraft at what were then extreme speeds (above 1,000 km/h) to prove the validity of the tailless concept and to keep the research program alive in view of the Air Ministry's ban on new aircraft development after Hitler's move into Poland.

Although Lippisch was just as interested in achieving forward motion as were his colleagues (and frequent competitors), Messerschmitt and Heinkel, he did not proceed with his work in a rash manner. That the Me 163, a rocket-powered research vehicle was converted into a gun platform angered Lippisch considerably. For a time he lobbied the Air Ministry to have a new delta-wing aircraft, the P.11, built especially for Walter's rocket drive. But his request was ignored, for the imagination of the Air Ministry had been fired by reports from Messerschmitt and others of the early flight performance of the Me 163. So buoyant was Udet after the October 1941 flight test, that he immediately ordered the Me 163 mass produced. Two weeks later Messerschmitt received a contract to construct seventy Me 163 prototype interceptors for an operational group by the spring of 1943. However, with Udet's suicide in November 1941, development of the Me 163 and a number of other aircraft he favored, was stopped on order of the man who succeeded him as chief of the Air Ministry, Erhard Milch.

When Me 163 production was halted, Lippisch and his design team used the time to redesign the prototype. Completed by April 1942, the new prototype was identified as the Me 163B V1. Lippisch left Messerschmitt in April 1943, about the time production of the Me 163B was beginning. A total of 237 Me 163Bs were produced by the end of 1944; another forty-two were built in January and February 1945. Then production came to a halt due to disruption of communications and inability to obtain components from subcontractors, which were dispersed throughout Germany. Of the 279 Me 163B interceptors built, an estimated 25% actually flew combat missions.

Between thirty and thirty-five Komets were brought out of Germany after the war. England took the bulk of them while the United States obtained five. Only ten are known to survive. Five remain in England at the Rocket Propulsion Establishment, Westcott, the Imperial

War Museum, London; the Science Museum; the Cranfield Institute of Technology; and RAF St. Athan, Wales. Two are in Canada, one at the Canadian War Museum, Ottawa, and one each remains in Germany at the Deutsches Museum, München, and the War Museum, Canberra, Australia, and the United States at the Smithsonian Institution National Air and Space Museum's warehouse in Silver Hill, Maryland, where although technically unrestored, it is on limited display and is in relatively good condition.

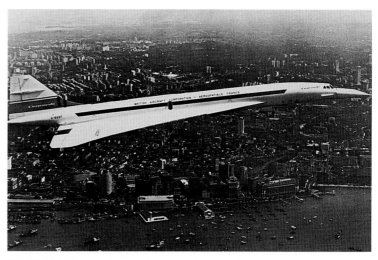

The British-French Concorde, Alexander Lippisch liked to point out, borrowed his concepts on delta-wing styling. In this photograph the Concorde is shown flying over Singapore.

Specifications: Project 163

Engine	1xWalter HWK 509A bi-fuel liquid-rocket engine of 3,748 lb (1,700 kg) thrust.
Wingspan	30.5 ft (9.3 m)
Wing Area	211 sq ft (19.6 sq m)
Length	18.7 ft (5.7 m)
Height	9.1 ft (2.8 m)
Weight, Empty	4,200 lb (1,905 kg)
Weight, Takeoff	9,061 lb (4,110 kg)
Crew	1
Speed, Cruise	435 mph (700 km/h)
Speed, Landing	160 mph (257 km/h)
Speed, Top	516 mph (830 kmh) at sea level and 596 (960 km/h) at 9,842 ft (3,000 km)
Radius of Operation	50 mi (80 km)
Service Ceiling	39,698 ft (12,100 m)
Armament	2x30 mm MK 108 cannon, each with 60 rounds of ammunition
Rate of Climb	3.4 min to 39,698 ft (12,100 m)
Bomb Load	None
Flight Duration	8 min

PROJECT DM-1

In wartime Germany, as in other countries, there were some scientists who believed sonic speed presented an unsurmountable barrier to human flight. Although it was thought not too difficult to develop an aircraft suitable for supersonic speeds only, it seemed difficult to design one that would perform as well in the subsonic range as in the supersonic range. Standard German aircraft during the 1940s were quite adequate at subsonic speeds, but their general configuration was considered impractical at supersonic speeds due to the formation of shock waves, with resultant adverse effects on stability and control. The challenge was to find a configuration that would work well in both speed ranges.

As for an engine, a propeller engine was ruled out immediately. Because of compression shocks, some aerodynamicists were even uncertain that the desired high-speed performance could be obtained with a turbojet power plant. Some believed that a form of ramjet propulsion system would be necessary to provide the acceleration required. Others thought the liquid-rocket Walter motor such as that used in the Me 163B would be best.

All the academic discussion stopped in late 1944 when Alexander Lippisch's call for a ramjet-powered interceptor was approved by the Air Ministry. What Lippisch proposed was an aircraft having a pure delta wing completely enclosing a centrally mounted ramjet duct and a triangular dorsal fin whose glazed leading edge section formed the cockpit. The proposed aircraft's propulsion system was to be complicated. A liquid-fuel rocket drive was to be used for takeoff and to boost the aircraft up to its ramjet operating velocity, about 149 mph (240 km/h). Once it reached 149 mph the ramjet's fuel of powdered coal mixed with oil would be ignited by a gas flame. Maximum estimated speed at high altitude was 1,025 mph (1,650 km/h). Flight duration would be about 45 minutes, during which time 1,760 lb (800 kg) of solid fuel, coal, would be consumed.

Lippisch's interest in ramjet power had remained active ever since his work at the Deutsches Forschungsinstitut für Segelflug (DFS) in the mid and late 1930s. When he left Messerschmitt AG and the Me 163 development program in 1942, he had joined the Luftfahrt Forschungs Anstalt Wien (LFA) in Vienna, where he led a new research program on high-speed aerodynamic problems and delta-wing aircraft. At the same time, DFS had continued to work on ramjet propulsion, and Lippisch continued to receive its data in exchange for his recent aerodynamic research findings.

It was Lippisch's delta-wing research at LFA and the ramjet-powered research at DFS which led to the possibility of a new, fast, fighter-interceptor aircraft. In November 1944, with DFS facilities at München and Darmstadt destroyed by bombing raids, the student group moved to Prien and began constructing an experimental test glider of form and dimensions similar to the projected fighter-interceptor. This test glider was designated DM-1 because it was developed by a small group of aeronautical engineering students from the universities in Darmstadt and München. These students worked without pay under the direction of Lippisch, who would commute between Vienna and the two universities supplementing their studies by building and flying aircraft. Together with Lippisch they planned to develop the aircraft in three phases.

Initially the DM-1 was to have no power plant and was intended only to indicate whether the chosen configuration would perform adequately during slow speed. Plans were for a bomber or a cargo aircraft to carry the DM-1 glider piggyback style to approximately 34,450 ft (10.5 km). It would then be released for free flight and go into a dive to

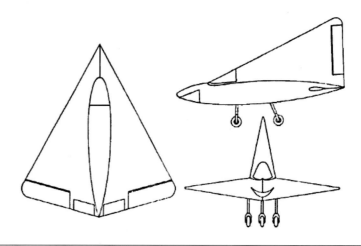

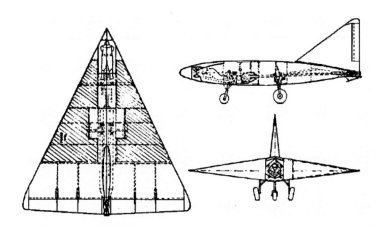

test high speed flight characteristics. After it pulled out of the dive, its performance at slow speeds could be investigated and its reactions in simulated stalls and sharp turns could be checked to determine what to expect during landing. A second model was to have been fitted with a Jumo 004B turbojet having 1,984 lb (900 kg) of thrust and would be used to ascertain flight characteristics at speeds of 497 to 746 mph (800 to 1,200 km/h).

It was the third model that would be fitted with the DFS ramjet propulsion system, with which speeds of 1,243 mph (2,000 km/h) were considered possible. The newly developed configuration could also be used for testing the potential of other power plants.

The most notable feature of the DM-1 was its large delta fin which extended almost to the apex of the wing. The fin had a wooden shell with conventional internal ribs and several thin stringers, and a rudder was hinged to it. The unusually large fin was necessary to accommodate the pilot. There was no fuselage inasmuch as all necessary equipment was incorporated in the wing and the fin.

The DM-1 had a tricycle landing gear which could be lowered manually during flight. Release of a knee-operated lever on the gear lowered it, and it locked into place by means of its weight. The two main wheels had conventional brakes, and either wheel could be braked separately for steering by tipping the rudder pedals.

The cockpit of the DM-1's had an adjustable seat whose framework, and the landing gear as well, was bolted to the main wing tips. A transparent (glazed) panel n the leading-edge of the fin and the bottom of the nose assured good visibility. The top hood opened readily for easy access and could be thrown off in an emergency. The steering mechanism consisted of a control stick and pedals which were connected to the surfaces with control rods.

At war's end in April 1945, the DM-1 was almost complete. When captured by the US Army, the students requested and were granted permission to complete the aircraft. Once completed it was towed behind a Douglas C-47 transport also found at Prien. The DM-1 made only one flight; then it was taken to the United States for evaluation in the National Advisory Committee for Aeronautic's (NACA) full-scale wind tunnel at Langley Field, Virginia.

Initial tests of the DM-1 were disappointing for they disclosed that its maximum lift coefficient was considerably lower than Lippisch had indicated. Believing that the unusually large fin/cockpit was adversely affecting lift, NACA staff removed the fin and lift improved dramatically. Then they sealed up the gaps between the wing and the elevator and the rudder. Lift improved even more, up to Lippisch's calculations. NACA then rebuilt the DM-1, adding a thinner, shorter vertical rudder and a small cockpit canopy, and making the leading edges sharper and more pointed. Altogether, NACA was able to obtain a maximum lift coefficient of 1.32. On the basis of all this research, NACA concluded that the DM-1, so modified, was a highly desirable form for supersonic flight, and also acceptable, perhaps even superior, at low speeds, and recommended that a turbojet-powered aircraft be built along the lines of the modified DM-1 glider. Convair was selected to build the powered version, which came to be known as the XF-92A. Later Convair came out with a series of delta-winged aircraft, all based on the DM 1 glider, including the F-102 Delta Dagger, the F-106 Delta Dart, and the B-58 Hustler bomber.

The original DM-1, the first delta planform, which ultimately found its design features incorporated into the British-French Concorde and the US space shuttle Columbia, now sits in storage at the Smithsonian Institution's National Air and Space Museum warehouse facilities in Silver Hill, Maryland, awaiting restoration and then display.

Specifications: Project DM-1

Engine	None. Towed to 26,246 ft (8 km) and released.
Wingspan	19.4 ft (5.9 m)
Wing Area	215 sq ft (20 sq m)
Length	21.7 ft (6.6 m)
Height	10.5 ft (3.2 m)
Weight, Empty	655 lb (297 kg)
Weight, Takeoff	1,014 lb (460 kg)
Crew	1
Speed, Cruise	NA
Speed, Landing	NA
Speed, Top	328mph (550 km/h)
Radius of Operation	NA
Service Ceiling	NA
Armament	None
Rate of Climb	NA
Bomb Load	None
Flight Duration	NA

PROJECT 11

Lippisch called the P.11, a craft designed to be powered by twin turbojets, his "tailless fighter-bomber." It was the first of the delta-wing designs he produced after leaving Messerschmitt in 1943, and it was the plane he wanted to build for the Luftwaffe instead of the Me 163 rocket plane. He believed his P.11 would be superior all-around to his Me 163 because of its lower drag coefficient, lighter weight, and better safety record. Lippisch also believed the P.11 would be less inclined to stalls and spins, thereby affording a far greater safety margin for inexperi-

US space shuttle Columbia as seen from above prior to its first space flight, landing at Edwards Air Force Base, California. Alexander Lippisch claimed that the wing planform of the US space shuttle Columbia, was based on his delta-wing research in the 1930s and 1940s.

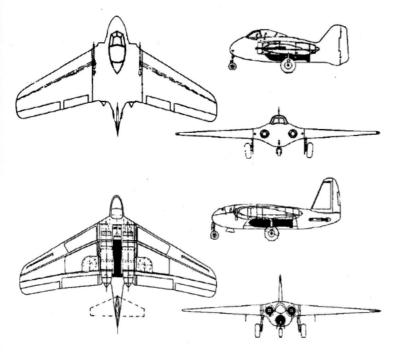

enced pilots. In addition, the P.11 was expected to be more economical to construct because the delta wing planform would require less manufacturing time. Finally, unlike the temperamental HWK 509A-1 bi-fuel liquid-rocket drive, the Jumo 004 turbojets powering the P.11 could be operated on relatively safe petroleum products. Indeed, many Me 262s at war's end were being fueled by unrefined petroleum straight out of the well, the only preparation being straining it through a filter to remove the dirt; if it could flow through the fuel lines, 004B turbojet could burn it.

Design drawings for the P.11 were started in November 1942. By late 1944 a full-scale glider had been constructed at workshops in Strobol, Austria and had been flown to test its aerodynamic qualities. Wind tunnel testing had to be bypassed because the facilities in nearby Vienna had been destroyed by Allied bombing raids. Despite the lack of wind tunnel data, Lippisch proceeded with construction of a flying prototype, and it was complete except for the wing tips when the Allies, along with the Red Army, captured the Lippisch-Strobol workshops in early April 1945. What happened to the P.11 after that remains a mystery. Shortly after the Allies moved into the Vienna area and Lippisch's abandoned workshops on Easter Sunday and only several hours prior to Allied takeover, the P.11 prototype with its twin Jumo's disappeared. Its whereabouts were never determined.

The P.11's twin Jumo 004B turbojets were mounted side-by-side in the center section of the delta wing. Each turbojet was to have its own air intake located along the leading edge of the wing close to the wing's root just below the cockpit. The center section was constructed of metal tubing and wood covering in a style and form similar to the Horten Ho 9. The P.11's landing gear was a tricycle type; the main gear folded up into the wing and the nose gear retracted backward up against the floor of the cockpit. Two huge vertical tail surfaces, rudders, were attached to the straight trailing edge of the wing. It is not known why Lippisch felt his P.11 had to have such large twin rudders.

The single-seat P.11 was expected to achieve speeds of 645 mph (1,038 km/h) at 19,500 ft (5,944 km). Lippisch's engineers believed its cruising speed would be about 528 mph (850 km/h), with a range of 1,860 mi (2,993 km). Armament was to be two 30 mm MK 103 cannon or one 7.6 cm anti tank gun. To increase its interceptor capabilities the P.11 would have been fitted with two solid-fuel rocket boosters for quicker takeoff and acceleration.

Specifications: Project 11

Engine	2xJunkers Jumo 004B turbojets each having 1,984 lb (900 kg) of thrust each, plus 2x2,205 lb (1,000 kg) thrust, solid-fuel rocket boosters for takeoff assistance.
Wingspan	41.7 ft (12.7 m)
Wing Area	NA
Length	NA
Height	NA
Weight, Empty	NA
Weight, Takeoff	NA
Crew	1
Speed, Cruise	528 mph (850 km/h)
Speed, Landing	93 mph (150 km/h)
Speed, Top	645 mph (1,038 km/h) at 19,500 ft (5,994 m)
Radius of Operation	1,367 mi (2,200 km)
Service Ceiling	NA
Armament	2x30 mm MK 108 cannon, or 1x7.6 cm anti tank gun
Rate of Climb	NA
Bomb Load	NA
Flight Duration	NA

PROJECT 12

The P.12 was the first supersonic ramjet-powered design proposed by Alexander Lippisch. He and his designers believed that the delta-wing aircraft powered by a Lorin liquid-fuel ramjet would reach speeds of up to 1,500 mph (2,414 km/h).

From the mid 1940s on, there was considerable interest in the Lorin engine, and tests had been conducted with them mounted on propeller-driven aircraft. German engineers felt that Lorin-propulsion had some distinct advantages, including a simple design concept and the possibility of economical flight at supersonic speeds. To gain experience with aircraft powered by a Lorin engine as the main propulsion unit without compromising the design of other power plants, Lippisch planned to build the P.12 as a test bed for the Lorin engine.

Lippisch's friend Professor Adolf Busemann at the Luftfahrt Forschungs Anstalt (LFA) at München, and others such as Hellmuth Walter of HWK, had experimented with "Lorin Ducts" from 1940 on. Although a great deal of research had been conducted with the goal of developing a practical ramjet-type engine, satisfactory results had still not been achieved by 1945. Some of he major problems included finding a satisfactory duct design, handling the extremely high temperatures generated by the ramjet, and building the units small enough to be placed in an aircraft. By 1945 no aircraft in Germany had flown on ramjet power alone.

Lippisch's design placed the P.12's pilot seat above the air intake for the combustion chamber. The undercarriage would have had a single central wheel with a skid projecting downward from each wingtip. The

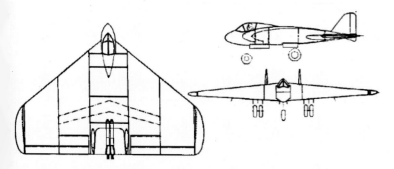

wing area was to be approximately 130 sq ft (12 sq m) with an aspect ratio of 1.33. All this was to be obtained by using a special elliptical-wing section profile. The P.12 never got further than the design stage. Lippisch abandoned the liquid-fuel ramjet Lorin engine for one that was to burn solid fuel, and scrapped his P.12 design in favor of his P.13A.

Specifications: Project 12

Engine	2xLorin liquid fuel ramjet engines.
Wingspan	NA
Wing Area	130 sq ft (12 sq m)
Length	NA
Height	NA
Weight, Empty	NA
Weight, Takeoff	NA
Crew	1
Speed, Cruise	NA
Speed, Landing	NA
Speed, Top	1,500 mph (2,400 km/h)
Rate of Climb	NA
Radius of Operation	NA
Service Ceiling	NA
Armament	None
Bomb Load	None
Flight Duration	NA

PROJECT 13A

Alexander Lippisch had long believed that flight in the supersonic range was possible. His conviction was based on wind tunnel tests of an aircraft that had a low aspect ratio and swept back wings, in short, a triangular planform. Lippisch's goal was to construct an experimental aircraft of this type, one that would demonstrate the advantages of a tailless design, namely, operational economy, a long range, a high service ceiling, and safety. Although he was motivated primarily by a desire to advance the state of the art in aircraft design, Lippisch also thought about future air traffic and the importance of improved economy and safety.

In wartime Germany, particularly during its desperate years beginning in 1943, political leaders had no time for fundamental scientific investigation. Any experimentation had to be closely related to practical application. Lippisch's way around this dilemma was to set up a partially experimental program similar to the Me 163. By pursuing what appeared to be an immediate military design goal, Lippisch was allowed to bring together theorists and practical engineers to work on a common problem. His experience with the Me 163 at DFS had convinced him that this was the best way to achieve milestones in designs and this is how Lippisch conducted fundamental scientific investigations on his new tailless delta-wing designs, too, known as the P.13A and the P.13B.

The P.13A was to be one of the powered versions in a long line of delta wing planform designs Lippisch was experimenting with. A mockup of the model was destroyed in a bombing raid at Vienna. However, the DM-1 glider designed to investigate the flight characteristics of the P.13A had been evacuated safely to an airfield at Prien.

The planform of the P.13A was to be almost triangular, with an aspect ratio of 2.0 and a 60 degree wing sweep. The aircraft was expected to have a maximum lift coefficient of 1.0 to 1.2, approximately the same ratio as the DM-1 glider, the testing vehicle.

Propulsion for the P.13A was to be provided by two Kronach Lorin-type ramjets installed on each side of the large central fin and fuselage. The ramjet tubes were to extend through the wing from the leading edge to the trailing edge in a straight line. Lippisch estimated that these ramjet engines would be self-supporting once a speed of 149 mph (240 km/h) had been reached. How the P.13A would accelerate to 149 mph is not known, but it is assumed that some form of takeoff assistance (solid-fuel rocket boosters) would have been necessary and provided for in the design.

The inlet duct for the air used in the Kronach Lorin ramjet was to extend forward of the apex of the leading edge. Originally, it was proposed that solid fuel in the form of small pieces of soft brown coal would be carried in a wire mesh container mounted in the duct at a small angle to the airstream, thus obstructing the free flow of air through the lower portion of the duct. Lippisch hoped to obtain a progressive reaction: oxygen would pass through the solid fuel, burning to CO, carbon monoxide, which in turn would combine with the oxygen in the air passing through the unobstructed upper portion of the duct to form CO_2, carbon dioxide. Combustion was to have been initiated by a gas flame; liquid fuel could be added to facilitate starting. It was estimated that about 1 ton of coal would provide an endurance of 45 minutes, enough to give the craft a range of 1,000 miles (621 km). Expected maximum speed was 403 mph (648 km/h).

Specifications: Project 13A

Engine	1xKronach Lorin coal-burning ramjet.
Wingspan	NA
Wing Area	NA
Length	NA
Height	NA
Weight, Empty	NA
Weight, Takeoff	5,060 lb (2,295 kg)
Crew	1
Speed Cruise	NA
Speed, Landing	NA
Speed, Top	403 mph (648 km/h)
Radius of Operation	1,000 mi (621 km)
Service Ceiling	NA
Armament	None
Rate of Climb	NA
Bomb Load	None
Flight Duration	45 min

PROJECT 13B

There is very little information available about the P.13B because it is believed that this Lippisch project existed only on the drawing board. Nevertheless, the P.13B is known to have included a number of changes from Lippisch's DM-1 and P.13A projects. Primarily, the P.13B had an entirely different delta-wing planform from the DM-1 and the P.13A, most notably a conventional cockpit and twin vertical rudders extending beyond the wing's trailing edges in place of the P.13A's huge dorsal

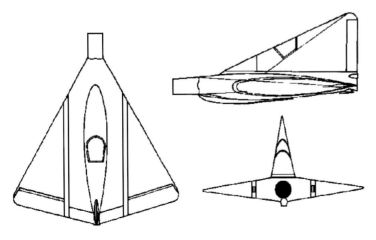

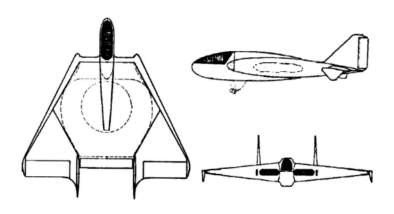
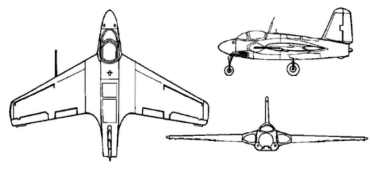

cockpit/fin. It appears that Lippisch was returning to his P.11 design for the P.13B.

Lippisch was not pleased with the coal-burning Kronach Lorin ramjet designed into the P.13A. The reasons for his displeasure are not known. So the Lorin ramjet's fuel arrangement was changed on his P.13B, the lump-coal arrangement given over in favor of a circular basket supported within the duct and positively rotated about its axis at some 60 revolutions per minute. Like the Kronach engine in the P.13A, combustion would be initiated by a gas flame and liquid fuel. Supplemental propulsion would have been required to help the craft obtain the approximately 149 mph (240 km/h) velocity at which the Lorin ramjet would operate. There are no details how the P.13B would have been propelled up to its self operating speed.

Although the P.13B appears to have been a conventional design, its landing arrangement was somewhat atypical, calling for wing tip skids and a retractable, steerable, nose wheel.

Specifications: Project 13B

Engine	1xKronach Lorin Duct ramjet of unknown size and thrust plus supplemental power in the form of solid-fuel rockets to bring the Lorin Duct up to self-operating speed.
Wingspan	NA
Wing Area	NA
Length	NA
Height	NA
Weight, Empty	NA
Weight, Takeoff	NA
Crew	1
Speed, Cruise	NA
Speed, Landing	NA
Speed, Top	NA
Rate of Climb	NA
Radius of Operation	NA
Service Ceiling	NA
Armament	None
Bomb Load	None
Flight Duration	NA

PROJECT 15

Alexander Lippisch was never happy about the Air Ministry's decision to turn his rocket-powered test bed, the DFS 194, into the series production Me 163B. It was not the design so much that displeased him, but the dangerous and unreliable HWK 509A-1 bi-fuel liquid rocket drive used to power it. The P.15 was Lippisch's attempt in early 1945 to turn his Me 163B into a turbojet-powered interceptor. To achieve the change, its bi-fuel liquid rocket drive would have been replaced by a single Jumo 004D turbojet having 3,960 lb (1,796 kg) of thrust at sea level.

With a Jumo 004D turbojet the P.15 was expected to have had a maximum speed of 621 mph (1,000 km/h). It would have been considerably smaller than the Me 163. Its overall length of 12.1 ft (3.7 m), compared with 17.7 ft (5.4 m) for the Me 163B but it would have used the same wings. Proposed armament were two 20 mm MK 103 cannon and two 30 mm MK 108 cannon.

Plans to convert the rocket-powered Me 163B into a turbojet-powered interceptor were not completed before the war ended, and no prototypes were constructed.

Specifications: Project 15

Engine	1xJunkers Jumo 004D turbojet having 3,960 lb (1,796 kg) thrust.
Wingspan	30.5 ft (9.3 m)
Wing Area	211 sq ft (19.6 sq m)
Length	12.1 ft (3.7 m)
Height	NA
Weight, Empty	5,333 lb (2,419 kg)
Weight, Takeoff	7,458 lb (3,383 kg)
Crew	1
Speed, Cruise	568 mph (915 km/h)
Speed, Landing	NA
Speed, Top	621 mph (1,000 km/h)
Radius of Operation	NA
Service Ceiling	56,761 ft (17.3 km)
Armament	2x30 mm MK 108 cannon, and 2x20 mm MK 103 cannon
Rate of Climb	NA
Bomb Load	NA
Flight Duration	NA

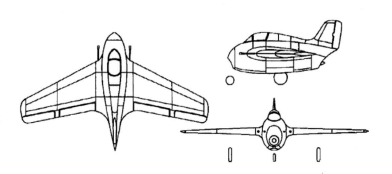

14

Messerschmitt AG - Augsburg, Germany

Secret Projects and Intended Purpose:
Project 262 - With a level flying speed of 540 mph (869 km/h) it had an 120 mph (193 km/h) speed advantage over the fastest propeller-driven fighter of the Allies.
Projet 262 HG-2 - This was an updated Me 262 to meet the Air Ministry's request for a high-altitude fuel efficient bomber interceptor. It retained the basic 262 fuselage, however, modified with more sweptback to the wing (18 degrees for the 262 and 43 degrees for the HG-2) and tail surface. It's twin turbojet engines were moved from under the wing and tucked into the wing-root at the side of the fuselage. It was believed that the P.262 HG-2 could reach 30,000 ft (9 km).
Project 1101 - A variable pitch, sweptback wing interceptor with a single seat and a single HeS 011 turbojet engine. This design would not "jell" and the Air Ministry rejected the project, but Voigt continued to work on it as a private research effort of Messerschmitt AG. The aircraft was to have been test flown in June 1945. It was captured at Oberammergau and was placed outdoors on static display for over a year prior to shipment to Bell Aircraft in the United States. Larry Bell later built a copy of the P.1101 calling it the Bell X-5. It, too, is considered a design failure.
Project 1106 - Similar to the P.1101, except that the cockpit was placed far to the rear, forming the leading edge of the T-shaped vertical stabilizer. This design was found to give very poor pilot visibility, and cockpit placement in the tail created a large amount of drag. Consequently, the project was abandoned.

Project 1107A - One of the Air Ministry's "Amerika Bomber" projects. This Messerschmitt version was to be fitted with four Jumo 004 turbojets in pairs beneath the wing. The design called for a new style of tailplane mounted at the top of a single rudder (fin) to clear the turbine's exhaust (later called the Multhopp "T-tail").
Project 1107B - A four HeS 011-powered 3x1,000 bomber with an anticipated range of 621 miles (1,000 km) at 621 mph (1,000 km/h) and carrying a bomb load of 2,205 lb (1,000 kg). This Messerschmitt project would have had a highly streamlined fuselage with a "butterfly" tail assembly. In keeping with a Messerschmitt AG design trend, the turbojet engines were to be placed in pairs and buried deep in the wing roots and as close as possible to the fuselage.
Project 1108 - Somewhat of a radical departure for Messerschmitt AG, the P.1108 was a proposed all-wing delta aircraft. Projected to be a 3x1,000 bomber, it would have been powered by four HeS 011 turbojets

Willy Messerschmitt in the early 1940s.

The Messerschmitt Bf 108.

The Messerschmitt Bf 109-G.

housed entirely inside the wing with their exhaust exiting out above the wing's trailing edge.

Project 1112 - A tailless fighter powered by a single HeS 011 turbine. The first tailless design by Messerschmitt AG since the Me 163 of Lippisch, the P.1112 is sometimes incorrectly referred to as the Me 263, which was a streamlined version of the 163. The P.1112 reached the stage of complete layout, and some component work had been started.

History:

The fastest fighters of World War II came from his factories - the twin turbojet-powered Me 262 and the sensational tailless HWK 509A bi-fuel liquid rocket-driven interceptor Me 163. Though the latter was not one of his designs, but that of a sometimes competitor, Alexander Lippisch, who had come to work with him, in the end it did not really matter, to Willy Messerschmitt it was performance that was everything

Speed had always been Messerschmitt's consuming interest, "The only thing that made airplanes worthwhile," he declared. But it was not Willy Messerschmitt's only passion. He was a pioneer in sailplaning, a professor of aeronautics, a designer of world wide repute, a champion air racer, and head of the world's largest airplane works during World War II. Although the United States was able to gain some measure of satisfaction by ending the war in the Pacific with a technological coup de maître, the atomic bomb, the war in Europe gave no such satisfaction, in large part because of Messerschmitt AG. His Bf 109 fighter plane was used to shoot down more aircraft in combat than any other in history. "A genius," they called him, "a genius born with a propeller in his brain."

In many ways Willy Messerschmitt was quite unlike any other German aircraft manufacturer, then or since. An unfailing believer in the totalitarian state, he was a conspicuous and important figure in the Third Reich. He was one of the very few German designer-builder-manufac-

The Messerschmitt Me 209-V1.

The Messerschmitt Me 209-V1 in flight.

turers who steadfastly refused to offer their expertise to the United States after the war. But above all else, Messerschmitt AG is one of the few whose names have become synonymous with an era of aviation. Messerschmitt played a role in every part of its spectrum: invention, design, manufacturing, management, politics, war, and rebuilding; and he did it better than any other aircraft manufacturer in Germany.

This remarkable self-taught engineer and businessman was born in Frankfurt am Main on 8 June 1889. He became interested in airplanes as a teenager while helping Freidrich Harth, a local inventor, build a foot-launched sailplane. Both men were drafted into the German Wehrmacht in 1917 and their sailplane activity came to a halt. They resumed their work again in 1919 after the Armistice of 1918 to form

A Messerschmitt Bf 109-E. The shield with the letter "S" forward of the cockpit, honors Albert Leo Schlägeter whom the French shot as an alleged terrorist in 1922.

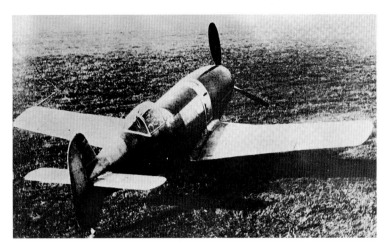

The Messerschmitt Me 209-V1 (D-INJR). On 26 April 1939 Fritz Wendel set a world record for propeller-driven airplanes which lasted for thirty years.

The remains of the Messerschmitt Me 209-V1 interned in a museum warehouse in Poland.

their own sailplane manufacturing firm: Segelflugzeugbau Harth und Messerschmitt. However Messerschmitt's 3 years in the army had shown him the possibilities of power-driven airplanes, and he found sailplanes much less interesting than before. Returning to school and working part-time at the sailplane works, he graduated in 1923 with an engineering degree from the Technical University of München. One of the first things he did after graduation was to buy out Harth (who had never fully recovered from head injuries suffered in a crash of a Pilotus sailplane in 1921) and changed the name of the sailplane firm to Messerschmitt Flugzeugbau GmbH. At his new factory, located in Bamberg, Messerschmitt translated into wood, metal, and fabric ideas for fast sport monoplanes which were then just beginning to replace the popular biplane aircraft of World War I.

In 1925 the Messerschmitt firm moved to Augsburg and was combined with the larger Bayerische Flugzeugwerke (BFW), a company founded by World War I flying ace, Ernst Udet. The firm prospered building sport and commercial aircraft. Messerschmitt's designs flew off with all the prizes in the Saxony Air Races of 1928. During the early 1930s, his planes captured many of the speed races on the European continent: London to Isle of Man, Berlin to Baghdad, Paris to Rome, around the Alps, and others. His sport aircraft, the Bf 108 Taifun (Typhoon), sold well in Switzerland, France, Britain, and the United States. In 1935 at the request of the German Air Ministry, he stopped producing sport monoplanes and concentrated on military aircraft. It was widely believed that when Messerschmitt emerged as sole owner of the merged Messerschmitt and BFW in 1932, he was able to do so through a secret 400-million-mark loan from the Air Ministry.

Messerschmitt AG's first military plane, the Bf 109, resembled in many respects his sport Taifun. Designed in 1935 and produced 1 year later, this single-seat fighter had all the trademarks that go with the name Messerschmitt. "If you look at my planes closely," he once remarked after the war, "you will see the hand of a man who can build a sailplane blindfolded." According to Messerschmitt, you could not design a good plane unless you knew how to build and fly sailplanes. More than 33,000 Bf 109s were produced between 1936 and 1945, making it the third most widely produced aircraft in the world, although the popular Cessna 172 Skyhawk may now have nudged it into fourth place. Only USSR's

Waldemar Voight in the late 1930s.

A late production model of the Messerschmitt Me 262 being flight tested in post-war United States. The metal seams show liberal amounts of putty and tape to cover up the poor fit by unskilled labor, possibly POWs and/or slaves..

A highly decorated Erhard Milch in the late 1940s.

"The Fat One," Hermann Göring.

Ilyushin series (I 11-2 and 11-10) and Yakovlev series (Yak-1, 3, 7, and 9) were produced in more units, 41,400 and 37,000 respectively.

Willy Messerschmitt's decision to go after turbojet-powered aircraft came on 26 April 1939, when Fritz Wendel, his chief test pilot, roared an Me 209 prototype along a triangular speed-timing course at Augsburg, establishing a new world speed record of 469.2 mph (755.1 km/h) for a piston-powered propeller-driven aircraft. This record remained Messerschmitt's until 1969.

On that evening of 26 April 1939 while celebrating over beer and skittles, Wendel told Messerschmitt, Woldemar Voigt, Messerschmitt's chief designer who had been working on airframe design concepts utilizing the promising new gas turbine, and other Messerschmitt AG engineers, "I felt I could have flown faster," said Wendel, "but the propeller seemed to be holding me back." Messerschmitt was wildly excited at this news. It confirmed what he and his colleagues, including Ernst Heinkel, had long suspected, that as a propeller-driven airplane approached 500 mph (805 km/h), the resistance (drag) created by the propeller became greater than the power developing in the engine. "Later that same evening," recalled Messerschmitt, "we began talking seriously about turbojets and rocket drives. We were on our way toward the speed of sound." Messerschmitt had to wait some time for that milestone to be reached, however; for the sound barrier was not broken until October 1947, and then by an American named Chuck Yaeger flying a US built Bell X-1.

To develop an airframe more suited to the higher speeds produced by the turbojet, Messerschmitt turned to his long-time airframe designer Woldemar Voigt. Messerschmitt's first turbojet-engined plane, the P.1065 (Me 262), as well as the HWK 509A rocket-driven Me 163, which colleague and competitor Alexander Lippisch had brought over from the Deutsches Forschungsinstitut für Segelflug (DFS) in January 1939, were

Lindburgh with Messerschmitt chief test pilot, Dr. Wurster in a Messerschmitt Me 108 "Taifun".

Udet, and Dr. Wurster, Messerschmitt's head test pilot, and others, Berlin.

Messerschmitt AG-Augsburg's control office building newly constructed, early 1940s.

Messerschmitt AG-Augsburg's control office building in 1944 - a burned out shell after American B-17 bombing raids. Spring 1944.

completed by 1941. The Me 262 airframe had been started late in 1938, at DFS and had first flown early in 1940 powered by a conventional, diesel aeroengine. Turbojet engines on order with the Bayerische Motoren Werke (BMW) did not arrive until November 1941 and then did not have sufficient thrust to power the P.1065 without the use of the nose-mounted diesel engine. The Junkers Jumo 004 turbojet engine Messerschmitt was waiting for was finally available in May 1942, and for the first time the 262 was flown under gas turbine power alone.

The first 262s had plenty of teething troubles. Four of the first five aircraft crashed during test flights and were destroyed completely. According to Messerschmitt:

"The (262) prototypes would roar down the runway, but we couldn't get the damned thing to take-off. We lengthened the runways to twelve hundred meters (3,937 ft), gave her a tricycle landing gear to keep her from nosing down, redesigned the tail rudder and placed outboard take-off rockets under the wings."

Finally, on 18 July 1942, the 262 showed what kind of plane it really was. In six successful test flights, the aircraft attained a speed of nearly 470 mph (755 km/h) and remained aloft more than an hour.

In 1939 and 1940 the German Luftwaffe was at the height of its power, basking in the glow of a succession of blitz victories, its glory dimmed only slightly by its check in the Battle of Britain. But, known to the outside world and recognized by only a few within the Reich, the Luftwaffe's demise was at hand. The future of the Luftwaffe, Messerschmitt felt, was in the hands of a fanatic triumvirate: two swaggering incompetents and a realist unable to oppose them. These three men were Hermann Göring, Erhard Milch, and Ernst Udet.

As the number two man in the Third Reich, Göring had come naturally to head the Luftwaffe. During its early years, in the 1930s he built the Luftwaffe from nothing. But in the mid 1930s he realized his own limitations in production matters and brought in as his adjutant his close friend Erhard Milch.

Few German aircraft manufacturers got along well with Milch. Messerschmitt's relationship with him was particularly sour; Milch looked on the professor from Augsburg as a prima donna and a braggart. The animosity between the two men dated back to 28 February 1928, when Milch was managing director for Lufthansa, the state-owned airline. Milch had ordered six new Messerschmitt M-20, ten seat commercial aircraft, for Lufthansa. The first three aircraft had crashed shortly after delivery due to a major design error, and a lifelong personal friend of Milch (Hans Hackmack), who was also chief of engineering at Lufthansa, had died in one of the crashes. Milch held Messerschmitt personally responsible and later canceled Lufthansa's remaining contract with Messerschmitt AG, an action that nearly forced Messerschmitt's struggling company into bankruptcy.

Messerschmitt AG-Augsburg inside the debris littered control office building, where designers were still attempting to initiate new designs!

Willy Messerschmitt postwar, building houses and home appliances instead of airplanes.

A Messerschmitt auto, post-war - the Mittelklasswagen P 511, 1953.

Another Messerschmitt auto, the Kabinenroller KR 200 - post-war.

Following the M-20 experience, the animosity continued and grew increasingly distasteful between Milch and Messerschmitt. It was Theo Croneiss, Messerschmitt AG's vice-chairman, the former brown-shirted SA General (the SA were the original body guards of Hitler) who started spreading rumors in early 1933 of Milch's Jewish blood. Croneiss even prepared a dossier including a photograph of a tombstone in a Jewish cemetery in Breslau bearing the family name "Milch" and sent it to Göring. Milch never forgave Messerschmitt, and although he gained the number two position in the Air Ministry on 15 May 1933 it required the full-weight of Göring to do it. Even though Milch was a fanatic Nazi he did in fact have a non-gentile father. To protect his good friend from harm, Göring had Milch legally declared a bastard under the Nürnberg Laws. "In the Third Reich," Hermann Göring declared publicly on the matter, "it is I who will decide who is and who is not a Jew."

From 1933 on Milch seldom missed an opportunity to embarrass Messerschmitt. Typical of Milch's actions aimed at making Messerschmitt's life and business operations a little uncomfortable was an event that occurred in June 1933. When Messerschmitt AG approached the newly organized Air Ministry for fresh contracts, Milch, then the Ministry's chief, required Messerschmitt's bankers to sign a two-million-mark bond, to be forfeited the day any new Messerschmitt aircraft crashed due to design failure. Messerschmitt was publicly embarrassed and personally pleaded with Milch to withdraw the bond requirement. Satisfied seeing Messerschmitt beaten down and begging, Milch waved the bond requirement.

Compounding Messerschmitt AG's problems with Milch was Messerschmitt's good friendship with Ernst Udet, the third-ranking

Willy Messerschmitt in retirement.

A Messerschmitt Zick-Zack automatic sewing machine. 1953.

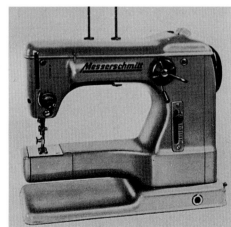

Willy Messerschmitt, post-war, in one of the prefabricated houses he was building and selling.

295

The Messerschmitt Me 262 as interpreted by US Army Intelligence reports. Early 1940s.

member of the Air Ministry. Udet, like Willy Messerschmitt, was cut from different cloth than Göring and Milch. A born flier who had flown in the von Richthofen Zirkus (circus) and who, like Göring, had achieved numerous combat victories (sixty-two confirmed kills for Udet making him the fourth highest scoring fighter pilot of World War I vs twenty-two for Göring). An extrovert, a lover of night-life, happy, dapper, smiling, Udet had spent the years after 1919 traveling the world, especially the United States, as a stunt and racing flier, a test pilot, and a movie maker. A man of the world who knew the world, a bon vivant with a caustic wit, a heavy drinker but a clear thinker, Udet regarded the Nazis as swine and often said as much. But as war approached, he had answered the call of patriotism and had returned to the Luftwaffe. On 10 June 1936, he came to head their entire Technical Department of the Air Ministry. His appointment came as quite a surprise, and not everyone liked the arrangement. Although Udet was an inspired pilot, his friends found him soft, vulnerable, and impressionable. The main reason for Udet's appointment was Göring's strategy of using Udet's name to attract bold, imaginative men who could really fly; Udet could help Göring build up his new Luftwaffe. "I don't understand anything about production or big airplanes," Udet once admitted. "The whole thing makes me uneasy. It's not in my line."

Consequently, men like Ernst Heinkel and Willy Messerschmitt thought they could call the shots, that they could specify the kind of airplanes the Third Reich should be using and that Udet would go along and authorize production. The appointment of Udet, and the way the Luftwaffe and the German aircraft manufacturers intended to use him, was the beginning of a human and technical tragedy for the Luftwaffe, for the aircraft manufacturers, and for Udet himself the tragedy of the right man in the wrong job.

German aircraft manufacturers, Heinkel AG and Messerschmitt AG among them, hoped Udet would be their conduit through which some of their favorite aircraft design projects would win Air Ministry approval and financial support. Heinkel had tried repeatedly to interest the Air Ministry in his turbine engine work, and Messerschmitt longed to see his Me 262 authorized for production. Udet was interested in the developing technology of turbojet-powered aircraft, but he was completely overruled by Hitler's General Staff, who couldn't see how these new designs would be currently needed. War with England was absolutely out of the question they said, since the Führer would never let the situation lead to war beyond the European continent; thus, any conflicts that might occur could be handled by medium-range bombers carrying a small bomb load. But not pursuing turbojet-powered fighters and heavy bombers was a gamble, and Udet felt it would mean catastrophe if Germany found itself fighting England and later America without these technologically-superior power plants. But Udet quickly learned that he was virtually powerless to change it.

On 22 June 1941, the week the invasion of Russia began, an important conference took place between Generalfieldmarshall Milch, Udet, and Messerschmitt at the Messerschmitt AG facilities in Augsburg. "I was very happy at the news of the invasion," Messerschmitt recalled, "for now it seemed that we could get the green light on producing turbine and rocket-powered engines we had been developing." He was backed by Udet, but Milch, in a fury, denounced the whole idea. Waving his Fieldmarshall's baton, he proclaimed that the war had already been won, that too many planes were being produced, and he categorically forbade Messerschmitt to plan production of any "radical" new plane (the 262). Milch accused Messerschmitt of "profiteering motives," and Udet of casting doubt on the Führer's strategic genius. Udet replied that he didn't know anything about the Führer's strategic genius, but he knew a lot about fighter planes, and that if the Luftwaffe did not get superior fighter planes rolling off the assembly lines quickly, the war Milch considered won in 1941 would be lost by 1943.

What happened later that fall of 1941 is still wrapped in mystery. But one September night in Berlin, Ernst Udet, after throwing one of his famous parties, retired to his apartment, loaded his old service pistol, and calmly fired five shots into a dartboard's bulls-eye at the far end of his study. The sixth he put through his brain.

Messerschmitt Me 262 prototype Number 2. Due to its tail wheel configuration, the aircraft was difficult to lift off its runway.

Throughout 1942, while Messerschmitt kept working to perfect the Me 262, Milch continued his bullying tactics against Heinkel, Messerschmitt, and anybody else who tried to oppose him on the issue of radical new designs for fighter and bomber aircraft. He particularly picked on Messerschmitt, reminding him of Udet's fate, over which he gloated, and threatening Messerschmitt with arrest if he continued to criticize the Air Ministry's fighter-construction program. But it appears that Messerschmitt's fate was already sealed, his influence diminished in part by Milch's and the Luftwaffe's lack of faith in him and in part by his own lack of attention to the details of production.

In the spring of 1943 the Messerschmitt AG's assignment was clear: produce the Me 209, the advanced piston-powered fighter, and the Me 410 bomber, both so desperately needed in the war against Britain. Although he had delivered on his successful Bf 109, Messerschmitt was not known for being totally reliable in matters of production. He had a reputation as a designer who emphasized performance over pilot safety, speed over armament, climb over tight turning, plane over pilot - and as one who moved on to ever more challenging design problems, rather than one who saw through production of perfected designs. Milch believed Messerschmitt AG should concentrate on its production task and leave the development of turbojet-powered fighter prototypes to someone else. Consequently, in 1943 the Me 163 and Me 262 projects were taken from Messerschmitt and were to be handed over to the rival Junkers firm in Dessau.

By 1943 Udet's prediction that Germany would need superior fighters had come to be; the Royal Air Force bombed at night, the American 8th Army Air Force bombed by day. In one momentous week in July 1943 the RAF dropped more than 7,000 tons of blockbusters on Hamburg. The German High Command and the Luftwaffe chiefs had completely failed to appreciate the significance of the United States' program of bomber manufacturing ability. For instance, at one of Göring's meetings with German aircraft manufacturers, Kurt Tank, of Focke-Wulf, dismissed a German intelligence report which stated that the United States was producing up to 2,000 heavy bombers per month as nonsense. "I know how much time and labor which goes into the construction of a single Condor (the large four-engined Focke-Wulf Fw 200 Condor passenger aircraft) and to think that the Americans have enough aircraft manufacturing facilities to build 2,000 each month, is pure fantasy." Although the figures showing the numbers of bombers the United States was turning out were known early on, little was done to insure that Germany's fighter defense was built up on anything approaching a large enough scale. The old fighter program continued unchanged, with a production of about 300 fighters a month, in the belief that Germany must maintain the air offensive, must drop bombs, must retaliate. According to Messerschmitt and others, any rational and far-sighted planning staff would have recognized the threat, the United States posed by 1942 and turned over all aircraft manufacturing to the production of day and night fighters such as Messerschmitt AG's Me 262.

By spring of 1944 the Allies had gained air supremacy over all of Europe. In March of that year the Messerschmitt AG and its entire production program and design development were taken over by a commissioner. The prime reason for this move was derived from Milch's long-term dislike of Messerschmitt. Another reason was the Air Ministry's decision in August 1943 (after Hitler had ordered the Me 262s produced) to pursue the production of fighters and interceptors seriously, combined with the belief that Messerschmitt could not be completely relied on. A third reason was Milch's and the Air Ministry's belief that Messerschmitt, because of his poor management, was unable to produce the quantity of weapons the Third Reich needed.

The argument that Messerschmitt AG was too poorly organized to fill a large order was not without merit. Milch constantly complained about Messerschmitt's difficulties and frequent failures to meet the small delivery quota his firm did have in 1943 and 1944. In fairness, it must be noted that some of Messerschmitt's difficulties stemmed from the devastating Allied bombing of his factories. Even so, leaders of the Air Ministry and Luftwaffe believed that Willy Messerschmitt continually made promises of delivery and of "miracle aircraft" he could not keep. For instance, Messerschmitt denied any responsibility for the delays in introducing the 262. Looking back after Germany's defeat, Messerschmitt blamed Milch. But the principal cause for the poor and tardy showing of the 262 was its unreadiness for operation and high-level disagreements within the Air Ministry over how it should be committed to battle. The cause for much of Messerschmitt AG's anger came from the fact that the aircraft which they had designed as a fighter, was turned into a high-speed bomber. Apart from this basic error the probability of success of the 262 as a high-speed bomber was doubtful, Messerschmitt claimed, for Germany had no bomb sight of the type that was needed in view of the particular qualities (speed) of the aircraft.

Perhaps it was a tragedy for the 262 to have come from Messerschmitt because he was a "personality" in whom Milch had little faith. "I know only too well how Messerschmitt likes to talk big," Milch would frequently remark. Defending himself, Milch noted that after

A Messerschmitt Me 262 post-war. The bulge on the nose indicated this aircraft was a reconnaissance version and under the bulge was the camera installation.

Messerschmitt Me 262-B-1A/U1 night fighter with the Hirschgeweih radar array mounted in the nose and two drop tanks of 79 gallons each. Aft of the cockpit are two 30 mm MK 108 cannons in a vertical "Schräge Musik" arrangement. Scale model by Reinhard Roeser.

Udet's suicide had left him in charge, he had seen the aircraft industry's immediate duty as providing enough of the existing aircraft, Bf 109s and Fw 190s, to prevent the Luftwaffe's complete collapse. Milch felt that the Junkers Jumo 004 turbojet engine, required for the 262, might develop the same maladies in mass production as generally plagued the introduction of any new aero-engine. Moreover, Milch reasoned, Messerschmitt, in all his arguing for production of a long-range fighter, was ignoring a basic tenet of strategy - that the home base from which all operations were launched must be defended first and foremost. Messerschmitt agreed with this. In fact, he and Milch were not very far apart on the issue of defending the home base. Milch wanted Messerschmitt to produce the Bf 109, while the designer felt the 109 was obsolete relative to the turbojet-powered 262.

The Messerschmitt AG managed to place only one turbine powered aircraft in the air - the Me 262. The HWK bi-fuel rocket-driven Me 163 cannot be counted since Messerschmitt AG merely provided production facilities. The only other Messerschmitt AG turbine-powered aircraft that made it to the prototype stage was the P.1101 and the P.262 HG-2. The P.1101 was destined never to fly. Although it won an Air Ministry competition in September 1944, the P.1101 was canceled by the Air Ministry the next month as an unworkable design. Instead the Air Ministry turned to Focke-Wulf's P.183. The P.262 HG-2 was a more streamlined, more fuel-efficient version of the standard 262 and a prototype was completed in May 1945. However, it was destroyed prior to its first test flight when another aircraft collided with it as it waited for takeoff clearance on the runway. Beyond the P.1101 Messerschmitt AG had precious little else in terms of turbine-powered designs being readied for eventual production. According to chief designer Voigt, Messerschmitt AG intended to proceed with development of only the models of the P.1101 series. At the time of the German surrender in May 1945, Messerschmitt AG had made no final decision about their future direction with turbojet-powered aircraft, whether they should go with a modified version of their P.1101 or with one of the tailless designs such as their P.1106 or P.1112 or with both.

Overall the Messerschmitt AG design team gradually lost its efficiency, creativity, and enthusiasm after completing their 262 in 1942. With the Allied bombing of the 262 production facilities in Augsburg in 1943, not much time remained for Voigt's design team to work on advanced project designs. Instead, they were seeking to put the 262 into the air with the modifications that the Air Ministry wanted in order to make it a fast bomber. By 1944 the P.1101 was considered a design failure. This was later confirmed by the failure in the early 1950's of Bell Aircraft's X-5, which was nearly identical to the P.1101. Voigt and other Messerschmitt AG designers subsequently sought a design solution to the problem-plagued P.1101, but time ran out before they could succeed.

Woldemar Voigt came to the United States under "Operation Paperclip" specifically to describe the P.1101's interesting variable-wing mechanism to the American aircraft design community. After working for a time at Wright Field at Dayton, Ohio, he went to work for Martin Aircraft in Baltimore, Maryland. After Martin merged with the Marietta Corporation in 1961 forming the Martin Marietta Aerospace Company, the Baltimore operations were closed. Voigt was then employed by the US Navy in Washington, DC. He died in obscurity in June 1980 at Annapolis, Maryland, thoroughly saddened by the fact that his career in America had never even remotely approached his expectations.

Willy Messerschmitt was probably the most famous builder of military aircraft in German history. Four times in his career he had built the world's fastest airplane. His Bf 109 fighter shot down more enemy planes in combat than any other in history. His Me 262 was the first turbine-powered fighter to prove itself in combat during World War II. Although the aircraft had been designed back in 1938 Messerschmitt AG was unable to achieve any further turbine design breakthroughs. The Me P.1101 was a bitter disappointment, and the firm won very few design competitions for proposed turbine-powered fighters and bombers held by the Air Ministry. Even though Messerschmitt AG was unable to repeat with turbines what they had accomplished with propeller aircraft, they have to be given credit for pushing turbine-plane development. "Plane building is both a science and an art," Messerschmitt once noted. "The design may look perfect. Then you build prototypes, or start mass production and you discover the worm."

The pursuit of the fast aircraft dominated Willy Messerschmitt's life. Early in his career he recognized that the key to advancing the field involved minimizing aerodynamic drag. Believing that speed was the only thing that made airplanes worthwhile, it was the designing and not so much the manufacturing that he enjoyed. He was almost fanatic in his attempts to discover sources of unnecessary drag and to reduce to the smallest area possible all surfaces exposed to the airstream. He once said that the designer sees not only the aircraft that is flying today, or even the aircraft that is being built or that which exists only as a project; any good designer must look much further into the future. Long before an aircraft is finished, Messerschmitt believed, a designer knows how it could have been improved. "Our work will never cease," he often said.

The Allied occupation of Oberammergau brought to a close the thing Willy Messerschmitt liked best: taking a design he believed in and

The Messerschmitt Me 262 flaps. The Junkers Jumo 004 exhaust cone is visible in the engine's tail pipe.

Messerschmitt Me 262-C1A. This prototype used an HWK 509A-1 bi-fuel rocket drive mounted in the aft fuselage to achieve a neck-snapping climb. The HWK 509A-1 provided 3,750 lb of thrust for 3 minutes enabling the 262 to reach 26,250 feet in 4 minutes. Scale model by Jamie Davies.

working however many hours it took to make it work. On 1 May 1945 Messerschmitt was arrested (under the automatic arrest feature of Pamphlet 31-110A). On 23 March 1945 the US War Department requested the US Army take into custody the dangerous Nazi Willy Messerschmitt. According to their Pamphlet 31-110A, Messerschmitt:

"Had thrived under National Socialism, welcomed it in the beginning, aided the Nazis to obtain power, supported them in office, shared the spoils of expropriation and conquest, or otherwise markedly benefitted in his career or fortune under the Nazis."

Unlike other German aircraft manufacturers, Messerschmitt was held as a war criminal at war's end because he had used forced labor (slaves) from concentration camps in his factories. He was detained until June 1946 when the charge (similar to the charge for which Albert Speer served twenty years) was dropped so he could testify at Nürnberg. Released from Nürnberg in 1947 and having never testified, Messerschmitt returned to his home in Augsburg. There a benevolent de-Nazification court on 26 May 1948 fined him 2,000 marks for being a Nazi follower. In 1948 the German mark was equivalent to 30 cents or a total fine of about $600. The court said Messerschmitt's claim that he opposed the Hitler regime had not been fully proven. Upon payment of the fine, Messerschmitt was free to go back into commercial business again, which he promptly did.

Messerschmitt claimed to have lost a personal fortune worth more than 50 million dollars when the Luftwaffe crashed along with Germany. Now, 2 years later, broke and out of work, he gathered many of his former staff in his old offices. He told his colleagues that he was a self-made man, and just in case there were ever any doubts about it, he was going to repeat the performance. Like Ernst Heinkel, he started out by manufacturing items needed by the German economy: prefabricated housing, mobile homes, little three-wheeled cars, small utility engines, and sewing machines. In 1952, 3 years before the aircraft manufacturing ban was lifted in Germany, Messerschmitt resumed his aircraft activities by constructing aircraft fuselages for the Spanish government. He also assisted them in aircraft design and production techniques.

Messerschmitt AG returned to aircraft production in Germany on a full-time basis after 1955 by producing a series of pilot-training aircraft for Egypt and Spain. This included the piston-powered Ha Me 100 and the turbine-powered Me 200 series. In 1957, after the state of Bavaria supported the reorganization of Messerschmitt AG by taking a 49% ownership, the renewed firm began the production of American F-104 Starfighters under license. During the 1960s, Messerschmitt AG, in cooperation with the Heinkel and Bölkow companies, was active in designing and constructing vertical takeoff-and-landing aircraft. For a while Willy Messerschmitt was considering the possibilities of manufacturing his famous Bf 108 Taifun sport plane but things did not work out and the plan was aborted before metal could be cut. The Bölkow and Messerschmitt AG groups were merged in the autumn of 1963 and then joined by a third party, the Hamburger Flugzeugbau (formerly Blohm and Voss) in June 1969, forming a consortium known as the Messerschmitt-Bölkow-Blohm (MBB) GmbH. Professor Messerschmitt took over as chairman of the board of directors becoming one of the forces in making MBB one of the three major firms in German aerospace activities. He died in Augsburg on 15 September 1979, at the age of 80.

Not one to express disillusion with the Third Reich, Messerschmitt never wavered in his belief that Germany could easily have won the war if only Hitler had been more accessible, Göring more competent, and Messerschmitt himself, more appreciated. He remained bitter about the 3-year edge in air superiority he felt Germany had wasted by waiting so long to produce the next stage of aircraft, including his Me 262; but he always blamed the shortsightedness of the Air Ministry's top officers, never the system itself. Harboring a basic respect for the dictatorial state, he retained a deep dislike for America until the day he died.

"Although you Anglo-Americans are undoubtedly ahead of the Soviets in aerodynamics," he taunted in 1950, "I don't think you will be for long. I am not so sure you will be in 5 years. And don't forget that a good part of the German aircraft and rocket industry and design brains have gone east (to the USSR)."

Messerschmitt loved to tell any Western reporter within earshot about the secret orders he allegedly received from the Führer's bunker just before Hitler's suicide. He said that he had been ordered to destroy all models of his company's top-secret projects, then to encase the blueprints in hermetically sealed chests and drop them into the depths of Lake Zell in Austria. This was done, and there they are today - waiting - Messerschmitt would say, waiting for what was never disclosed.

PROJECT 262

It is considered the most formidable combat aircraft to emerge from World War II. Equipped with four 30 mm cannon in the nose and 24 wing-mounted air-to-air rockets firing 1 pound 37 mm warheads, it was the most heavily armed fighter of the war. Unofficially it was called the Schwalbe (Swallow), a bird regarded as the harbinger of spring. But to the Allied pilots who first witnessed this new aircraft in July 1944 in the skies over Germany, it appeared like a shark out of water. To the men of the US 8th Army Air Force, the Me 262's other unofficial name, Sturmvögel (Stormbird), was more appropriate because it seemed to come "straight at you from out of the sun." With a level flying speed of

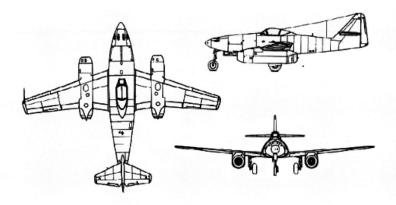

540 mph (869 km/h), it had a 120 mph (193 km/h) speed advantage over the fastest propeller-driven Allied fighter.

Turbojet-powered aircraft had been given a fair amount of thought in Germany prior to Messerschmitt's decision to get involved. The Air Ministry's interest might well be said to have started in America when a young man named Helmut Schelp was working for a master's degree at Stevens Institute in Hoboken, New Jersey. Schelp returned to Germany in 1936, and soon afterward went to work for the Air Ministry in Berlin. The following year he began work on gas turbines, and in 1938 he was made turbine-engine inspector for the Air Ministry. By the end of 1938 he had the Bayerische Motoren Werke (BMW) and Junkers Jumo working on what later became the BMW 003 (at that time designated the P.3302) and the Jumo 004. He did not succeed in getting Ernst Heinkel to even cooperate, for Heinkel preferred to work alone, and it was not until 1941 that Schelp succeeded in persuading Daimler-Benz in Stuttgart to enter the turbine-engine field. That firm's first design, the DB 007, proved extremely complicated, and was abandoned when the company was switched to development of a propeller-driven version of the Heinkel-Hirth HeS 011 known as the HeS 021.

On the basis of what appeared to be a promising future for the turbine, the Technisches Amt (Technical Office) of the Air Ministry contacted Messerschmitt AG in late autumn of 1938 to see if he had any ideas for a suitable airframe capable of accommodating two TL Strahltriebwerke (axial-flow turbines) that BMW had in the preliminary design development stage. The Air Ministry anticipated that these two BMW P.3302 turbines, each of which would develop 1,320 lb (599 kg) of thrust, would be available for delivery by December 1939. For Messerschmitt the timing could not have been better; all he had to do was pull out Voigt's preliminary design for a twin-turbojet-powered fighter. The Air Ministry was elated, and on 7 June 1939, Messerschmitt submitted a complete set of design drawings for an aircraft called the P.1065.

Project 1065 called for an all-metal, low-wing, cantilever monoplane with a fully retractable tail wheel undercarriage and two Air Ministry sponsored BMW P.3302 turbine engines buried in the wing roots (similar to Voigt's P.262 HG-2 design of 1945). The P.1065 was expected to have a level flight speed of 560 mph (901 km/h). On the strength of this proposal, Messerschmitt was ordered to proceed with construction of a mockup. Technisches Amt officials inspected the mockup in January 1940, and on 1 March 1940, Messerschmitt was awarded a contract for construction of three airframes for flight testing. The aircraft was officially designated the Me 262.

BMW had been unduly optimistic in its estimate of the time necessary to develop a turbine-engine of sufficient reliability and thrust for flight testing. The P.3302 turbine itself, which by now had been officially designated the BMW 003, had a substantially larger diameter than originally anticipated. Consequently, its installation in the 262's wing roots, as planned by Voigt, was no longer practical. A complete redesign of the project was undertaken and submitted to the Technisches Amt on 15 May 1940. The fuselage was still an all-metal semi-monocoque of a triangular form with rounded corners; the wing, which passed through the wide fuselage base, was also an all-metal structure, with a single built-up I-section main spar and flush-riveted stressed skinning. The main spar had a few degrees of sweep back outboard of the turbine nacelles and carried automatic leading-edge slots. The main wheels retracted inward into the underside of the fuselage, and the tail wheel retracted aft.

Messerschmitt AG's revised proposal was accepted by the Technisches Amt at the beginning of July 1940, and in August the first metal was cut at Augsburg on the three 262 prototypes. Meanwhile, Messerschmitt waited on the BMW 003 turbines of 1,015 lb (460 kg) thrust to be delivered. But delays continued to plague BMW engineers. Impatient to get his design into the air, Messerschmitt placed a 700-horsepower Jumo piston engine in the nose of one of his prototypes and on 4 April 1941, test flew the airframe for the first time.

In November 1941 the first BMW 003 turbine engines finally arrived at Augsburg, and Messerschmitt AG engineers attached them to the 262's wings. But these engines proved to be unsatisfactory because of their low thrust and poor operating efficiency, and Messerschmitt was forced to wait again now for the more promising Jumo 004 with their 1,848 lb (840 kg) of thrust. These turbines arrived in mid 1942, and on 18 July 1942, the P.262 flew for the first time on turbine power alone.

Messerschmitt presented a report on the progress of their 262 and asked for further development funds from the Air Ministry, but Milch refused. Consumed with plans to increase production of conventional fighters and bombers in the face of growing Allied air power in Western Europe and the demands of the Russian and North African Fronts, Milch was in no mood to support experimental turbine-powered aircraft. But Willy Messerschmitt was also in no mood to be put off, especially in view of the time and money his firm had invested in this project, and he lined up a number of individuals to fly his new turbine-powered fighter prototype One of these people was General der Flieger (General of the Fighter Group) Adolf Galland. On 22 May 1943, Galland flew the Me 262 and later reported to Reichsmarshal Hermann Göring:

"This model is a tremendous stroke of fortune for us. It puts us way out in front provided the enemy continues to utilize piston engines. As far as I can tell, the fuselage seems to be entirely satisfactory, and the engines are everything that has been claimed for them, except for their performance during takeoff and landing. This aircraft opens up entirely new possibilities insofar as tactics are concerned."

Galland went on to suggest that the single-engine fighter production henceforth be restricted to the Focke-Wulf Fw 190 and that the production capacity thus freed be switched to the 262 program. Galland was the single, most important factor in bringing the 262 into operational use, beginning with his own test flight and later through his glowing report to Göring.

On 2 June 1943, the Air Ministry's Chief of Procurement and Supply cut an order to begin construction of the pre-production Me 262A-O because of its superior speed as well as its many other qualities. Still, Messerschmitt felt that the Air Ministry was dragging its feet over full deployment of his 262. In frustration he appealed directly to Hitler and when the Führer saw the 262 perform for the first time on 26 November 1943, he declared that it was just the thing for carrying a 1,102 lb (500 kg) bomb to England. Somewhat dismayed, Messerschmitt admitted that the 262 did have sufficient power to carry bombs, but cautioned Göring that such a load would drastically reduce the aircraft's performance. Nevertheless, it was becoming apparent that Hitler was worried about an Allied invasion somewhere along the Eastern Coast. Göring told Messerschmitt that the Führer envisioned a turbine-powered bomber with the power to drive the Allied armies into the sea before they could establish a beachhead. Messerschmitt went back to the drawing board, wasting precious months creating a "Blitz Bomber," while officials believed what they could have was the world's best fighter. Adolf Galland remembered the Luftwaffe's costly and humiliating defeat in the Battle of Britain, when its bombers were rendered ineffective because support fighters could not establish a protective air umbrella. Many felt the situation would be no different over Western Europe; but Hitler could think only in terms of offense, never defense, and he allowed 262 production to come to a virtual halt while Messerschmitt AG designers worked day and night to convert the aircraft into a bomber.

It quickly became apparent that the offshoot bomber was completely unsuited for tactical bombing operations. The enormous amount of fuel consumed during low-altitude flight made ground-level bombing runs impractical and operational ranges unprofitably low. As a dive bomber, the converted 262 fighter also was shackled by its limited gliding and diving speeds, and because of some high-speed instability. Only a small number of Me 262 bombers ever saw combat service, and no units went into operation before August 1944, 2 months after D-Day. The 262 as a bomber project was a disaster.

Finally, in October 1944 Hitler relented and allowed limited production of the 262 as a fighter. The aircraft had a takeoff weight of 15,720 lb (7,130 kg) and was powered by two Jumo 004B-1 turbines, each rated at 1,984 lb (900 kg) thrust. Using a kerosene-based fuel, the Me 262 A-1A fighter had a range of 652 miles (1,050 km) at 29,530 ft (9 km). Service ceiling was 36,080 ft (11 km).

Overall, the Me 262 A-1A was a very good package. It had one of Germany's most advanced turbojet engines, a good airframe, and the heaviest armament package yet seen in a fighter. The aircraft was designed with a thought for subcomponent production at diverse locations, although such manufacturing sites had never been envisioned by its designer, Voigt. It could be assembled with relative ease in the primitive forest and cave factories to which the Reich was reduced in the last months of the war. The fit of the aluminum skin was rough, but Messerschmitt personnel solved this problem with generous amounts of tape and body putty.

Like other Messerschmitt designs, the 262 was reported to be a pilot's airplane, one which had to be flown, not just heaved, into the air. Basically underpowered and fitted with engines sufficiently lacking in reliability to keep the adrenaline flowing, its pilots claimed that it was thoroughly exciting to fly, particularly in view of its lack of an ejector seat. The 262 had a tendency to snake through the air, a problem which, it had been determined, resulted from lack of an effective rudder fin area. Voigt believed that since the destabilizing effect of the airscrew (propeller) was not present, the vertical rudder could be made smaller than that found on many piston-powered aircraft, and some of the snaking had been reduced by removing at least a third of the vertical rudder. But this change also threatened aircraft safety in the event that the aircraft had to fly on only one engine. In such a situation, it was found, the aircraft could fly no slower than 200 mph (320 km/h), which was unacceptable. The problem lay in the aircraft's lengthy nose. However, shortening the nose in 1944 was out of the question, for Germany was in desperate straits and Messerschmitt AG workers already were struggling to get the aircraft into operational status the way it was. Despite its problems, the 262 proved itself to be very docile and responsive in flight, virtually all of its pilots, Allied as well as Luftwaffe, liked to comment that it was a first-class combat aircraft for both fighter and ground-attack roles.

A substantial number of the estimated 1,400 262's produced survived the war and were flight tested, taken apart, and studied in numerous countries throughout the world. It is estimated that fewer than ten survive today, on static display at military bases and in aircraft museums.

Specifications: Project 262

Engine	2xJunkers Jumo 004B-1 turbines, each having 1,984 lb (900 kg) thrust.
Wingspan	41 ft (12.5 m)
Wing Area	233.6 sq ft (21.7 sq m)
Length	34.9 ft (10.6 m)
Height	NA
Weight, Empty	9,742 lb (4,420 kg)
Weight, Takeoff	15,720 lb (7,130 kg)
Crew	1
Speed, Cruise	334 mph (538 km/h) at 65% throttle
Speed, Landing	109 mph (175 km/h)
Speed, Top	514 mph (827 km/h) at 14,264 ft (6.4 m)
Rate of Climb	32,810 ft (10 km) in 26 minutes
Radius of Operation	652 miles (1,050 km) at 29,530 ft (9 km)
Service Ceiling	36,080 ft (11 km)
Armament	4x30 mm MK 108 cannon, the upper pair firing 100 rounds per minute and the lower pair firing 80 rounds per minute, plus 24 wing-mounted air-to-air rockets.
Bomb Load	1x2,205 lb (1x1,000 kg) bomb or 1x1,102 (2x500 kg) bombs
Flight Duration	NA

PROJECT 262 HG-2

The Me 262 HG-2 was a streamlined, more fuel-efficient version of the standard 262. Designed by Woldemar Voigt of Messerschmitt's Advanced Design Office at Oberammergau in 1944, the plane was in fact an entirely new aircraft although it carried the designation of 262. Voigt had designed the original 262 in 1938. With its twin Jumo 004s, it theoretically had a service ceiling of approximately 36,080 ft (11 km). In operation, however, the Jumo's had a tendency to flame out at that altitude and to suffer compressor stall at high speeds and altitude. Consequently, the 262 had an imposed limit of about 26,240 ft (8 km), making it ineffective against a fleet of B-17 heavy bombers flying above 29,525 ft (9 km). In addition to the poor performance of the Jumos, the Me 262 consumed fuel at a high rate. With declining fuel supplies, the Air Ministry felt that the Me 262 as a fighter design was becoming obsolete. Voigt and his engineers sought to meet the Air Ministry's need for a high-altitude, fuel-efficient interceptor by replacing their aging 262 with something more modern.

In his search for better performance and greater fuel economy, Voigt retained only the fuselage and the tail assembly from the 262; and even then the fuselage was changed somewhat to accommodate a second seat for a radar operator. In addition, the radar scanner disk was placed in-

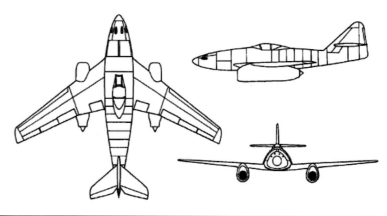

Messerschmitt Me 262 HG IV. (Hochgeschwindikeit high speed flight). This configuration was intended to achieve the greatest speed possible from an Me 262 and perhaps its sole purpose was to set up a world's speed record maybe even breaking the speed of sound in level flight. Scale model by Steve Malikoff.

Messerschmitt Me 262 HG IV. Scale model by Steve Malikoff.

Messerschmitt Me 262 HG IV. Scale model by Steve Malikoff.

side the nose cone. The moderately (18 degree) sweptback wing of their standard 262 was replaced with one of 43 degree sweep, and the new twin HeS 011 turbojets with a combined thrust of 5,732 lbs (1,600 kg) thrust were to be brought in as close as possible to the fuselage side. No performance data are available regarding the anticipated speed of the P.262 HG-2. The original 262, with its standard Jumo 004Bs of 1,980 lb (900 kg) thrust, achieved a level flight speed of 536 mph (868 km/h) at 22,880 ft (7 km). With all the aerodynamic improvement and the 33% increase in power afforded by the HeS 011 turbines, the P.262 HG-2 should have been able to reach 621 mph (1,000 km/h). Air intakes for the buried turbines were located in the leading edges of the wings, and no loss of thrust was expected because no long intake duct was required.

Armament on the P.262 HG-2 was to consist of four 30 mm MK 108 cannon able to fire about 500 rounds of ammunition per minute, to be placed in the nose of the aircraft. The MK 108 cannon was becoming standard armament for Luftwaffe fighters in 1944 and 1945 because of its small size and compact shape allowed for ane extremely neat installation.

At least one prototype of the P.262 HG-2 had been completed with Jumo 004B turbine engines prior to the end of the war; reportedly it was destroyed in a ground accident. The story goes that in May 1945, as the prototype waited near the edge of the runway for takeoff clearance for its first flight test, another aircraft attempting an emergency landing missed the center line of the runway and clipped the 262 HG-2 as it came down. The prototype was completely destroyed in the ensuing fire, and Voigt and his engineers had no time to complete another prototype before the war ended. It is believed, too, that Soviet aviation experts obtained the plans for the P.262 HG-2, the Russian "Neuer," was flown in the late 1940s.

Very few specifications or performance data on either the P.262 HG-2 or the Russian version are available.

Proposed Messerschmitt Me 262 Schnellbomber Ia. February 1944. Scale model by Jamie Davies

Specifications: Project 262 HG-2

Engine	2xHeinkel-Hirth HeS 011 turbojets each having 2,866 lb (1,300 kg) of thrust.
Wingspan	NA
Wing Area	NA
Length	NA
Height	NA
Weight, Empty	NA
Weight, Takeoff	NA
Crew	NA
Speed, Cruise	NA
Speed, Landing	NA
Speed, Top	NA
Rate of Climb	NA
Radius of Operation	NA
Service Ceiling	NA
Armament	NA
Bomb Load	NA
Flight Duration	NA

Proposed Messerschmitt Me 262 Schnellbomber Ia. Scale model by Jamie Davies.

Proposed Messerschmitt ME 262 Schnellbomber Ia. Scale model by Jamie Davies.

Proposed Messerschmitt Me 262 Schnellbomber Ia. Scale model by Jamie Davies.

Messerschmitt AG-Augsburg, 1944 raid destroyed the facilities. Here we see nose sections of former Me 262s lying about.

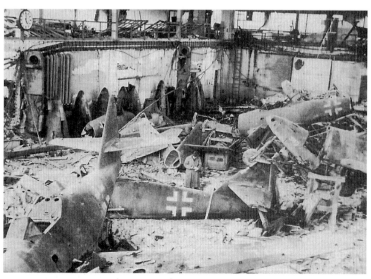

Messerschmitt AG-Augsburg. The remains of several Me 262s inside a bombed out assembly hall.

Secret Aircraft Designs of the Third Reich

A Messerschmitt Me 262 forest factory, suffering the results of an Allied fighter attack.

Another Messerschmitt Me 262 destroyed at the forest factory near Rengensburg.

PROJECT 1101

The Me P.1101 was Willy Messerschmitt's labor of love, his favorite research project. He anticipated that it would allow Messerschmitt AG to exceed Mach 1 (speed of sound) before any other company. The aircraft's secondary purpose was to test the sweptwing concept in flight. It had a variable-backsweep wing which could be set at different angles (up to 45 degrees) while the aircraft was on the ground. When the P.1101 was discovered by American troops as they overran the Tyrol mountain area around Oberammergau, Messerschmitt engineers, headed by chief designer Woldemar Voigt, were working on a mechanism that would allow the pilot to change wingsweep during flight. The P.1101's wings were unmounted when it was found. They were hastily reattached by US Army engineers, and the aircraft was pulled out on the front lawn to be displayed along with other aircraft found in Messerschmitt's semi-private research workshop. There it stayed, falling back on its tail section, and its port main wheel gear collapsing as it sank into the grass.

When it was relieved of its display duties 12 months later, it was missing parts, seriously deteriorated from the weather, and full of dents, the result of soldiers walking, sitting, and sliding on the aircraft while taking pictures for the folks back home

To see his pet project treated by the Americans as a piece of playground equipment, then shipped to the United States for study and, when experts had finished, discarded in a scrap heap, angered Messerschmitt the rest of his life. Nor did he ever forgive chief designer Voigt for accompanying the P.1101 to America in 1946, and then deciding to remain. Messerschmitt blamed Voigt for all that happened to the P.1101, calling him a traitor.

As research projects go, Messerschmitt AG's P.1101 had a sort of rugged simplicity about it. American aviation experts found the little craft rather ugly, and since it was originally designed as a research aircraft, with no provisions for guns, radar, and the combat-related equipment, they initially regarded the P.1101 as having no immediate intelligence value.

A typical supply store house at a forest factory where production for Messerschmitt Me 262s had moved.

Production huts at a forest factory for Messerschmitt Me 262, nose and tail pieces.

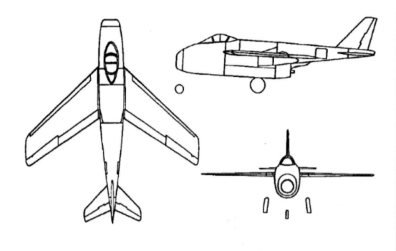

A Messerschmitt Me 262 freshly built in a forest factory, sits by the Autobahn near war's end, 1945. The Autobahn served as the 262's take-off and landing strip for forest-based 262s.

As early as 1935 Voigt had been interested in the theoretical work on swept wings done by Dr. Adolf Busemann, one of Germany's most respected aerodynamicists, who had presented a paper on the theoretical advantages of sweeping wing leading edges. Later Dr. Albert Betz of the Deutsches Versuchsanstalt für Luftfahrt (DVL), or German Experimental Institute for Aviation, began some research into the relationship between compressibility and wing sweepback. This research had been brought to Voigt's attention early on, and he continued to follow it with great interest. As the work at DVL showed increasing promise, Voigt persuaded Messerschmitt to sponsor a comprehensive wind tunnel program to test the soundness of Busemann's theories and some wind tunnel work at DVL. In July 1942, Voigt started primary design work on a sweptwing aircraft known internally as Project 1101. Project 1101 was conceived as a research venture to investigate in flight the characteristic of wings swept in increments of 30 to 40 degrees. The intention was to lock the wings at a certain angle, fly the aircraft throughout its entire performance envelope, and then reset the wing at a different angle for comparative testing.

Voigt made up a model of the P.1101's proposed design, one having a 6.1 ft (2 m) wing span. The model was exhaustively tested in the Berlin-Adlershof wind tunnel of DVL, with encouraging results. However, the Me 262's first flight solely on turbine power had indicated that a great deal of additional thrust would be necessary for a single-engine aircraft such as the P.1101 to fly well. So, in 1942 project work slowed considerably for lack of an adequate turbojet engine. Moreover, since the Me 262 was requiring more work than had been anticipated (and because of efforts to convert it to a bomber, then back to a fighter), Voigt and other members of the Messerschmitt AG design team began devoting their full attention to their long suffering 262.

The P.1101 remained on Messerschmitt's drawing boards in an inactive status until September 1944 (almost 2 years later), when the Air Ministry issued a requirement for a single-seat fighter to be powered by a single turbojet engine, one that would achieve greater speed, altitude, and fuel economy than the fuel-hungry twin-turbine Me 262. Power for

The general office of a Messerschmitt Me 262 forest factory assembly facility.

Another factory fresh Messerschmitt Me 262 abandoned by its constructors, May 1945.

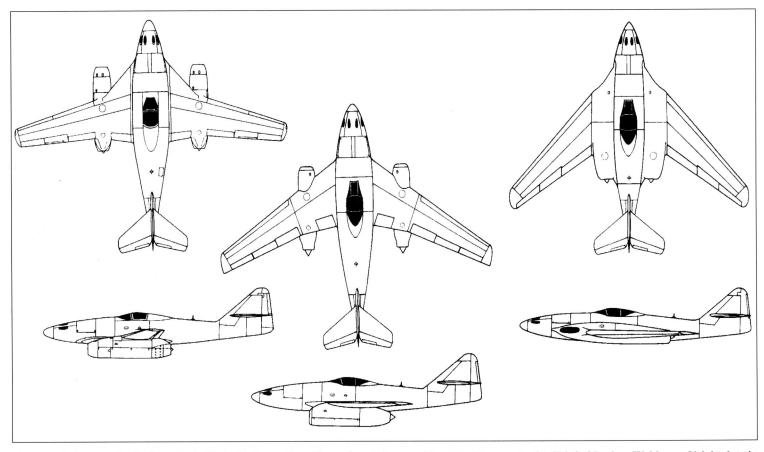

Three variations on the Hochgeschwindikeit (high speed) configurations being considered by Messerschmitt Chief of Design, Woldemar Voight, late in the war.

the proposed aircraft was to be provided by a single HeS 011 turbojet with 2,866 lb (1,300 kg) thrust. The HeS 011 turbine being readied for mass production in early 1945.

Messerschmitt AG offered their P.1101 to the Air Ministry. It was in competition with Blohm and Voss' P.215, Focke-Wulf's P.183, Heinkel's P.1078A, and Junkers' P.128. No award was made, but several firms were encouraged to construct a prototype of their entries. Messerschmitt AG among them. The Air Ministry provided some limited funds to Messerschmitt to help them build a prototype P.1101.

As the prototype was taking shape at Oberammergau under bomberless skies, for the Allies did not learn of this Messerschmitt research facility existence until after the war, the Air Ministry found it suffered from considerable problems. It seemed to Voigt and the others that the little P.1101 just did not want to "jell." The principle difficulties were many. First its cannon installation around the HeS 011's air intake duct in the aircraft's nose was very crowded and the ammunition boxes and spent shell exit chutes did not work well. Second the P.1101's low thrust axis necessitated large changes of trim with changes in thrust. Third the motion of the main wheel upon retracting into the fuselage presented a serious gear door closing problem. Fourth the wing, turbine engine, and landing gear loads were grouped in such a way that an excessive number of strong points throughout the fuselage were required. Fifth the aircraft was becoming heavier with each new modification. Sixth anticipated performance of 612 mph (985 km/h), was appearing much less than desired.

Wind tunnel model of the Messerschmitt Me 262 HG-3.

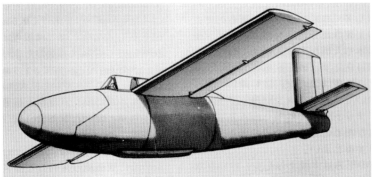

Messerschmitt Me P.1104 proposed interceptor powered by a HWK bi-fuel or solid fuel rocket. The RLM chose the "Natter" over the P.1104.

The Messerschmitt Me P.1106. Scale model by Dan Johnson.

The Messerschmitt Me P.1106. Scale model by Dan Johnson.

The Messerschmitt Me P.1106. Scale model by Frank Henriquez.

The Messerschmitt Me P.1106. Scale model by Dan Johnson.

Home of Messerschmitt's secret advanced design group headed by Woldemar Voight. Oberammergau, south of München. Spring 1945.

Three members of the US Army Field Intelligence standing in front of one of the design departments of Messerschmitt AG-Oberammergau.

Secret Aircraft Designs of the Third Reich

A view of Messerschmitt AG-Oberammergau's advanced design facilities looking down from the mountains.

The Messerschmitt Me P.1101 accommodates an American soldier sitting by its starboard wing root as the aircraft itself slowly sinks into the mud after being on public display for almost a year.

The side view of the Messerschmitt Me P.1101. Scale model by Ed Bailey.

Overhead view of the Messerschmitt Me P.1101 showing its highly swept back wings and tail section. Scale model by Ed Bailey.

The Messerschmitt Me P.1101 side view, as it might have looked on the front lines with its air-to-air missiles. Scale model by Dan Johnson.

Messerschmitt Me P.1101, seen from side and rear. Scale model by Dan Johnson.

308

Rear view of the Messerschmitt Me P.1101. Scale model by Dan Johnson.

Overhead rear view of the Messerschmitt Me P.1101. Scale model by Dan Johnson.

For these reasons Air Ministry officials concluded that the P.1101 could not meet design specifications, and they decided to withdraw their financial support of the prototype in favor of Focke-Wulf's P.183, designed by Hans Multhopp.

Voigt eventually tried to complete the P.1101 on is own. He returned to his original idea of using it to reach Mach 1; he would also use it as a turbine engine test bed, and to evaluate the characteristics of sharply sweptback wings. During the early months of 1945 work on the P.1101 continued at a slow pace, and Voigt anticipated that flight trials would start about mid-June 1945. However, Allied troops arrived in the area in mid-April, when the project was about 80% flight-ready and a BMW 003 installed. Prior to the American troops appearance in Oberammergau, Messerschmitt had instructed his personnel to microfilm all the P.1101's engineering drawings and hide them away at a site near Oberammergau. Although the American troops found the P.1101 and later used the facilities for intelligence gathering operations in Germany, it was the French who found the P.1101's microfilmed design and engineering drawings and took them back to Paris. Despite continued efforts by Americans to obtain the microfilm, the French never did cooperate, and the Americans did not receive any documents or microfilm.

Had it not been for its increasing weight problem, the P.1101 would have been a relatively easy aircraft to maintain and service. Voigt's fighter design called for the wing to be swept back 40 degrees at quarter chord. The wings would have been shoulder-mounted with steel spars and wooden ribs and skin. The craft would have had wing leading-edge slots, plain camber-changing flaps, and conventional aileron, elevator, and rudder controls. All tail surfaces were to be swept. The barrel-like fuselage was to have a slender tailboom aft of the turbine's exhaust nozzle, a pressurized cockpit set well forward, and a nose air intake. The nose wheel was to retract with a 90 degree turn into the space beneath the air intake and the main undercarriage wheels, rearward and inward. Three fuselage fuel tanks, with a total capacity of 345 gallons (1,565 liters), were to be installed behind the cockpit above the wing spar. Armor plate would give the pilot protection against 12.7 mm cannon fire from the front and 20 mm from the rear. Nose-mounted armament was to consist of either two or four MK 108 cannons.

Allied intelligence officials and their civilian counterparts were eager to talk with Voigt about the P.1101 as well as other proposed Messerschmitt AG projects under consideration at Oberammergau. In July 1945 the Americans talked at length with Voigt and found him an articulate man with a good command of English. It was during these "interrogation" sessions that American officials learned the truth about the ugly P.1101 and heard a litany of its design faults and how the aircraft seemed to defy attempts of the best minds within Messerschmitt AG to mold it into a workable design. These discussions virtually ended American interest in the P.1101, and the pride and job of Willy Messerschmitt was pulled out of its hangar at Oberammergau and put on display along with an Me 262 and other "interesting" German aircraft. There it probably would have remained but for its "discovery" by the late Robert Woods, Bell Aircraft Corporation's chief designer, who saw the aircraft on display while visiting Allied intelligence headquarters at Oberammergau in 1946. He wanted the aircraft despite the fact that Voigt considered it a design failure. The US Army complied with Woods' request and shipped the badly deteriorating P.1101 (its steel spars were rusting, the wooden wing covering was coming loose, its main

Head on view of the Messerschmitt Me P.1101. Scale model by Dan Johnson.

Bell Aircraft's identical copy of the Messerschmitt Me P.1101, known as the X-5, this aircraft prototype was never put into production. It is considered a design failure.

Another view of the Bell Aircraft's X-5 lifting off the runway at Buffalo, New York.

Another view of the Messerschmitt Me 264 V1.

View of the Messerschmitt Me 264 V1 cockpit.

gear had collapsed, and its fuselage boom had been bent) to Wright Patterson Air Force Base outside Dayton, Ohio. Woods also interested Voigt in coming to America to help explain the P.1101's variable wing geometry to Woods and other members of Bell's engineering staff. Therefore, in 1946, under "Operation Paperclip," Voigt and his family came to America along with numerous other German aircraft designers and engineers and scientists. In the autumn of that year, the P.1101 arrived at Wright-Patterson, where it sat outdoors for 2 years before finally being moved to Bell's headquarters at Buffalo, New York, in 1948. Ultimately it would become the nonflying mockup for two identical aircraft test bed built by Bell and known as the X-5. The X-5 first flew in June 1951, and it became Bell's offer to the US Air Force as an interceptor. It was rejected after a series of exhaustive tests by the Air Force.

If the P.1101 did not create any problems for the Allies during the war, it certainly was the cause of a number of confrontations in America, particularly within Bell. Trouble began after Larry Bell, president of the company, had been persuaded by his friend, Robert Woods, that the P.1101 was a design and engineering achievement so great that its potential should not be ignored. Bell assigned Robert Stanley, the firm's chief engineer, to study the P.1101 with an eye to its potential as a salable military aircraft.

Stanley put his best designers and engineers on the project. After a thorough review, the design team reported that the P.1101 was inferior to American turbine-powered aircraft already in the air, the F-86 for example. Woods could not accept this conclusion and continued to champion the merits of the aircraft, finally convincing Bell to proceed with the construction of a prototype.

The decision to proceed with the prototype X-5 created many difficulties within the company. Feelings of the design review team against the X-5 were so strong that several engineers left in a rage. Stanley quit his position as chief engineer. Chief designer Woods, apparently unswayed, took over Stanley's duties and saw through the Americanization of the P.1101.

Stanley and the other engineers who lost out at Bell later learned that their calculations had been correct. Most of their fears - that the

The Messerschmitt Me 264 V1 (P.1107A). One of the first attempts at designing and building an "Amerika Bomber" in 1942/43.

Another view of the Messerschmitt Me 264 V1.

American version of the P.1101 would lack endurance, speed, range, and a weapons-carrying capability - were confirmed in testing by the Air Force. In addition the design was considered inappropriate for the plane's intended purpose. Because of its size, the P.1101, with its HeS 011 power plant, was considered underpowered. The swing wing seemed a superfluous option on an interceptor intended to serve as a last line of defense. Furthermore, armament on the swing-wing fighter had to be carried almost entirely inside the fuselage, as the General Dynamics F-111 fighter later demonstrated, and wing loading was nearly impossible. Indeed, although the US Navy believed that a ship-based interceptor might be needed, the US Air Force did not need an aircraft with home interception capabilities during the 1950s.

Few people, then or now, consider the P.1101 very pleasing aerodynamically. However, the P.1101 generally receives high marks for inventiveness in terms of systems layout, state-of-the-art engineering, and airframe design. The simple layout of the turbine engine would have made maintenance easy for Luftwaffe field-based mechanics. Thus, field servicing could have been accomplished quickly. It might have served Germany's needs nicely, especially in late 1944 and 1945, when B-17s and B-25s were bombing freely nearly everything of any value. But whether it could have made a real difference in Germany's war effort is doubtful due to its limits as a gun platform.

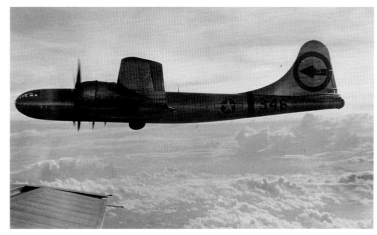

Boeing B-29. The Messerschmitt Me 264 V1 would have been three quarters the size of this B-29.

The sole example of the P.1101 was junked at the Bell factory in the early 1950s. It was used for structural test, and the remains were scrapped after completion of the two X-5 prototypes. Only one of the two X-5s survives. It is on permanent outdoor display at the US Air Force Museum, Wright-Patterson AFB, Dayton, Ohio.

A side view of the Messerschmitt Me 264 V1 in flight.

Secret Aircraft Designs of the Third Reich

Overhead view of the Messerschmitt Me 264 V1.

Specifications: Project 1101

Engine	1xHeinkel-Hirth HeS 011 turbine of 2,866 lb (1,300 kg) thrust.
Wingspan	26.9 ft (8.2 m)
Wing Area	171.1 sq ft (15.9 sq m)
Length	29.0 ft (9.1 m)
Height	NA
Weight, Empty	5,719 lb (2,594 kg)
Weight, Takeoff	8,966 lb (4,067 kg)
Crew	1
Speed, Cruise	357 mph (575 km/h)
Speed, Landing	107 mph (172 km/h)
Speed, Top	550 mph (885 km/h) at 22,967 ft (7 km)
Rate of Climb	4,370 ft/min (1,520 m/min)
Radius of Operation	932 miles (1,500 km)
Service Ceiling	45,931 ft (14 km)
Armament	5x30 mm MK 108 forward-firing cannon
Bomb Load	None
Flight Duration	40 minutes

PROJECT 1106

With the rejection of Messerschmitt AG's P.1101 in the design competition of September 1944, Woldemar Voigt and his design team immediately took steps to correct its obvious deficiencies. Shortly after awarding the project to Focke-Wulf, with its P.183, the Air Ministry announced that it wanted another fighter-interceptor design with substantially greater range and longer flight duration. The P.1101 had a projected flight duration of only 40 minutes. Voigt and his engineers felt that they could modify the aircraft to eliminate its airframe problems and could give it greater range and duration by providing greater fuel-carrying capacity. The modified P.1101 came to be called the Me P.1106.

One of the major requirements of the Air Ministry's new design request was that the interceptor have a flight duration of at least 60 minutes. To be competitive, Messerschmitt AG would have to devise a larger fuel tank than the one on the P.1101. Eventually the P.1106 became a rearrangement of the P.1101 in order to accommodate additional fuel. The only area where additional fuel tanks could be placed, Voigt and the other designers found, was in the cockpit area. They could do this if the pilot's area were moved back to the base of the vertical rudder. No other new parts or components would be required.

The P.1106 was to retain the P.1101's 40 degree sweptback wing, but it would not have the complicated and heavy variable-pitch mechanism. With the cockpit to be located back along the rear fuselage above the turbine exhaust duct at the base of the vertical stabilizer, fuel was to be located in the fuselage above the turbine, forward of the pilot. The P.1106 became a mid-wing design, with the wing structure passing under the fuel tanks and above the engine. To accommodate the pilot at the base of the rudder, the designers had to scrap the P.1101's conventional tail (empennage) and convert it to the Multhopp T-tail found on Focke-Wulf's design-winning P.183. The landing gear was to remain unchanged from the P.1101.

As the design work progressed into October 1944, it became apparent to Messerschmitt and his designers that this design was not going to "jell" either. Its principal difficulties were the same as those experienced on the P.1101. Moving the pilot back to the base of the vertical

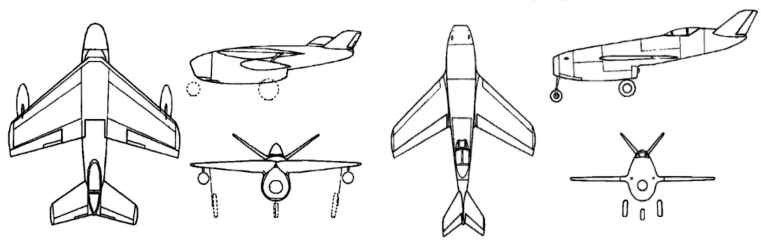

Ilyushin IL-16 looks like the turbine powered Me 264. The IL-16 may, in fact, have been based on the Arado Ar 234 variants.

The proposed Messerschmitt Me P.1107.2. Illustration by Hugh W. Cowin.

The proposed Messerschmitt P.1107.2 incorporated air intakes similar to the Handley-Page B-35/46 "Victor 8," post-war bomber.

rudder presented another problem, one that virtually doomed the proposed design as far as the Air Ministry was concerned: with the pilot placed well aft, looking down the long fuselage, his vision was badly obscured by the wing.

Before abandoning the P.1106 as unworkable, Messerschmitt AG studied the possibility of using the airframe for supersonic research in order to test its sweptback wings at high speeds. Calling their project the P.1106R (for rocket propelled), Messerschmitt engineers would have taken out its single HeS 011 turbojet and replaced it with two Walter HWK 509A bi-fuel liquid rocket drives of 3,968 lb (1,800 kg) thrust each. Only a few modifications to the airframe would have been required in order to turn it into a true supersonic research aircraft. These changes included closing the air intake and making it more pointed, extending the aft section of the fuselage a small amount to accommodate the HWK 509A bi-fuel liquid rocket drive, and removing the Multhopp T-tail and replacing it with a V-type controlling surface, or butterfly-style tail. The pilot's cockpit was to remain at the rear of the aircraft but would not have been built into either one of the V-type controlling surfaces. Instead, it was enclosed in a one-piece bubble canopy.

Specifications: Project 1106

Engine	1xHeinkel-Hirth HeS 011 turbojet of 2,866 lb (1,300 kg) thrust.
Wingspan	21.9 ft (6.7 m)
Wing Area	140 sq ft (13 sq m)
Length	26.6 ft (8 m)
Height	NA
Weight, Empty	5,181 lb (2,350 kg)
Weight, Takeoff	8,362 lb (3,793 kg)
Crew	1
Speed, Cruise	541 mph (871 km/h)
Speed, Landing	108 mph (174 km/h)
Speed, Top	677 mph (1,090 km/h) at 19,686 ft (6 km)
Rate of Climb	4.5 minutes to 19,685 ft (6 km), and 12 minutes to 39,372 ft (12 km)
Radius of Operation	994 miles (1,600 km) at 39,372 ft (12 km)
Service Ceiling	45,934 ft (14 km)
Armament	4x30 mm MK 108 cannon, rockets, and an oblique-firing gun
Bomb Load	NA
Flight Duration	1.8 hour at 39,372 ft (12 km)

PROJECT 1107A

Entered into the Air Ministry's "Amerika Bomber" project, the P.1107A was a proposed four-turbojet-engined version of Messerschmitt AG's exceptional aerodynamically smooth Me 264 prototype piston-powered heavy bomber. The Me 264's origin goes back to the early 1940s, when the Air Ministry requested feasibility studies from German aircraft manufacturers for a bomber capable of flying nonstop to the United States and back. It was not known at the time whether the United States would join in the war against Germany, but the Air Ministry wanted to prepare for what they considered a remote possibility. Therefore the airplane was scheduled to be used for propaganda purposes by dropping leaflets in the United States before any such decision was made. Its use for spotting conveys for submarines was also contemplated. Of the designs submitted - Focke-Wulf's Fw 300, Junkers' Ju 290, and Messerschmitt AG's Me 264 - only Messerschmitt's proposal was new and fresh, the others being modified versions of existing aircraft. Messerschmitt AG was able to obtain a contract for several prototypes, the others did not,

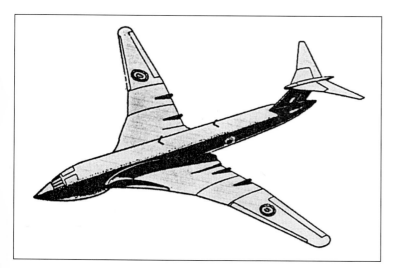

A pen and ink drawing of the Handley-Page "Victor 8" bomber showing its wing root air intakes and crescent wing.

Secret Aircraft Designs of the Third Reich

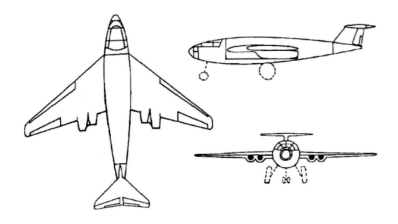

Siegfried Knemeyer, chief of technical development at the RLM (Udet's old job), was told by Göring in late 1944, to have the aircraft companies resubmit for a second "Amerika Bomber" competition in order to reach the 11,000 km needed.

and work began in 1941 on the Me 264 V1. In December 1942 it made its first flight.

The Me 264 V1 was about 3/4 that of the B-29 in size and it was one of the most aerodynamically clean all-metal aircraft to be designed in Germany during World War II. Cigar-shaped, with a circular-section monocoque fuselage and sweptback wings having a high aspect ratio of 14.6, the Me 264 had a rounded, glazed nose enclosing the crew.

After America entered World War II in December 1941, the Air Ministry commissioned various studies for an even more advanced version of the Me 264. However, it was concluded that a conventional piston-powered bomber attack on the United States, even though it might be successful, would not materially harm American's ability to produce weapons of war. Some thought was given to force a negotiation settlement of the war by dropping one of Germany's atomic bombs on one or more of the major metropolitan cities of the United States, such as New York City, Detroit or Washington, DC.

To achieve the greater speed and distance required to carry out such a mission, Messerschmitt AG suggested to the Air Ministry that the basic shape of the Me 264 be retained, but that the aircraft be converted to turbine power. Four BMW 003D or two BMW 018 turbojet engines would be paired and hung beneath the main spar on each wing. Messerschmitt now began calling their turbojet-powered Me 264, the P.1107A. With a turbojet-powered P.1107A, the Air Ministry hoped for better range, speed, and fuel economy. Because of its projected great speed of 621 mph (1,000 km/h), the aircraft could dispense with much of the armament intended for use on the prototype Me 264 such as engine protection and top, side, rear, and dorsal turrets; the de-icing systems for the propeller engines also would not be needed. A new Multhopp-style tailplane was to be mounted at the top of a single vertical rudder fin to clear the turbine's exhaust. On the Me 264 the tail surfaces were of the twin vertical tail type somewhat reminiscent of the B-24 arrangement.

"The Fat One," Hermann Göring, had hoped that airplane designs such as the Me P.1107.2, and the Junkers Ju 290 would be able to achieve the specifications set out for the "Amerika Bomber" project, namely 11,000 km (6,835 miles) non-stop.

With the failure of the second "Amerika Bomber" competition, Oberst Knemeyer spoke to his friend Walter Horten who had not been invited to submit bids in either competition, about designing an all-wing airplane which could achieve the 11,000 km (6,835 mi) range. The Horten's first attempt at the "Amerika Bomber" project was the Ho 18-A, which was able to achieve the range required by Göring.

314

Messerschmitt AG - Augsburg, Germany

The P.1107A was never built. The Air Ministry determined that the P.1107A, although showing improvement in speed and range over the Me 264 bomber, was still not quite fast enough and would not be able to obtain the altitude needed to reach the United States safely without being shot down by existing US fighters. The Air Ministry felt that an all-wing aircraft might be a better choice and ultimately chose the Horten P.18B for further consideration.

Specifications: Project 1107A

Engine	4xBMW 003 turbine engines, each having 2,425 lb (1,100 kg) of thrust, or 2xBMW 018 turbojets, each having 7,500 lb (3,400 kg) of thrust.
Wingspan	141 ft (43 m)
Wing Area	1,370 sq ft (127sq m)
Length	69 ft (21 m)
Height	NA
Weight, Empty	51,509 lb (23,360 kg)
Weight, Takeoff	95,700 lb (43,400 kg)
Crew	
Speed, Cruise	NA
Speed, Landing	NA
Speed, Top	621 mph (1,000 km/h)
Rate of Climb	NA
Radius of Operation	9,300 miles (15,000 km)
Service Ceiling	49,872 ft (15.2 km)
Armament	None
Bomb Load	8,818 lb (4,000 kg)
Flight Duration	24 hours

Rear view of the Junkers Ju EF 140 showing its six turbines mounted in two clusters of three engines each.

Reimar returned to Göttingen, re-designed his Horten Ho 18A and took the design to Göring. Göring accepted and instructed Knemeyer to find acceptable facilities and labor for immediate construction. This aircraft was to be known as the Horten Ho 18B.

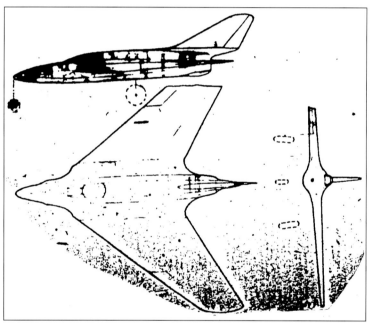
The proposed Messerschmitt Me P.1111 was to have been tailless with a large vertical rudder sweeping out aft of the turbine exhaust. The cockpit was to end in a form of a projectile nose.

Göring accepted the Horten Ho 18A for immediate construction and instructed the Horten brothers to build the aircraft in conjunction with Messerschmitt and Junkers. In the process of discussions, Junkers wanted Reimar to place a large vertical fin on his all-wing bomber with an attached hinged rudder. Junkers was now calling it the Junkers Ju EF 140. Angered, Reimar walked out of the committee.

Secret Aircraft Designs of the Third Reich

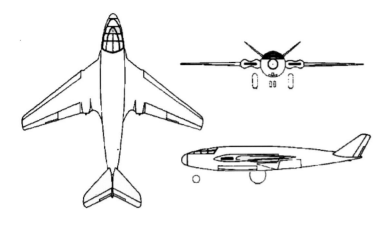

PROJECT 1107B

The P.1107B was the second of two early designs for turbine-powered heavy bombers conceived by Messerschmitt AG for use in raiding cities along the United States' east (Atlantic) coast. It was proposed in November 1944. The Air Ministry wanted a highly fuel-efficient, long-range bomber with round-trip capability of at lest 4,350 miles (7,000 km). Plans called for the Luftwaffe to make air raids on New York City, Detroit, and Washington, DC. Other projected targets included other northeastern industrial cities and some inland cities such as Chicago and the steel mills around Pittsburgh. To achieve the required range, the turbine-powered bombers were to start their flights from occupied France and make refueling stops at the Azores in the north Atlantic.

Plans for raiding American cities had been drafted as early as April 1942. At that time the Air Ministry envisioned using Me 264s powered by BMW 801 E (in powerpack 801 TC form) piston engines. These plans languished after Germany became bogged down in heavy fighting on the Russian Front. Then as the United States continued to supply war material to England and the USSR, the Air Ministry reconsidered its earlier plans for attacking the heavy industrial areas of America's northeastern seaboard. Messerschmitt AG proposed that the task be undertaken by their P.1107B.

To obtain the required range and fuel economy, the P.1107B was to come with circular-section, cigar-shaped fuselage with no unbroken lines. The cockpit was to be blended into the nose, and the entire frontal area glassed over. To obtain better aerodynamic efficiency, Messerschmitt AG designers planned to mount four turbine engines as close to the fuselage sides as possible.

Air intakes were designed into the wing leading edges like those on their P.262 HG-2. Overll, the P.1107B design gives evidence of Messerschmitt AG's designers to achieve greater aerodynamic efficiency, better fuel economy, and greater range by reducing air resistance and drag through greater streamliing. Control devices were to be standard ailerons and flaps, but a butterfly V-tail was to be used to clear the exhaust line of the turbines and to achieve improved aerodynamic efficiency. The entire fuselge of the P.1107B was to be used for carrying its cargo of 8,818 lb (4,000 kg) of high-explosive bombs. A simple tricycle undercarriage composed of two large single wheel main wheels and twin nose wheels completed the design.

Messerschmitt AG's Fernbomber (long-range bomber) project did not make it beyond the design state; it was dropped from consideration by the Air Ministry in February 1945 in favor of proposed all-wing designs as the Horten P.18B which was expected to have superior performance. Design of the P.1107B was handled by Messerschmitt AG's Paris design operations and was headed by Ludwig Bölkow. With the liberation of France in August 1944, the Messerschmitt AG works came to be occupied by the French, who captured design plans for the proposed P.1107B. Although the French did not appear to incorporate any of the P.1107B's design features in their post-war bomber or passenger designs, this Messerschmitt project looks a great deal like the bombers in England's 1960-era Handley Page "Victor" series, which had the turbines in pairs in the wing roots. A similar placement of the turbines also was seen in the Hawker Siddeley Comet transport series of aircraft in the 1950s.

Specifications: Project 1107B

Engine	4xHeinkel-Hirth HeS 011 turbine engines, each having 2,866 lb (1,300 kg) of thrust.
Wingspan	57 ft (17.4 m)
Wing Area	646 sq ft (60 sq m)
Length	56 ft (17 m)
Height	NA
Weight, Empty	30,100 lb (13,653 kg)
Weight, Takeoff	67,693 lb (30,703 kg)
Crew	NA
Speed, Cruise	355 mph (572 km/h)
Speed, Landing	NA
Speed, Top	547 mph (880 km/h)
Rate of Climb	NA
Radius of Operation	4,350 miles (7,000 km)
Service Ceiling	45,931 ft (14 km)
Armament	NA
Bomb Load	8,818 lb (4,000 kg)
Flight Duration	NA

PROJECT 1108

The Messerschmitt AG P.1108 was a proposed 800 to 1,000 mph (1,287 to 1,609 km/h), all-wing. Schnellbomber or fast bomber with a projected range of 10,000 miles (16,093 km).

By late 1944 the tailless designs of Alexander Lippisch, the all-wing planform of the Horten brothers, and the delta all-wing projects of Heinrich Hertel of Junkers were being viewed by the Air Ministry as the aerodynamically efficient shapes of the future. With support from Hermann Göring, the Air Ministry was now selecting tailless, all-wing, and delta planforms time and again over conventional designs submitted in various design competitions. The P.1108 reflects the revolution Messerschmitt AG's design team went through in less than a year's time in response to the Air Ministry's call for a fast "Amerika Bomber."

Initially, Messerschmitt AG designers had responded with a turbojet-powered version of their Me 264, updated with sweptback wings and a Multhopp T-tail to keep the controls out of the turbines' exhaust. This was their P.1107A. When this design failed to interest the Air Ministry, Messerschmitt's designers developed an even more modified version: the P.1107B with its four turbines buried in the wing roots as close as possible to the fuselage sides (the T-tail had been changed to butterfly tail believing that the butterfly tail was more efficient). The P.1108

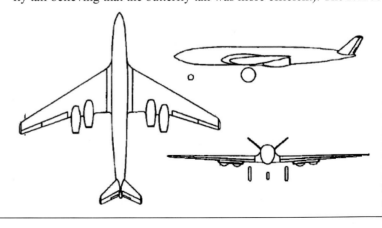

was as aerodynamically clean as any in the world at that time. But again, Messerschmitt AG failed to capture the Air Ministry's interest. Göring said the failure had as much to do with Messerschmitt's personality as anything else, and he ordered the Air Ministry to award prototype construction to the Horten brothers for their P.18B. At that point Messerschmitt AG engineers came back with the P.1108, in essence an all-wing delta design that represented an attempt to incorporate the best of Lippisch's and the Hortens' design ideas.

The P.1108 was to have one of the thinnest wing cross-sections of any aircraft built up to that time - only 6.6 ft (2 m). The wing was to have no protrusions (cockpit canopy, turbines, or cannon) of any kind. Although the cockpit was to extend beyond the wing's leading edge by about 9.8 ft (3 m), its small diameter of only 6.6 ft would have allowed it to merge right in to the wing's leading edge. No vertical control surfaces such as a rudder was to be employed; control was to be achieved through devices built into the wing. Even the four BMW 018 turbines, each with an anticipated 7,500 lb (3,400 kg) of thrust each, were to be completely housed in the P.1108's wing. Turbine exhaust was to exit through openings in the wing's trailing edge. No humps or bulges were visible, and no protrusions or drag-creating air intakes were planned. Instead, air intakes were to be located in the wing's upper surface in the form of openings, or drains, leading directly to the turbine engine beneath the surface. An alternative was to locate similar sculptured openings on the wing's undersurface. A tricycle landing gear was to be used; the wheel's were to turn 90 degrees while retracting, so that they could rest flat inside the wing.

Messerschmitt AG officials had hoped to begin work on a prototype by mid 1946, but this is doubtful because design drawings had only begun when Germany surrendered in early May 1945.

Specifications: Project 1108

Engine	4xBMW 018 turbojet engines, each having 7,500 lb (3,400 kg) of thrust.
Wingspan	71.2 ft (21.7 m)
Wing Area	NA
Length	51 ft (15.5 m)
Height	6.6 ft (2 m)
Weight, Empty	NA
Weight, Takeoff	35 to 40 ton (3,175 to 3,629 kg)
Crew	NA
Speed, Cruise	640 to 800 mph (1,030 to 1,287 km/h)
Speed, Landing	NA
Speed, Top	800 to 1,000 mph (1,287 to 1,609 km/h)
Rate of Climb	NA
Radius of Operation	10,000 miles (16,093 km)
Service Ceiling	49,871 ft (15.2 km)
Armament	NA
Bomb Load	4 to 5 ton
Flight Duration	NA

PROJECT 1112

The Messerschmitt AG P.1112 was the last turbine-powered fighter design to come out of Woldemar Voigt's advanced design office at Oberammergau prior to the end of World War II. It was a proposed tailless aircraft with a sweptback vertical rudder. With this tailless fighter planform and along with their P.1108 Schnellbomber design, Messerschmitt AG was chasing the growing trend by the Air Ministry in selecting more and more tailless, all-wing and delta turbojet-powered aircraft for the Luftwaffe.

Willy Messerschmitt had never been particularly interested in tailless or all-wing aircraft, although he did show some interest in hiring

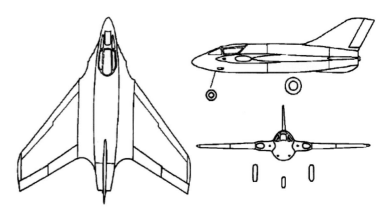

the Horten brothers in 1938. For a time in 1944, Messerschmitt AG sought to take over the Horten Flugzeugbau but that idea was quickly stopped by Milch. Messerschmitt then hoped that his P.1101 and its two subsequent modifications, the P.1106 and P.1110, would work out. However, the P.1101 turned out to be a design failure. For all practical purposes Messerschmitt AG never had a successful turbojet-powered aircraft after their Me 262, which had been drafted up in 1938. With the proposed tailless P.1112, Messerschmitt hoped that he might again come up with the right combination of elements that would make for a successful fighter.

Design studies for the P.1112 began in January 1945. It was Willy Messerschmitt who instructed his design team to make the aircraft tailless. His chief designer, Woldemar Voigt, opposed the tailless configuration, thinking that a tailless fighter would not be able to adjust to the violent changes of trim it would experience as it approached the speed of sound. But Messerschmitt believed that since the Air Ministry was picking tailless aircraft planforms over conventional types, such as his P.1101, P.1106, and P.1110, he had to join the trend in order to be competitive.

The first tailless study to take shape, the P.1111, had wings swept back 45 degrees at the quarter chord line. The wings were made wide of chord to form a near-delta planform, but had considerable trailing edge sweepback. A single sweptback vertical fin and rudder extended beyond the rear of the fuselage. Wing controls consisted of elevons, inboard split flaps, and outboard leading-edge slats. There were air intakes in the wing roots, and slightly curved ducts passed either side of the cockpit to the turbojet engine.

The fuselage arrangement had the main battery of two 30 mm MK 108 cannon in the extreme nose, with the nose wheel beneath, retracting aft. The pilot's cockpit followed, and behind this was the ammunition for an auxiliary battery of two additional 30 mm MK 108 30 cannon.

The wind tunnel model of the Me P.1111/1112 proposed series. April, 1945.

Secret Aircraft Designs of the Third Reich

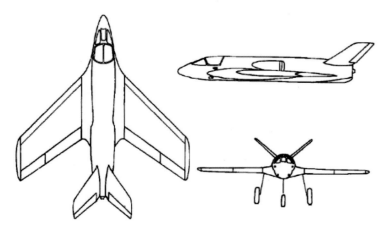

The Northrop X-4.

Pen and ink sketch of the proposed Messerschmitt Me P.1111.

The proposed Messerschmitt Me P.1112 with a "V" tail. Air intakes for the single turbojet are located each side of the aft fuselage. *Courtesy of Luftwaffe Secret Projects: Fighters, 1939-1945.*

The single Heinkel-Hirth HeS 011 turbine of 2,866 (1,300 kg) thrust occupied the aft portion of the fuselage. The air intakes were elliptical in their shape, located well outboard from the fuselage for better efficiencies during high-speed flight. The fuel system had not been entirely worked out, but it was anticipated that all fuel tanks would be located in the wing, principally in the outer panels.

Messerschmitt AG submitted the P.1111 design to the Air Ministry in March 1945. Several objectionable features were noted during the Air Ministry's review, and Messerschmitt immediately redesigned it coming to call the change the P.1112. In the redesign the pilot's cockpit was moved to the extreme nose of the aircraft so that the nose profile blended into the windshield, with no re-entry. The main battery of two MK 108 cannon was moved out into the center sections of the wing. Overall wing area was reduced to 22 sq m (from 301 ft 2 or 28 sq m), mainly to save weight and to obtain an increase in speed. As much fuel as possible was to be placed inside the fuselage in order to obtain better

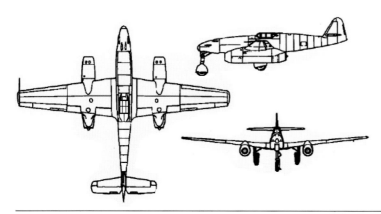

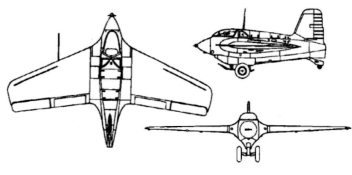

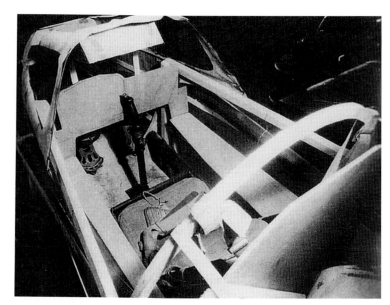

The proposed Messerschmitt Me P.1112 wooden mockup found at Oberammergau by US Army Intelligence, April 1945, shown is the cockpit area.

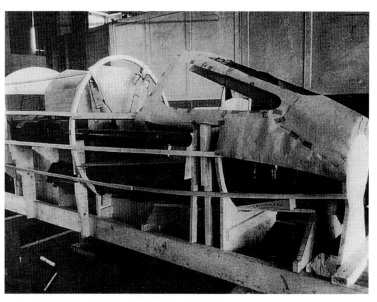

The proposed Messerschmitt Me P.1112 wooden mockup found at Oberammergau by US Army Intelligence, April 1945, shown is the forward fuselage.

protection for the self-sealing fuel tanks from enemy cannon fire.

It is not known how the Air Ministry responded to Messerschmitt's P.1112 proposal. According to Voigt, Messerschmitt AG had intended to proceed privately with the P.1112 unless further tests and calculations showed the speed gain due to the reduction in wing area to be negligible; in that case the lower stalling speed and better takeoff of the P.1111 would probably make the P.1111 preferable. Complete design studies and structural calculations on the P.1112 had not been completed prior to the end of the war. The specifications presented below are for its sister aircraft, the P.1111.

Specifications: Project 1111

Engine	1xHeinkel-Hirth HeS 011 turbojet having 2,866 lb (1,300 kg) of thrust.
Wingspan	29.9 ft (9.1 m)
Wing Area	301 sq ft (28 sq m)
Length	21.3 ft (6.5 m)
Height	NA
Weight, Empty	6,041 lb (2,740 kg)
Weight, Takeoff	9,440 lb (4,282 kg)
Crew	1
Speed, Cruise	507 mph (816 km/h)
Speed, Landing	102 mph (165 km/h)
Speed, Top	634 mph (1,020 km/h)
Rate of Climb	4,688 ft/min (1,423 m/min)
Radius of Operation	932 miles (1,500 km)
Service Ceiling	45,934 ft (14 km)
Armament	4x30 mm MK 108 cannon
Bomb Load	None
Flight Duration	1.8 hours

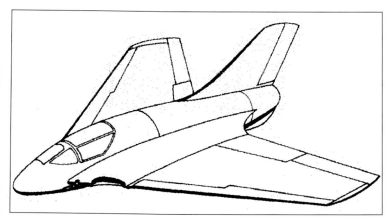

Proposed Messerschmitt Me P.1112. This variation was tailless incorporating a tall vertical rudder with leading edge wing root air intakes. *Courtesy of Luftwaffe Secret Project: Fighters 1939-1945.*

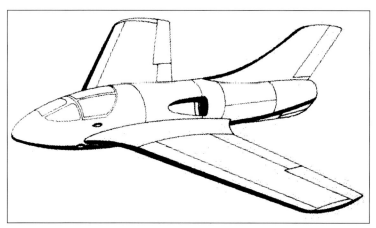

Proposed Messerschmitt Me P.1112, tailless with tall vertical rudder. Fuselage side openings are air intakes for the turbine. *Courtesy of Luftwaffe Secret Projects: Fighters 1939-1945.*

15

Eugen Sänger LFA - Ainring, Germany

Secret Project and Intended Purpose:
Project Sänger Orbital Bomber - Manned liquid rocket drive and/or Lorin Duct-powered 2,284 mph Intercontinental bomber.

History:
Had this rocket-drive orbital bomber proposed by Dr.Ing. Eugen Sänger in 1944 been built, it would have made Germany the first country in the world to send pilots to the edge of space and back down again over an intended target on the opposite side of the globe. Its pilots would have been the first astronauts and this project the first intercontinental bomber.

Eugen Sänger was a specialist on ramjets, also known as Lorin Ducts, at the Luftfahrt Forschungs Anstalt (LFA, or Aircraft Research Establishment) located at Ainring. It was Sänger who along with Adolf Busemann, Otto Lutz, and others at LFA's München facilities, in the early 1940s carried out considerable research on the so-called "stove pipe," or ramjet engines. In principle, the ramjet was a less complex aircraft propulsion system than either the turbine or the bi-fuel liquid rocket drive. Inducted air becomes compressed when it meets a central contraction in the "stove pipe;" at the point of contraction, fuel was ignited - and the resulting thrust was very powerful. Although the ramjet was very economical to construct and operate at this stage of development, it had a very major drawback. The aircraft had to be moving at a forward velocity of about 149 mph (240 km/h) before there was sufficient pressure inside the ramjet for the system to operate. Thus, an auxiliary engine was required to get the ramjet up to its operating speed.

As the war continued the Air Ministry looked for engines suitable for high-altitude flight able to reach ultrasonic speeds of Mach 3 (2,284

Eugen Sänger, left, and Ferdinand Brandner. Brandner was the former Chief of Engine Gas Piston Development at Junkers Jumo, Dessau. Post war he was arrested by the Soviets and charged as a war criminal. In prison for two years, he was released and worked on the Jumo 022K turboprop at Kuibyshyev. Photo taken in France in the late 1950s.

Dr.Ing. Irene Sänger-Bredt.

The flight path of Eugen Sänger's proposed stratospheric bomber. Upon lift off, the bomber would climb to a height of 990 mi (145 km) and begin circling the earth to reach its target (the United States) on the opposite side of the globe. Sänger believed his rocket-powered bomber would have a range of 14,600 mi (23,500 km).

View of the Sänger orbital bomber wind tunnel model. 1944-45.

Wind tunnel model of Sänger's proposed orbital bomber. 1944.

Artist's rendering of the proposed Sänger orbital bomber in flight high over the earth.

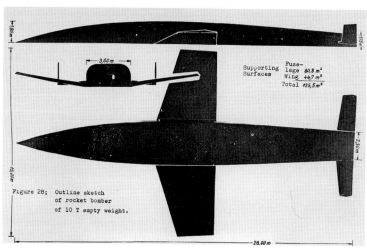

A pen and ink sketch of Sänger's proposed orbital bomber. 1944.

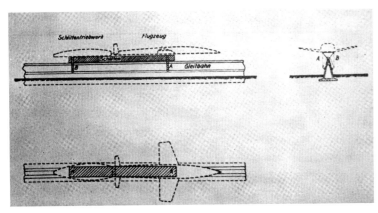

A drawing showing the general arrangement for launching one of Sänger's orbital bombers. The propulsive units behind the bomber (vehicle) consist of V2 rocket engines strapped together forming essentially a single unit.

Eugen Sänger's orbital bomber would have been launched on a ramp similar to France's SFECMAS Ars. 5.5 pilotless target aircraft shown in this photograph. Sänger went to work for SFECMAS after World War II and continued his research work on ramjet-powered aircraft.

mph or 3,675 km/h) or more. Turbine engines and ramjets were out of the question for the lack of sufficient thrust, and the Air Ministry officials were looking to LFA and scientists such as Sänger for solutions. Sänger's research by 1945 was centering on chemical fuels that would allow liquid rocket engines to operate with peak efficiency at altitudes up to 20 miles (32 km) and over 12,428 miles (20,000 km) in range.

The ultimate aim of the Air Ministry in Sänger's research and that of his assistant Irene Bredt whom he married after the war, was development of a liquid rocket propulsion system for a bomber aircraft that could strike the United States (which is situated at Germany's antipode, that is, exactly opposite Germany on the globe), drop its bomb, presumably an atomic bomb, and return to the point of departure by circling the earth. One of the most secret of the Air Ministry projects, the so-called "Orbital Bomber Project" was to be tried and only with existing ramjet engines and with a minimum of expense. Sänger believed the project would succeed if the aircraft sort of "ricocheting in and out of the dense

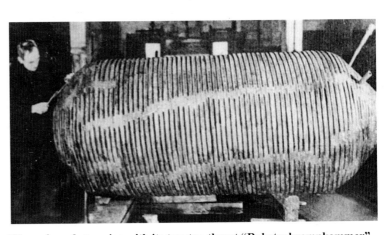

Sänger's rocket engine with its two ton thrust "Raketenbramnkammer".

Cut-away scale model of Sänger's rocket engine.

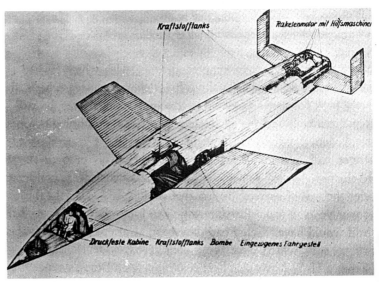

A cutaway of the Sänger-type orbital bomber, showing the pilot's compartment, the propellant tanks, and the rocket engine. The bomb load for the Sänger bomber would have been on the order of ten tons.

Werner von Braun in the early 1940's Germany.

Fully assembled V2 liquid rocket showing its engine and plumbing, found by US Army troops in an underground factory. An engine like this one was to be used by Sänger for his orbital bomber.

A Soviet version of the orbital bomber by Sänger in the early 1950s. State of the art technology was not available for this ambitious effort.

layers of the atmosphere above the earth," and he spent considerable time experimenting with various ramjet configurations and the fuels to power it at an altitude of 20 miles.

By 1944 Sänger had developed a shape for the proposed orbital bomber ramjet plane. Taking into account the ultrasonic speeds the bomber would be required to sustain, the nose of the aircraft was to be very slender. Overall the proposed bomber design took on the appearance of a projectile. The wings were to be relatively small, their only real purpose being to assure stability of the aircraft in flight and upon landing. Nor was the plane's tail really needed during a flight of Mach 3 and at an altitude of 20 miles above the earth. Its only purpose was to give the aircraft stability at subsonic speeds and during landings and Sänger placed a vertical surface on each end of a relatively straight horizontal elevator. A wide separation of the vertical surfaces (rudders) was necessary, Sänger believed, due to the high temperature generated by the aircraft's twin ramjet-drive engines. Separation also was called for because of the highly corrosive nature of the chemicals consumed in the ramjet engines; engineers wanted to avoid the chemicals coming in contact with the tail if at all possible.

The proposed bomber, expected to weight as much as 100 tons, was to be launched by use of a catapult. Plans called for the bomber to be placed on a monorail-type launcher and propelled on this device, beginning with a gradual incline and ending with the craft leaving the monorail at nearly a vertical assent. For landing, the aircraft would use built-in skids or a proposed retractable landing gear.

It is known that Sänger's orbital bomber was to have been powered by liquid chemical fuels, but no details on the proposed engine or its propellants are available. Sänger once mentioned the possibility of developing an atomic-powered engine for his intercontinental bomber aircraft, but it is unlikely that the technology was available at the time the aircraft was being considered. Sänger believed that after the war the rocket-drive airplane might be used to transport people and cargo in much the same way the USA space shuttle is planned for in the years ahead. An avid enthusiast over the possibilities of space travel, Sänger believed that his rocket-driven bomber could also form the basis for interplanetary travel. Wernher von Braun's V-2 liquid fueled rockets would be the vehicle for travel to the stars, Sänger fantasied, while his

Troisiéme Congres, Aeronautique Europeen 1958, Bruxelles, 22-27 September 1958. Left to right: Professor Baade, Dr. Cordes, Dip.Ing. Schneider, and Dip.Ing. Gerlach.

Horst Schneider, Covina, California, 1989.

Sänger experimental ramjet mounted atop a Dornier Do 217 leaves a trail of fire during one of its test flights. 1942.

own rocket ship could be used to ferry people and cargo between von Braun's V2 rocket ship and the earth.

Sänger's proposed orbital bomber project did not get beyond the design study stage. After the war Sänger and his assistant, later wife, Irene Bredt went to France, where they worked on conventional turbine-powered aircraft.

The following is an excerpt from an interview between the author and Horst Schneider in June 1989, Covina, California. "Eugen Sänger's orbital bomber had little support during World War II. This was because it was just too far out on the horizon of possibility. Sänger's efforts, however, were mostly paperwork with some fundamental experimentation related to expected problems with body heating during reentry, and friction heat on the skids during landing. On project drawings I saw in Russia, the vehicle provided for the installation of two of von Braun's V2 liquid rocket engines. The Soviets also considered the use of solid propellent rockets. Eugen Sänger's wife, Irene Sänger-Bredt, performed research work with metal dispersions for solid propellent rockets to increase specific impulse. In Russia I saw German GEKADOS (geheime Kommamdosache) reports on the Sänger Project translated into Russian which had been issued by ZIAM and ZAGI. To my knowledge the Russians never picked up on Sänger's work. However, the Soviets offered to make any German a very rich man if they knew where Sänger could be found post-war. Sänger probably was aware that there was a

An operational Sänger experimental ramjet mounted on a truck undergoing tests in 1942.

Eugen Sänger's experimental ramjet mounted atop a Dornier Do 217 in flight out of Fassberg in 1942. These tests were satisfying enough for the Air Ministry to continue its financial support right up to the end of the war.

Sänger's experimental ramjet testing in one of the DFS' laboratories.

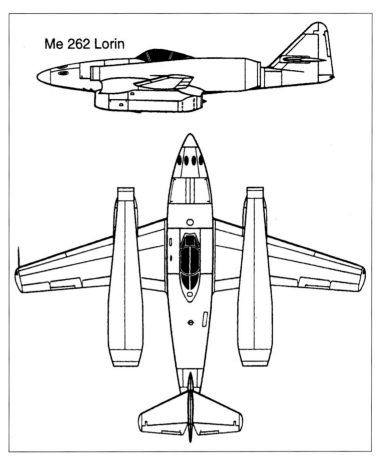

Sänger's experimental Lorin ramjet was even envisioned for the Messerschmitt Me 262. It would have been mounted on top of the standard Jumo 004 nacelles. *Courtesy of WARBIRDTECH Series Volume 6 - Messerschmitt Me 262 Sturmvogel.*

price on his head and left immediately after the surrender for France and was welcomed in the aero-space research community. Also, the General Electric company in the United States picked up on Sänger's work with a similar project to serve as a reusable space booster. That was in the 1950s. I saw a report relating to Sänger's project in Huntsville, Alabama in 1958. The author was a German by the name of Kappus, who had emigrated to the United States shortly after the war with Dr. A. Franz, the former head of the Junkers Jumo 004 jet engine. Dr. Franz was many years the head of the gas turbine development at Lycoming, Connecticut. He knew Sänger."

Specifications: Project Sänger Orbital/Intercontinental Bomber

Engine	multiple liquid fueled rocket and/or ramjet engines. Type and thrust rating unknown.
Wingspan	49.2 ft (150 m)
Wing Area	NA
Length	91.9 ft (280 m)
Height	5.9 ft (18 m)
Weight Empty	11 ton (10,000 kg)
Weight Takeoff	100 ton
Crew	NA
Speed Cruise	963.2 mph (1,550 km/h)
Speed Landing	93.2 mph (150 km/h)
Speed Top	Mach 3 or 2,284 mph (3,675 km/h)
Rate of Climb	NA
Radius of Operation	12,428 miles (20,000 km)
Service Ceiling	104,987 ft (32 km)
Armament	NA
Bomb Load	NA
Flight Duration	NA

16

Heniz G. Sombold (Bley Segelflugzeugwerke) - Naumburg, Germany

Secret Profect and Intended Purpose:
Project Sombold 344 - A manned, single bi-fuel HWK liquid rocket drive aircraft carrying a high-explosive aerial flak bomb up to the altitude of a passing B-24 or B-17 bomber formation, hurling it into the bomber pack, and detonating it by remote control.

History:
Another suggestion for countering USAAF B-24 and B-17 heavy bomber formations, which many Luftwaffe pilots were calling "Feuerspeidenden Berg" (fire-spitting mountains), was offered to the Air Ministry by Heinz Sombold. In January 1944 Sombold, who worked in the engineering office of Bley Segelflugzeugwerke at Naumburg/Saale, proposed a plan for breaking apart these "Fire-spitting mountains" by having an interceptor hurl an explosive projectile into the middle of the bomber pack. The nonpowered projectile, falling downward in a 45 degree arc would self-detonate when it reached the same altitude as the bombers, scattering the formation so that Luftwaffe pilots could fight them one-on-one.

B-24's especially B-17s were particularly hard to bring down when flying in close formation due to their massive defensive firepower as a group. Luftwaffe pilots were finding it harder and harder to take on a tightly grouped formation with any hope of success. Individual B-17s could be shot out of formation, but the rest of the pack moved on to their targets; and taking out one or two B-17s often meant heavy losses for the attackers. The Luftwaffe was learning that there would never be enough planes or pilots to stop the bombing runs unless a way could be found to break apart the bomber formations, thereby destroying their

This reconnaissance photograph shows the airstrip at Peenemünde where jet and rocket-powered aircraft were first sighted by the British. Note bomb craters among buildings. Since the airfield was primarily for research and development and not an active Luftwaffe base, the facilities were considered more important as a target than the runway itself.

A B-17's twin 50 caliber tail gun.

A hard-working B-17 proud of its bomber runs over Germany as indicated on its nose. Nine fighters shot down in addition to its bomb loads.

A B-17's twin 50 caliber top turret gun.

The nose of a B-17 showing its four 50 caliber machine guns.

group defense. Several Me Bf 109s or Fw 190s could take on a lone B-17 with good results, but how could the formations be broken up to set up a one or two-on-one situation?

Sombold felt his P.344 was the answer. The designation 344 had been previously assigned to Ruhrstahl AG for their X4 air-to-air missile project, however, the Air Ministry canceled it in late 1944. Sombold proposed that his interceptor powered by a HWK 509 bi-fuel, liquid rocket drive could carry aloft in its nose a 1,102 lb (500 kg) high explosive bomb. The interceptor would hurl the missile into the middle of the bomber pack. The projectile would detonate itself through the use of a time-fuse when it reached the altitude of the B-17 bomber formation. After the detonation, Sombold believed, B-17s would be scattered all over the sky and the ones still flying could be picked off by Luftwaffe fighters. This interceptor aircraft would not be involved in the fighting after releasing its projectile, but would return to its take off location, land on a built-in skid, refuel, and then rearm.

However, Sombold's P.344 differed from Bachem's Natter project. The Natter was to be launched vertically, then attack the bomber forma-

LEFT: USAAF bombers could be brought down, Heinz Sombold knew, but it generally required a direct hit by flak guns. Bomber formations were too heavily armed, and Luftwaffe fighters were generally no match for B-17s if they strayed to close to a tight formation. For example, a formation of fifty B-17's with twelve 50 caliber guns each meant a total fire power of 600 guns. RIGHT: A B-17 losing its starboard wing from a direct hit by flak.

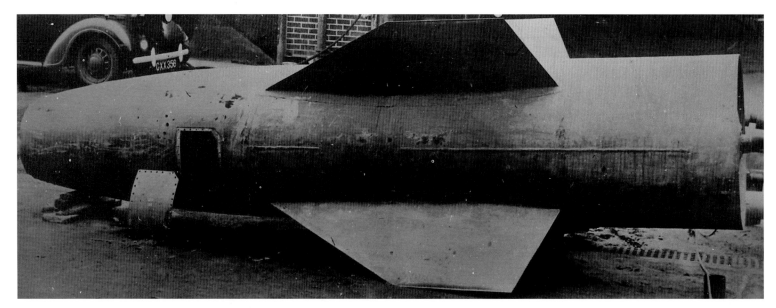

The warhead which Sombold wanted his pilots to hurl into bomber packs would work by scattering them all over thr sky to be picked off by individual fighters.

The "Fallbobe" also known as SD-1400-X, potent air-to-air missile. A popular name among the Allies was "Fritz X."

tions, for example, with a volley of 24 73-mm Föhn RZ-73 anti-aircraft rocket missiles fired from their containers in the nose of the aircraft. The P.344 would be towed by a "mother-aircraft" into a position that would be a suitable starting point for the attack. Sombold believed that a suitable altitude would range between 6,562 ft (2 km) and 13,123 ft (4 km), depending on the altitude of the attacking bomber formation that particular day.

By January 1944, when Sombold presented his P.344, which really was an air-launched flak weapon, the Air Ministry had a variety of air-to-air missiles and projectiles that could have been adapted easily to Sombold's interceptor. Although few details are available on the P.344's specifications or performance or on the projectile it was intended to carry, the projectile was probably an advanced version of the "Fritz Z" designed and manufactured by Ruhrstahl AG in Westfalen.

German efforts in the field of rockets dated back to the 1920s and 1930s. The Treaty of Versailles had prohibited the manufacture and development of artillery and powered aircraft, but not rockets. Numerous individuals carried out experiments with rockets, but it was not until Hauptmann Walter Dornberger in 1929 began sponsoring the work of

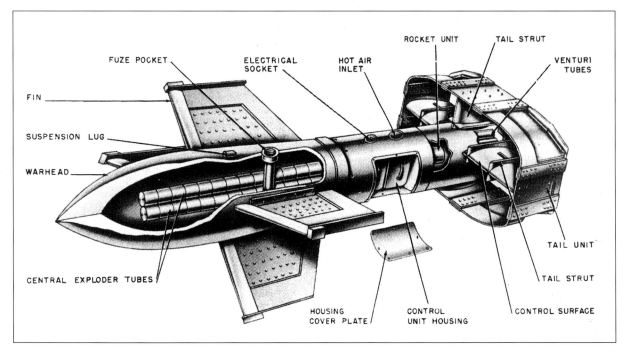

The internal components of the "Fritz X" are shown in this illustration.

Heniz G. Sombold (Bley Segelflugzeugwerke) - Naumburg, Germany

Customary position of a "Fritz X" when being carried by a twin engine Luftwaffe bomber, such as a Dornier Do 217.

Heinz Sombold believed that air-launched missiles had reached an advanced state of readiness by late 1943, especially "Fritz X" guided missiles when launched by a Dornier Do 217. In this photograph a Do 217 has launched a "Fritz X" missile, which is on its way to attack the Italian battleship "Roma" on 9 September 1943. After being hit, the "Roma" caught fire and later sank between Corsica and Sardinia.

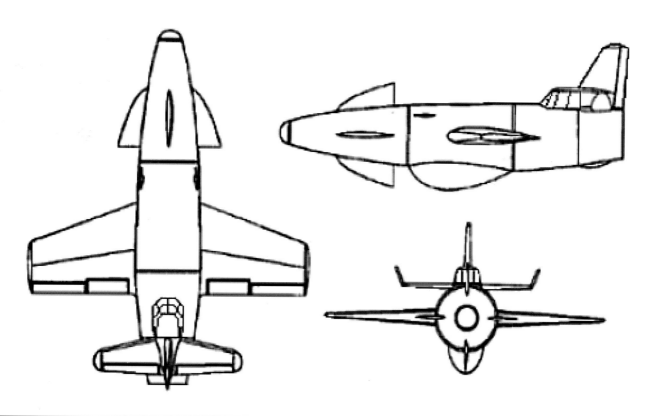

Secret Aircraft Designs of the Third Reich

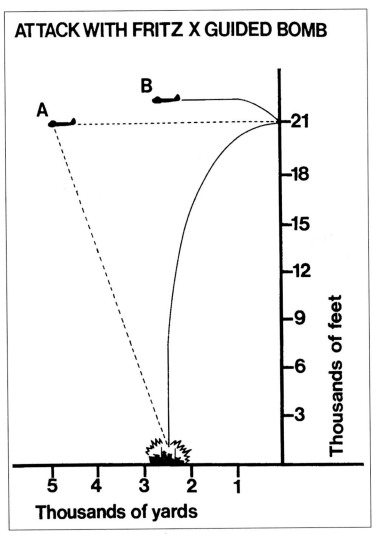

The preferred approach when attacking with a "Fritz X" guided bomb. The carrier aircraft at approximately 21,000 ft (64 km), would drop its bomb about 3,000 yds (2,743 m) behind the intended target. As the bomb lost its forward momentum, it would be right on target.

A ship receives a mortal blow from the attack of just one "Fritz X" guided bomb.

X-4 - a fin-stabilized guided missile with a proximity fuse warhead developed specifically for use by fighter planes against enemy bomber formations.

promising experimenters that rocket work was conducted in a coordinated and systematic manner. In 1932 Dornberger set up a test facility at the Wehrmacht's Kummersdorf-West proving ground outside Berlin. It was here that individuals such as Wernher von Braun began work on what would later become the V-2 rocket program.

Then, in early 1939, the Henschel Flugzeugwerke in suburban Berlin established a group for engineers to study the possibilities of developing air-to-ground missiles which could be guided onto targets from some distance away. One of the individuals Henschel hired was the brilliant turbine development engineer at Junkers, Dr.Ing. Herbert Wagner. About the same time Dr.Ing. Max Kramer of the Deutsches Versuchsanstalt für Luftfahrt (DVL) began development of bombs that could be controlled during free-fall to attack armored ships. This led in turn to a series of controlled-trajectory bombs, code-named "Fritz Z;" when these projectiles were dropped from altitudes of up to 19,685 ft (6 km), they could be guided to their targets and then exploded.

By 1942 the Luftwaffe's flak regiments were becoming increasingly effective against the high-flying B-17 bombers. But there were not enough fighters to go after every bomber, and the bombers' combined anti-aircraft fire was so awesome (up to 12 pair of 50-caliber guns

Heniz G. Sombold (Bley Segelflugzeugwerke) - Naumburg, Germany

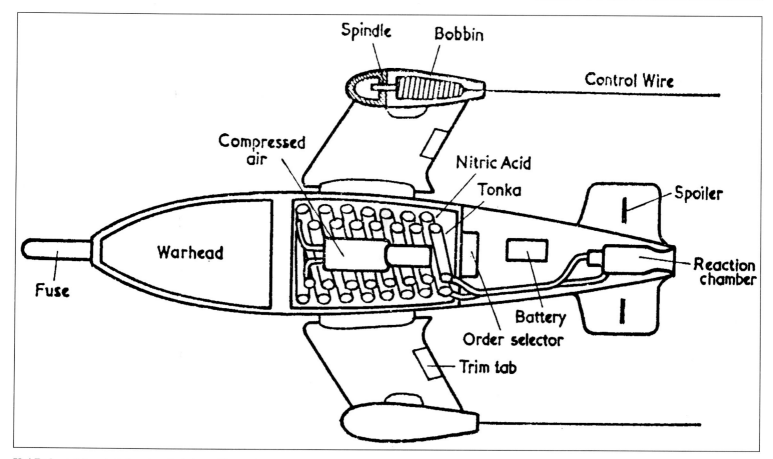

X-4 Ruhrstahl rocket showing its internal layout.

each) that frequently more Luftwaffe fighters were lost than Allied bombers. It was during this time that official interest was concentrated on developing some ground-to-air missiles and other flak-type rockets.

Early in 1943 work began on air-to-air guided missiles that could be carried by the several turbine-powered aircraft under development such as the Me 262. The projectile to be used on the P.344 may have been a Fritz Z-type projectile or something similar or an extension to the work being conducted on ground-to-air missiles, air-to-ground missiles, air-to-air missiles, and aerial-launched flak-type explosive bombs.

Specifications: Project Sombold 344

Engine	1xWalter HWK 509A bi-fuel liquid rocket drive providing 3,748 lb (1,700 kg) of thrust.
Wingspan	NA
Wing Area	NA
Length	NA
Height	NA
Weight, Empty	NA
Weight, Takeoff	NA
Crew	1
Speed, Cruise	NA
Speed, Landing	NA
Speed, Top	NA
Rate of Climb	NA
Radius of Operation	NA
Service Ceiling	NA
Armament	NA
Bomb Load	1xFritz Z-type aerial-launched high explosive projectile
Flight Duration	NA

17

Zeppelin Abteilung Flugzeugbau GmbH - Fredrichshafen, Germany

Secret Project and Intended Purpose:
Project "Rammer/Sideswiper"- Manned, solid fueled rocket-powered interceptor for the purpose of ramming or sideswiping individual B-17 heavy bombers with the purpose of cutting off their tail assembly, wing sections and so causing the bomber to go out of control and crash down.

History:
Despite the fact that the Air Ministry had allowed Focke-Wulf to develop some design ideas for a self-destroying aircraft and a mother-ship to carry them to in the vicinity of their intended victims: RAF Lancasters and USAAF B-24's and B-17s , the attitude throughout Germany was that such kamikaze-style raids would never really be permitted. Yet there

Ferdinand Graf von Zeppelin, 1838-1917. The namesake of the Zeppelin Abteilung Flugzeugbau. The original Zeppelin Airplane Works had changed its name to Dornier.

A Fieseler Fi 103 manned flying bomb of the type Hanna Reitsch wanted to use for suicide missions.

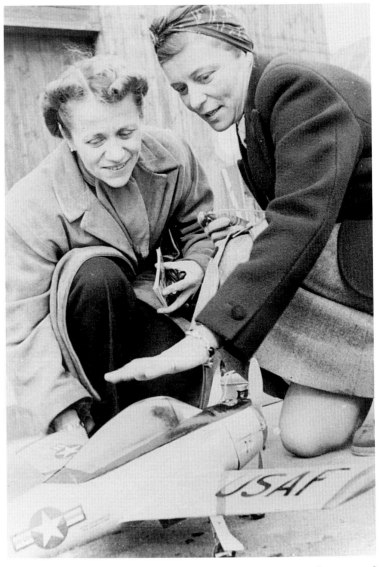

Hanna Reitsch (left), with friend talking over the features of a powered P.51 scale model.

Zeppelin Abteilung Flugzeugbau GmbH - Fredrichshafen, Germany

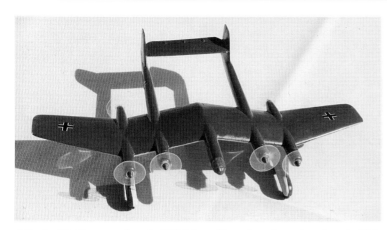

A Focke-Wulf proposal to the RLM for a "kamikaze" carrier aircraft. This example became Focke-Wulf's proposed design for a mother carrier. Scale model by Reinhard Roeser.

The "kamikaze" aircraft would have selected as targets, aircraft fuel tankers.

The Zeppelin "Rammer" as proposed. Scale model by Steve Malikoff.

This B-17 bomber suffered damage from a collision. It was hoped that the Zeppelin "Rammer" could inflict such mortal damage.

were individuals within the Luftwaffe who were willing to volunteer for suicide attack missions. In late 1944 with defeatist thinking growing stronger day after day, a few so-called "true believers" in Nazism began signing pledges stating that they were willing to sacrifice their lives in order to destroy large, important targets with the hope that such action late in the war might have enough impact to stop the conflict through a negotiated settlement.

Kamikaze activities were never taken seriously in war-time Germany and that angered advocates such as famed Deutsches Forschungsinstitute für Segelflug (DFS) pilot Hanna Reitsch. She firmly believed that suicide flights against specific targets, such as some of the Allies' mightiest warships, with a 2-ton bomb might bring the war to an end through some type of negotiated truce, thereby saving German Nazism and hundreds of thousands of German lives an Allied invasion would cost Germany. Reitsch continued to argue the merits of kamikaze action right up to war's end.

While kamikaze-type actions were receiving a cool reception in official Air Ministry and Luftwaffe circles, work went on to find an alternative means of destroying enemy ships at sea and major facilities such as industrial centers, water reservoirs, and electric power generating sites. Heinz Sombold's Project 344 seemed a better way of taking out ships at sea as well as major infrastructure facilities without requiring the pilot to lose his or her life in the process, but the Air Ministry

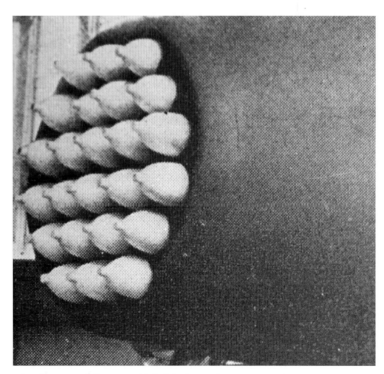

The "Rammer's" nose held R4M air-to-air rocket missiles shown here installed in the nose of a "Natter."

The Zeppelin "Rammer" also had a stinger in its long proboscis. Scale model by Steve Malikoff.

A rear view of the Zeppelin "Rammer." Notice the small hole in the tail below the vertical rudder, was for the exhaust/thrust of its Schmidding solid-fuel rocket drive. Scale model by Steve Malikoff.

still had to deal with the massive Allied bomber squadrons. Since suicide attacks against heavily defended B-17 bombers, for instance, were ruled out by the Air Ministry, the Zeppelin Abteilung Flugzeugbau at Friedrichshafen in 1945 suggested an alternative, a solid fuel-rocket propelled aircraft which could break up the bomber packs and possibly destroy one or two in the process by actually flying right into one or more of them. The idea was to build a small aircraft so strong that could ram or sideswipe a B-17 and cut off its tail, for instance, sending the stricken bomber falling to earth out of control. After this feat was accomplished maybe it could easily find another victim. Afterward this Zeppelin "rammer" or "Sideswiper" would land on a built-in skid, refuel and rearm, then be sent aloft again to hurl its load of 14 R4M rockets into another B-17 heavy bomber formation.

The Zeppelin Rammer/Sideswiper was to be relatively small aircraft a constant chord wingspan of 16.4 ft (5.4 m) and a length of 16.1 ft (4.9 m). It was to have been so solidly built that it could be expected to hit a bomber and slice cleanly through its fuselage or tail section, thereby sending the enemy plane into an uncontrollable and fatal dive to earth. In essence, Zeppelin's proposed aircraft was not a self-destroyer, but one that would ram or touch enemy bomber aircraft in flight and in doing so inflict mortal damage. The designers at Zeppelin suggested that their little Sideswiper would be towed or carried to an altitude above that of the incoming bomber formation. Upon reaching the appropriate altitude, it would release itself, glide down to the bomber formation, and begin its attack by firing its 14 nose-mounted R4M air-to-air rockets into the advancing bombers. A container holding the R4M anti-aircraft projectiles was fitted to the nose in an arrangement similar to that of the Ba 349 Natter. Following the firing of its projectiles, the pilot was to select a single bomber and attempt to slice through its fuselage or cut off its tail assembly through a sideswiping maneuver. Power for making the sideswiping was to come from a single Schmidding 2,205 lb (1,000 kg) solid-fuel rocket tube. This rocket tube, which was attached to the wing's rear spar, terminated in a venturi. Overall it was 5.4 inches in diameter and 9 feet (2.7 m) long.

The Schmidding solid-fuel rocket was to be used to accelerate the aircraft and help it achieve the altitude it needed in order to sideswipe as many bombers as possible before the rocket drive had spent all its fuel. Immediately after touching the enemy bomber and cutting through it, the Zeppelin aircraft was supposed to fly away.

On completing its mission, the Zeppelin pilot who flew the aircraft in a prone position, was to immediately reduce its altitude, lower the aircraft's retractable landing skid, then put down on any convenient short stretch of flat ground. Later the Zeppelin would be retrieved by a flatbed truck and carried back to its takeoff base, Since the aircraft was to have such a short wingspan, no transporting problems were anticipated even though it would be public road network. The reason was that by 1944-1945 due to lack of gasoline, there were no personal automobiles on German roads.

To enable the Zeppelin to remain intact after hitting a B-17 bomber, this proposed rocket-propelled fighter was to be strongly built. Steel

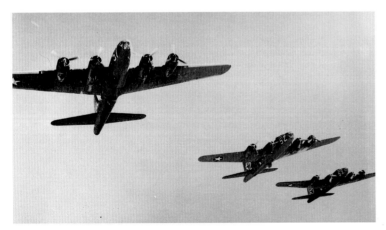
The "Rammer's" R4M air-to-air rocket missiles would be fired into a formation of B-17s, not necessarily to destroy them in flight, but to disable and later be picked off by fighters.

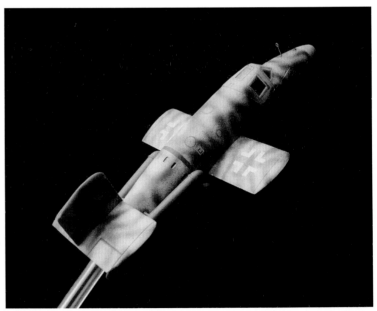
The "Rammer's" competition for acceptance at RLM was the Bachem "Natter." Scale model by Jamie Davies.

Also in competition the Heinkel He P.1077 "Julia." Scale model by Jamie Davies.

The ultimate goal of the Zeppelin "Rammer" was total destruction of the bomber. Although pictured here is a B-29, Zeppelin wished the same results for the current force of B-17s over Germany as well as when B-29s were sent, too.

was to be used throughout its airframe. The wings were to be especially strong; three successive tubular spars were to run the entire length of the wing, from tip to tip. This solid construction would, the Zeppelin engineers believed, enable the aircraft's wing leading edge to slice cleanly through the more delicately built B-17's fuselage and tail section. To protect the pilot from the massive firepower of a B-17 and other aircraft, the cockpit windscreen was to be formed of 80-mm thick plate glass, and the side window panels of 40-mm thick plate glass. The Zeppelin's nose and its wings' leading edges were to be fabricated of 20-mm and 30-mm cannon-proof steel plate. With its small overall size, high speed while making sideswiping passes, the aircraft would have been a difficult target for B-17 gunners.

Time ran out before Air Ministry officials could make up their minds on proceeding with Zeppelin's proposed "Ramming" or "Sideswiping" B-17 destroyer. No prototypes or mock-ups are known to have been built.

Specifications: Project Rammer

Engine	1xSchmidding 533 solid-fuel rocket drive engine with 2,205 lb (1,000 kg) of thrust.
Wingspan	16.4 ft (5.4 m)
Wing Area	132 sq ft (12.3 sq m)
Length	16.1 ft (4.9 m)
Height	NA
Weight Empty	NA
Weight Takeoff	NA
Crew	1
Speed Cruise	NA
Speed Landing	NA
Speed Top	600 mph (970 km/h)
Rate of Climb	NA
Radius of Operation	NA
Service Ceiling	NA
Armament	14 R4M nose-mounted 55-mm anti-aircraft rocket projectiles
Bomb Load	None
Flight Duration	NA

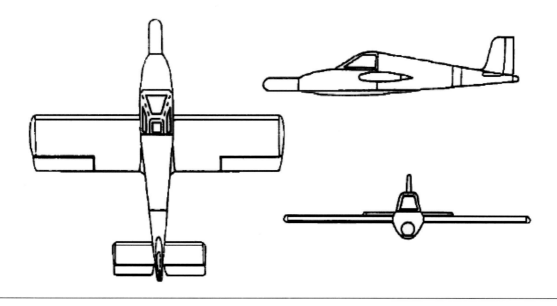

Epilogue:
The Legacy

When Ernst Heinkel demonstrated the world's first turbine-powered airplane in August 1939, Luftwaffe and Air Ministry officials smiled at one another. They were impressed, yet for all the trouble and expense Heinkel had gone through to put that aircraft into the air, all he received was a scolding. "Why are you wasting your time and private resources on a turbine engine and that noisy rocket with running boards?" they asked. Undaunted, Heinkel pleaded with Air Ministry officials that Germany was close to being the first country in the world to have a superior air weapon, a turbine-powered fighter aircraft. "That may very well be," they told Heinkel, "but we'll have to wait until the Führer takes care of the Polish issue. Then there will be sufficient time to develop your promising turbine, just in time for the next war in 1942-43." As events unfolded, the issues set in motion by the "Polish matter" would not be settled until Germany's unconditional surrender in May 1945 and Heinkel would not be called on to produce a turbine powered fighter until late in 1944, when his firm designed and built the He 162 "People's Fighter" in the record time of 90 days.

During the six years between Heinkel's first turbine powered flight in August 1939 and Germany's surrender in May 1945, politics at the highest level mitigated against rational exploitation of a truly advanced resource of war. If anyone should bear the blame, it should be Hitler, with his mistaken belief that any war over Poland would be short, so short that it would be over before science could provide anything that would really make a difference. Hitler took little interest in airplanes and much less interest in promising turbine projects, beginning with Heinkel's He 280 twin-turbine fighter and later Messerschmitt's superb Me 262 turbine-powered fighter until the declining fortunes of war forced him to do so in 1944. By then it was too late. Hitler was desperate and looking for, indeed, demanding, miracle air weapons to save him and the country from impending defeat. Aircraft manufacturers responded to his belated official interest and support with an unending flood of proposals for turbine-powered fighters, bombers, and rocket-driven interceptor planes, each hopeful that one of his designs might turn out to be a war-winning weapon.

But most aircraft designers and engineers knew it was impossible at that point to introduce new aircraft that would make any difference to the war effort. They knew development work took at least 2-1/2 to 5 years, sometimes longer, and they had learned from experience (Heinkel, for instance, with his He 111 and He 162) that trying to speed up the process only led to greater difficulties. Then why did the aircraft engineers and designers continue to work on the new designs? Hans Wocke, famed Junkers aerodynamicist gives us an idea of how German aircraft design people approached what they knew to be a hopeless task. He recalls working on proposed turbine-powered aircraft designs at Junkers' Dessau facilities in 1944 and 1945. Just before lunch time, work in the entire factory would stop and the employees would file out along the Elbe river into the countryside, away from the factory. As they ate their lunches, they would see groups of USAAF B-17 bombers overhead, usually on their way to a bombing target somewhere beyond Dessau (though the Junkers' facilities were also a frequent target). An hour later the B-17s would again pass over Dessau on their way back to England. At this point, Junkers' personnel would file back into the sprawling factory and resume their work. From time to time Air Ministry and Luftwaffe officials would visit Dessau and meet with the project design group and inquire into their progress. The visitors invariably would bring along several brand new turbine-powered aircraft designs and request that the Junkers group commence work on the project immediately. Wocke and his colleagues knew the war would be over before these proposed fighters, interceptors, and bombers could be built. Whether the people at the Air Ministry recognized this, Wocke never knew; but he and his colleagues throughout the German aircraft industry kept on designing because it was their profession - designing and building superior air machines to the best of their ability, regardless of what was going on around them.

In most cases the dedication of these men had nothing to do with Nazism. They, like their colleagues in other aircraft factories in other nations, were motivated mainly by a desire to advance the so-called "state-of-the-art" in aircraft design. Yet each one had his own reasons. Some were passionate scientists and very conscious of a duty to science and the Reich. Others, such as Ernst Heinkel and Willy Messerschmitt, were motivated by their massive egos; both loved fast planes.

In contrast, Kurt Tank, Walter Blume, and the Junkers team of Ernst Zindel, Hans Wocke, and Heinrich Hertel were not so sure that racing improved the breed. These men wanted more than anything else to create aircraft that would carry their loads from one place to another in the most economical and safest possible manner. A fourth group of designers cared neither about race planes nor about mass-produce work a day aircraft; they spent their time working on aerodynamic problems that would face all aircraft in the years ahead as turbine engines allowed greater speeds. Such men as Alexander Lippisch, Woldemar Voigt, and Hans Multhopp, for example, addressed airflow problems. They and other German designers believed there was a hurdle to be over come, the phenomenon of "compressibility," if aircraft were to achieve supersonic speeds. They frequently used the analogy of a snowplow to explain the problem of compressibility and the nature of their task ahead. When plowing a street covered with heavy snow, a snowplow may encounter a place where the snow is so packed that no amount of power

Epilogue: The Legacy

can push it aside. Air pressure over the wing at sonic speeds also tends to pack. Although it does not impede the forward motion of a wing such as packed snow does a plow, it does build up detrimental shock waves. These waves destroy the smooth flow of air around the wing. The breakdown of this so-called "laminar flow" destroys the lifting power of the wing and sets up serious vibrations in the structure.

One solution to the problem of compressibility was thought to lie in developing a wing so thin that compressibility was minimized as forward velocity was increased. Another course was to find ways to remove the layer of air closest to the moving wing (the boundary layer) during high-speed flight, thereby minimizing compressibility problems. Aerodynamicists sought to control the boundary layer of air by experimenting with different types of wingsweep and different amounts of wing thickness.

In addition to trying to solve the problem of compressibility, these designers and engineers were working to determine the best placement of the turbine on the aircraft. Hanging the turbine in a nacelle under the wings created a large amount of drag on the plane's forward motion, yet putting it inside the fuselage called for a great deal of duct work, which in turn cut down on the turbine's operating efficiency. The difficulties Heinkel had encountered with his pioneering He 178 when he placed his turbine engine inside the fuselage were well known, the duct work had robbed the engine of much of its efficiency, and prevailing opinion was that turbines should be hung beneath the wings. Thus, on later designs such as the Arado Ar 234 reconnaissance bomber, the Messerschmitt Me 262 fighter bomber, the Heinkel He 280 fighter prototype, and the Junkers Ju 287 heavy bomber prototype, turbines were placed under the wings. Engineers promised more thrust and better operating efficiency on the second generation of turbine engines, such as the HeS 011, and aircraft designers once more began thinking about placing the turbines, if not inside the fuselage, at least in the wing root and as close to the side of the fuselage as possible. Messerschmitt advanced project designer Woldemar Voigt had redesigned their Me 262 by placing the turbines in the wing roots and were anticipating a big increase in speed and overall performance. In addition, Voigt's proposed heavy bomber project, the Me P.1107B, was scheduled to have its four turbines placed in the wing roots (in the 1950s the British Vickers "Valiant" MK.1 bomber would be designed with an overall configuration similar to the proposed Messerschmitt Me P.1107B aircraft).

In anticipation of the time when turbines would be so powerful that only one would be required to power a fighter, German engineers designed an original single-engine configuration. For instance, Voigt, with his P.1101, established the barrel fuselage with a nose inlet for the turbine. Other features of what came to be a classic German configuration were a short jet-pipe, a fuselage-mounted main landing gear, a nose cockpit placed above the inlet duct, and a standard tail carried on a boom extending out above the turbine's exhaust nozzle. Other manufacturers adopted this configuration for their single-engine fighters and interceptors, including Focke-Wulf in its early turbine designs and Heinkel in its P.1078 series aircraft.

Some engineers and designers were certain that a more efficient shape could be found and began exploring the possibilities of delta wing, tailless, and even all-wing aircraft. Alexander Lippisch had determined through his considerable research that the delta wing might be the most

A monument to the victims of the air war in Europe, is located in Dresden, Germany. It is a memorial to those civilians who perished in the two-day fire bombing of Dresden in 1945. In German it reads: "To these hundreds of thousands without names buried here who died in the fiery hell of 13-14 February 1945. A testimonial of man's inhumanity to man."

efficient planform for high-speed flight. His experiments had shown that compressibility could be minimized by using a very thin wing. However, making wings very thin was a problem because to be strong enough they would have to be almost solid metal or solid wood). Lippisch found that instead of actually making the wing thinner, he could achieve the same effect by increasing its chord (that is, the distance from the leading edge to the trailing edge). When this was done, the wing took on a delta, or triangular shape. Eventually the wings got to be so wide that the trailing edge ended up right at the rear of the aircraft. Lippisch found that the tailplane could be removed and that pitch/roll could be controlled by hinged trailing edge surfaces, called elevons, which acted as elevators. Since elevators were now unnecessary, all of Lippisch's designs came to be tailless. This was a turning point in aerodynamic circles throughout Germany. Lippisch's research and experiments in wind tunnels and with prototypes showed that his aircraft had superior aerodynamic qualities. They had almost no tendency to stall and suffered no loss of control at extreme angles of attack. Moreover, structure weight was held to a minimum, greater fuel economy was possible, the aircraft experienced less flutter, thinning of the wing, and fewer trim changes were required while passing from transonic to supersonic speeds.

While aerodynamic engineers throughout the aircraft community were looking with favor on Lippisch's delta-wing configuration as a means of controlling compressibility, the Horten brothers, Walter and Reimar, advocates of the all-wing aircraft, were struggling to keep up with the fast-paced aircraft research. By late 1944 they, too, were abandoning the all-wing as a configuration for military aircraft and were about to place vertical fins/rudders on all their fighter designs. The difficulties they had encountered with their all-wing Horten Ho 9V2 in terms of directional stability showed that not enough was known about how to make this configuration a stable aircraft during combat conditions. Although the all-wing remained the ideal configuration in the minds of most aerodynamicists, people recognized in 1944 that suitable flight controls were, perhaps, decades away, and the entire design concept in Germany was shelved.

Although in 1944 and 1945 many aircraft manufacturing firms began proposing turbine-powered airplanes full of features suggested by Lippisch's work on tailless design, it was not a style embraced by everyone. Hans Multhopp of Focke-Wulf and Woldemar Voigt of Messerschmitt tended to avoid the head-long rush of designers to imitate Lippisch's tailless designs. Their own extensive research had led them to believe that a conventional aircraft could be developed which would lack the disadvantages of the delta wing, poor visibility and poor low-speed maneuverability. Multhopp and Voigt suggested using highly sweptback wings, and automatic wing slots in the wing's leading edge, to improve the stability and control, with all surfaces swept at 35 degrees. Voigt had applied his ideas on his Messerschmitt Me 262; he extended them in his design for the Me P.1101, which was destined not to fly, being in the "fitting out" stage when American army troops captured Messerschmitt's Oberammergau facilities in April 1945. Multhopp, working independently yet along the same lines as Voigt, extended the state-of-the-art beyond the level established by anyone else in the world, and his ideas were embodied in the radically new Focke-Wulf Ta 183. Time ran out before Focke-Wulf could complete the Ta 183 but the Soviet Union did virtually duplicate many of the aircraft's design features (with the exception of its T-tail, although prototype models of the MiG-15 did carry it). Experiments with the MiG-15 in the Korean War proved the viability of Multhopp's ideas; it handled superbly as a fighter and needed only a few refinements from Multhopp's original design calculations.

The Ta 183 is only one of many German turbine-powered designs that left a legacy to the world of aircraft design. German influence on postwar aircraft has been particularly strong in the United States, England, and the Soviet Union, primarily because these three nations had the largest aircraft manufacturing industries at the end of the war, but also because these countries sought out German aircraft designers and engineers after the war and collected virtually all their research material, documents, and interesting airplanes.

A great many turbine-powered postwar aircraft owe their ancestry to the designs of Alexander Lippisch. It is in the United States that Lippisch's influence is most evident. His delta-wing DM-1 glider was the direct basis for all of Convair's delta-wing designs, including the XF-92, the XF2Y-1 Sea Dart, the YF-102 Delta Dagger, and the B-58 Hustler bomber. Lippisch's ideas on delta-wing designs also lived on in the Swedish SAAB 210 Draken, the British Boultan Paul P.111, and the American Douglas F3D Sky Ray. These aircraft, in particular, are descendants of Lippisch's proposed P-12 design. Two of the postwar aircraft Lippisch always pointed to with pride as a continuation of his delta-wing ideas were the British-French Concorde and the United States' space shuttle Columbia.

Woldemar Voigt, Messerschmitt's brilliant design chief, lived to see his design ideas, especially those embodied in his prototype P.1101, take shape in the SAAB J-29, Bell aircraft's X-5, and North American's XF-86. Voigt's P.1111 survived in the form of the De Havilland 108, and the Northrop X-4. Messerschmitt's proposed P.1107B, a large multi-turbined bomber, influenced Handley Page's "Victor" bomber, design specifications which were issued in 1948.

Aircraft in several countries were patterned after Walter Blume's Arado Ar 234 turbine-powered reconnaissance/bomber, including, in the United States, the Glenn Martin Aircraft Company's XB-48 bomber, Convair's XB-46 bomber, and North American's B-45C reconnaissance/bomber. Several bomber projects in the USSR also had their roots in the Ar 234B, including the Ilyushin 26 and 28 bombers. The British Avro 707A fighter shows the unmistakable influence of Arado's Nachtjäger NJ-1 proposed fighter.

The work of Hans Wocke, Ernst Zindel, and Heinrich Hertel at Junkers was not widely influential throughout the world, primarily because until very late in the war Junkers manufactured aircraft designed by other firms. Junkers' six-turbine swept forward-wing bomber, the Ju 287, did influence several Soviet Union prototype bomber aircraft such as the Type 150 and the Type 151, for awhile, but neither of these aircraft progressed beyond the prototype stage.

The aircraft of the Horten brothers left no lasting legacy among aircraft designers after the war. The French firm of Payen did fly a turbine-powered prototype design, the Pa-49, that resembled the Horten's Ho 9B and the Ho 13B, combining a delta-type wing with a large dorsal fin/rudder. After the war, while Reimar Horten was working for the Argentinian Instituto Aerotècnico, he designed and built an aircraft (the IAe 37) patterned after the brothers' proposed wartime, single-seat, delta-wing supersonic fighter with twin vertical tails, the Ho 10C. However, it did not go beyond the prototype.

The designs from the creative mind of Richard Vogt of Blohm and Voss left no immediate legacy to postwar aircraft designers and engineers, but several of his proposed designs from the 1940s have found their way into prototype designs in the 1980s. These late-bearing Vogt designs include his proposed P.202 "Swiveling Wing." Its distinctive feature was a wing that would rotate 35 degrees horizontally on its axis, giving a backsweep to the starboard wing and a forward sweep to the port wing. The National Aeronautics and Space Administration (NASA)

Epilogue: The Legacy

built a prototype, turbine-powered model of an aircraft with a similar pivoting wing, the AD-1. NASA officials were impressed and suggested that the concept be explored further, for possible application to commercial aircraft. Vogt's forward-sweeping wing, a feature he patented and later proposed on his Bv P.209 fighter, has been proposed for combat aircraft by the North American Aviation Company.

Perhaps Hans Multhopp, the design genius from Focke-Wulf, had more influence on postwar aircraft design than any other German designer, with the possible exception of Alexander Lippisch. It was the Soviets who took Multhopp's Project 183 project and translated its revolutionary ideas into a whole generation of Mikoyan and Gurevich designs which came to be known as the MiG-15 through MiG-19 turbine-powered fighter aircraft. When Kurt Tank went to work for Argentinean dictator Juan Peròn in 1946, Tank took all Multhopp's Project 183 calculations and he and other Focke-Wulf designers built Argentina's second turbine-powered fighter prototype, the Pulqui Dos. It was a virtual copy of the P.183. Multhopp continued his aircraft design work after the war in England for the Royal Aircraft Establishment (RAE), drafting plans for a proposed RAE supersonic project along with fellow Focke-Wulf design colleague Martin Winter. Afterward, in 1949 Multhopp's ideas concerning lifting surface for swept wings were applied to English Electric's P-1, the first supersonic fighter aircraft in the world. Multhopp's lifting surface theories also contributed to NASA's initial thinking in the late 1950s when designers began laying down ideas for the proposed space shuttle Columbia.

There are no monuments to the many German aviation designers of World War II, as there are for the pilots who flew their aircraft. But the designers did leave a legacy for their colleagues in other countries to follow. In some cases, their aircraft survived (Walter Blume's Arado Ar 234, Walter and Reimar Horten's Ho 229, Woldemar Voigt's Messerschmitt Me 262, for example) - though none are flying. But the real legacy has come in the form of design features - Hans Multhopp's T-tail, Richard Voigt's "pivoting wing," Alexander Lippisch's delta wings, to name a few - that are commonplace on today's aircraft and, indeed, on tomorrow's aircraft as well. These are visual reminders that there was a time in the not too distant past when German airframe designers and developers of turbine engines had the imagination and skill to conceive and produce aircraft which were the most advanced in the world.

Glossary

A

A-Stoff - Liquid oxygen.
Aero - Of or for aeronautics or aircraft.
Aerodynamics - The science dealing with air and other gaseous fluids, and with the forces acting on bodies when they move through such fluids, or when fluids move against or around the bodies.
Aerodynamicist - A person who is trained in the science of aerodynamics.
Aeronautics. A general term applied to everything associated with or used in any way in the study or design, construction, and operation of an aircraft.
Aerospace - Of or pertaining to, both the earth's atmosphere and space.
Afterburner - A device for augmenting the thrust of a turbojet engine by burning additional fuel in the exhaust gases of the turbojet.
AG - Aktiengesellschaft (Joint Stock Company).
Ailerons - Pairs of control surfaces, normally hinged along the wingspan, designed to control an aircraft in a roll by their different movements.
Air brake - Any device used primarily to increase the drag or air resistance of an aircraft at will.
Aircraft - Any vehicle designed to be supported by the air, being borne up either by the dynamic action of the air on the surfaces of the structure or object, or by its own buoyancy.
Air drag - The drag exerted by air particles on a moving object.
Air frame - The structural framework and covering of an airplane, rocket, etc., not including the engine and its related accessories.
Airfoil - A part having a flat or curved surface, as a wing, rudder, or elevator, designed to keep an aircraft up or to control its movement by reacting in certain specific ways to the air through which it moves.
Air force - The aviation branch of the armed forces of a country.
Airplane - An aircraft, heavier than air, which is kept aloft by the aerodynamic forces of air acting on its wings and is driven forward by a screw propeller or by other means, such as turbojet propulsion. (Altered from earlier Aeroplane).
Airport - An area of land or water, including buildings and facilities, that is used or intended to be used for the landing and takeoff of aircraft.
Air scoop - A hood or open end of an air duct or similar structure, projecting into the airstream about a vehicle in such a way as to use the motion of the vehicle in capturing air to be conducted to an engine.
Airspeed - The speed of an aircraft relative to its surrounding air mass.
Airstrip - An unimproved surface adapted for takeoff and landing of aircraft, usually having minimum facilities.
Air-to-air missiles - (AAM) A missile launched from an airborne carrier to impact a surface target.
Airworthiness certificate - A certificate issued by a representative of a government's aviation administration after an aircraft has been inspected and found to meet the requirements of the aviation administration's regulations to be in a condition for safe operation.
All-weather fighter - A fighter aircraft having radar devices and other special equipment that enables it to intercept its target in dark or daylight weather conditions that do not permit visual interception.
Altitude - Height as measured above sea level.
Amerika Bomber - German engineer Dr.Ing. Eugen Sänger's idea to develop a winged rocket that glided back to Earth at the completion of its mission: to deliver bombs on New York City and Washington, DC. Never developed, however, the idea re-emerged in February 1945 to use a Horten Ho 18B all-wing flying nonstop round trip carrying the German Atomic Bomb. Never developed.
Angle of attack - The angle between the leading edge of a wing and the direction of air flow approach. That part of an airplane flight prior to landing.
Arrow wing - A wing planform consisting of a highly swept (60 degree) forward wing section combined with an aft wing section having a sweepback of approximately 45 degrees. The basic arrow-wing configuration consists of the wing as described and a long fuselage including a tail. The complete configuration is therefore in the shape of an arrow with an arrowhead, a slender shaft, and tail feathers.
Aspect ratio - The ratio of a wing's length to its width. The usual aspect ratio of pre-World War II aircraft was about 6 or 7 to 1 (with width measured along the chord). In soaring planes the aspect ratio (span to chord) may be anywhere from 10 to 15 to 1. Wingspans of 60 ft (18.3 m) with a chord of approximately 3.3 ft (1 m) have been noted in some German sailplanes.
Athodyd - Also known as a ramjet. A type of turbojet engine consisting essentially of duct or tube of varying diameter and open at both ends, which admits air at one end, compresses it by the forward motion of the engine, adds heat to it by the combustion of fuel, and discharges the resulting gases at the other end to produce thrust.
Auxiliary landing gear - That part or parts of a landing gear, as an outboard wheel, which is intended to stabilize the craft on the ground but which bears no significant part of the weight.
AVG - Aerodynamische Versuchsanstalt Göttingen (Göttingen Aerodynamic Experimental Establishment).
Axes of an aircraft - Three fixed lines of reference in relation to an aircraft: (1) the longitudinal axis in the plane of symmetry, usually parallel to the axis of the propeller or turbojet and called the thrust line; (2) the vertical axis, about which the plane rotates in yawing; (3) the lateral axis which is perpendicular to the other two. In mathematical discussions the first of these axes, drawn from front to rear, is called the X axis; the second, drawn upward, the Z axis; and the third, running from left to right, the Y axis.
Axial-flow compressor - A rotary compressor having interdigitated rows of stages of rotary and stationary blades through which the flow of air is substantially parallel to the rotor's axis of rotating.

B

B-Stoff - Hydrazine hydrate (hydrazinhydrat).
Base - A locality from which operations are projected or supported. A locality containing installations that provide logistic or other support. A home airfield.
Base leg - A flight path in the traffic pattern at a right angle to the landing runway, off its approach end and extending from the downward leg of the extended runway centerline.
Besatzung - Crew, seating.
Bench check - A functional test of an item to determine condition made at a workshop or servicing bay.
BHP - Brake horsepower, or power delivered at the output shaft of an engine.
BK - Bordkanone (fixed aircraft cannon).
BMW - Bayerische Motoren Werke-München (Bavarian Engine Works-Munich).
Body - The main part or main control portion of an airplane, airship, rocket, or the like, a fuselage or hull.

Glossary

Bombardier - A member of a bomber crew who operates the bomb sight and the release mechanism.

Bomber - An aircraft designed to deliver bombs: light, a bomber designed for a tactical operating radius of less than 1,000 mi (1,609 km) at design gross weight and design bomb load; medium a bomber designed for a tactical operating radius of between 1,000 and 2,500 mi (1,609 km and 4,023 km); heavy, a bomber designed for a tactical operating radius of more than 2,500 mi (4,023 km).

Booster engine - An engine, especially a booster rocket, that adds its thrust to the thrust of the sustaining engine.

Boundary layer - The layer of turbulent air (as in aerodynamics) in the immediate vicinity of a bounding surface such as the surface of a wing.

Boundary layer control - Efforts to minimize the layer of turbulent air immediately above the surface of a wing by withdrawing it from the boundary layer by surface suction or by injecting air into the boundary layer in order to achieve greater lift coefficients.

Bramo - Brandenburgische Motoren Werke (Brandenbug Engine Works).

Buffeting - The beating of an aerodynamic structure or surface due to unsteady flow of air, gusts, and so on, or the irregular shaking or oscillation of a vehicle component due to turbulent air or separated flow.

Bulkhead - A wall, partition, or similar member in an aircraft fuselage, rocket, or similar structure that helps give shape and strength to the structure.

Bv - Blohm und Voss Schiffswerke - Abteilung Flugzeugbau GmbH - Hamburg.

C

C-Stoff - Catalyst (30% hydrazine hydrate, 57% methanol, 13% water) used in the HWK 109-509A bi-fuel liquid rocket-drive engine.

Cabin - In an aircraft, all the compartments used for the carriage of passengers or cargo.

Camouflage - The use of concealment and disguise to minimize the possibility of detection and identification, especially by blending in with the natural environment.

Canard - When used on an aircraft, the horizontal surfaces used for trim and control is forward of the main lifting surface of the wing.

Canopy - A transparent, bubble-like enclosure for the flight crew of an aircraft.

Cantilever wing - A wing built on the principle of a cantilever beam with no external struts or bracing.

Catapult - A power-actuated machine or device for hurling forth an object at a high speed (see also "launch").

Ceiling - The maximum altitude an aircraft is capable of attaining under standard conditions.

Center of gravity - The point within an aircraft around which its weight is evenly distributed or balanced.

Center section - The middle or center section of a wing, to which the outer wing panels are attached.

Centrifugal compressor - A compressor having one or more vaned rotary impellers which accelerate the incoming air radially outward into a diffuser.

Chemical fuel - A fuel that depends on an oxidizer for combustion or for development of thrust, such as liquid or solid rocket fuel. Also, a fuel that uses special chemicals.

Chord - In aeronautics, the distance in length between the leading edge and the trailing edge of an airfoil. Frequently measured in percentages or factions of the chord.

Close air support - Air attacks against hostile targets that are in close proximity to friendly forces on the ground.

Cockpit - A compartment housing the pilot(s) in an aircraft.

Coefficient - A number indicating the amount of some change under certain conditions, often expressed as a ratio.

Compressibility - Effects that occur when local velocities on parts of an aircraft such as wings and tail surfaces exceed the speed of sound, producing shock waves that result in a sharp increase in drag and other effects. Compressibility may be encountered at a Mach number as low as 0.75 (570 mph or 919 km/h, at sea level, and 495 mph or 797 km/h at 35,107 ft or 10.7 km).

Condensation trail - A visible trail of condensed water vapor or ice particles left behind an aircraft in motion through the air. Sometimes called a contrail or vapor trail.

Configuration - The arrangement of parts to form a plan or outline.

Contra-rotating propellers - Two propellers mounted on concentric shafts having a common drive and rotating in opposite directions.

Control surface - An airfoil or part thereof that moves to produce changes in the forces or moments acting on an aircraft in order to control it.

Co-pilot - A pilot, responsible for assisting the first pilot in flying the aircraft.

Coupling - A device for joining adjacent ends or parts of anything.

Cowling - A cover surrounding all or part of a power unit when installed in an aircraft.

Craft - An aircraft or machine designed to fly through air or space.

Crew member - A person assigned to perform duties in an aircraft during flight.

Cruise speed - The velocity required to maintain sustained economical flight.

D

Damp - To level out or retard. Also, to reduce vibration.

Datum line - A baseline or reference line from which calculations or measurements can be made.

DB - Daimler-Benz.

Dead reckoning - Finding one's position from an unknown position by means of a compass and calculations base on speed, elapsed time, effect of wing, and direction.

Dead stick landing - A landing of an airplane without engine power.

Debug - To isolate and remove malfunctions from an aircraft.

Defense classification - One of several possible categories assigned to defense information or material that denote the degree of danger to national security that would result from its unauthorized disclosure. The grades include: confidential, secret, and top secret.

Delta-wing - A triangularly shaped wing on an aircraft.

Design studies - Studies conducted to determine the characteristics of a system needed to satisfy a particular requirement.

DFS - Deutsches Forschungsinstitut für Segelflug (German Research Institute for Gliding Flight).

Dienstgibfellhohe - Operational ceiling.

Dihedral angle - The angle created by two intersecting planes such as where two wings are joined.

Ditching - Controlled landing of a distressed aircraft on water.

Dive - A steep descent, with or without power, in which the air speed is greater than the maximum speed in level flight.

Dive flap - A flap-type air brake used to reduce the velocity of an aircraft.

DLH - Deutsche Lufthansa (German State Airline).

DLV - Deutsche Luftsportverband (German Aviation Sport Union).

Dorsal fin - A vertical surface or fin located on the aft and upper side of a fuselage.

Downward leg - A flight path in the traffic pattern parallel to the landing runway in the direction opposite to landing.

Drag - A retarding force acting on a body in motion through the air.

Drop tank - An external tank designed to be dropped in flight after the contents have been consumed, for example, gasoline drop tanks.

DSV - Deutsche Schiffsbau-Versuchsanstalt (German Shipbuilding Experimental Institute).

Duct - Specifically, a tube or passage that confines and conducts air to the compressor of a turbojet engine, or a pipe leading air to an engine.

Ducted fan engine - An aircraft engine incorporating a fan or propeller enclosed in a duct which ingests ambient air to augment the gases of combustion in the airstream.

Dutch roll - A lateral oscillation with a pronounced rolling component.

DVL Deutsche Versuchsanstalt für Luftfahrt (German Experimental Institute for Aviation).

E

Entwurg - Project.

EF - Erprobungsflugzeug (experimental aircraft).

Ejection seat - A seat capable of being ejected in any emergency to carry the occupant and his equipment clear of the aircraft.

Elevator. A hinged, horizontal control surface used to raise or lower the tail in flight, thereby producing a pitching moment (force) to the airplane. The elevator is usually hinged to the horizontal stabilizers and connected to the pilot's control stick.

Empennage. The rear part of an airplane, usually consisting of a group of stabilizing planes (horizontal stabilizers and vertical fin) to which are attached the control surfaces (elevators and rudders). Sometimes referred to as the airplane tail assembly or tail unit.

Endurance - The length of time an aircraft can continue flying under given conditions without refueling.

Engine - A machine or apparatus that converts energy, especially heat energy, into work. Also called a motor.

Engine pod - The housing for each externally mounted engine on a multi-engine aircraft.

Erector - A vehicle used to support a rocket for flight.
E-STELLE - Erprobungsstelle (proving or test facility).
Experimental model - A model of the complete equipment to demonstrate the technical soundness of the basic idea, also known as a prototype.
External load - A load that is carried, or extends outside the aircraft fuselage, including fuel tanks, armament, missiles, and cargo carriers.

F

Fairing - An auxiliary member or structure whose primary function is to reduce drag on the part to which it is fitted.
Feathering propeller - A propeller having blades that can be rotated in the hub so that its leading and trailing edges are parallel, or nearly parallel, with the line of flight; in case of engine failure, propellers are rotated to decrease drag on the aircraft and also to prevent the propeller from being rotated by the air and possibly damaging the engine.
Fighter - An aircraft designed to intercept and destroy enemy aircraft and missiles.
Fin - A fixed or adjustable airfoil or vane attached longitudinally to an aircraft, rocket, or similar body to provide a stabilizing effect.
Flachenbelastung - Wing loading.
Flak - Fliegerabwehrkanone (anti-aircraft gun).
Flugelflache - Wing area.
Flugelstreckung - Aspect ratio.
Fluggewicht - Flying weight.
Flying - The movement of an aircraft through the air.
Flying Boat - A seaplane supported, when resting on the surface of the water, by a hull or hulls which provide flotation and serve as a fuselage.
Forschungsanstalt - Research establishment.
FuG - Funkgerat (radio or radar set).
Fuselage - The structure of an aircraft which houses the crew and cargo and to which is attached the wings, the tail unit, and the engine mounts. In some designs the tail unit is attached to outriggers or tail booms and the fuselage is simply a streamlined structure for housing the crew, etc., in this case the fuselage generally is called a nacelle.

G

Gasoline, German Aviation: Aviation Grade A3 - light blue color and rated at 80 octane. Aviation Grade B4 - dark blue color and rated 87 octane. Aviation Grade C3 - dark green color and rated at 100 octane.
Gas-turbine engine - An engine incorporating as its chief element a turbine rotated by expanding gases.
Geschwader - Group.
GmbH - Gesellschaft mit beschrankter Haftung (limited-liability company).
Glazing - Glass windows or cockpit canopy.
Gross thrust - The total thrust of a turbojet engine without deduction of the drag due to the momentum of the incoming air (ram drag).
Gross weight - The total weight of an aircraft when loaded, including fuel and crew. Also called takeoff weight.
Ground loop - An uncontrollable violent turn of an aircraft while taxiing, or during a landing or takeoff run.
Guided air-to-surface missile - An air-launched guided missile used against surface targets.
Guided missile - Any missile that is subject to some degree of guidance or direction after launch.
Gun - A cannon having a relatively long barrel and operating with a high muzzle velocity.

K

HWK - Hellmut Walter-Kiel. (Werke).
Hard landing - An impact of an aircraft on the surface, possibly destroying the landing gear and the engines in the process.
Heavy water - (D20) - Water containing hydrogen (deuterium) atoms, used as a moderator in some nuclear reactors because it slows down neutrons effectively.
Height - Vertical distance above some reference point or plane.
High altitude - An altitude above 33,000 ft (10 k).
High-lift device - Devices employed to increase the lift of a basic wing. The most common forms are wing flaps and wing slots.
Höhe - Height.
Höchstegeshwindigkeit - Maximum speed.
Hull - The main structural and flotation body of a flying boat.

Hydrofoil - A surface on a seaplane similar in form to an airfoil, used to facilitate takeoff by providing hydrodynamic lift.
Hypersonic - Pertaining to speeds of Mach 5 (3,800 mph or 6,115 km/h at sea level) or greater.

I

Inlet - An entrance or orifice for the admission of air to an engine.
Instability - The condition of an aircraft if, when displaced form a state of equilibrium, it continues, or tends to continue, to depart from the original condition.
Instrument flight - Flight in which the path and altitude of the aircraft are controlled solely by reference to instruments.
Interceptor - A manned fighter aircraft used for identification and engagement of enemy airborne objects.
Ion engine - An engine that provides thrust by expelling accelerated or high velocity ions. Ion engines use energy provided by nuclear reactors.

J

JATO - Jet-assisted take off. A takeoff utilizing auxiliary jet-producing units, usually rockets, for additional thrust.
Jet-engine - Any engine that ejects a jet or stream of gas, obtaining all or most of its thrust by reaction to the ejection.
Jumo - Junkers Motorenbau (Junkers engine works).

K

KG - Kommandit-Gesellschaft (limited-partnership company).
Knot - One nautical mile (1.1508 statute miles) per hour.
Kraftstoff - Fuel.
Kriegsmarine - German Navy (after 1935).
KWK - Kampfwagenkanone (tank gun, e.g. KWK 39).

L

LT - Lufttorpedo (aerial torpedo).
Laminar boundary layer - In regard to air, the layer of air next to a fixed boundary such as a wing.
Laminar flow - In a smooth flow of air in which there is no crossflow of air particles between adjacent stream lines, hence, a flow conceived as made up of layers; commonly distinguished from turbulent flow.
Laminar flow control system - A technology for reducing drag on an airplane by maintaining laminar boundary layers. Laminization is accomplished by sucking a small amount of the external boundary layer flow through the aircraft's skin. The system requires a perforated or slotted skin and a compressor for expelling the sucked air.
Landing - The act of terminating flight and bringing the airplane to rest.
Landing gear - The apparatus comprising the components of an aircraft that support and provide mobility for the craft on land, water, or other surfaces. The landing gear consists of wheels, floats, skis, bogies, and treads, or other devices, together with all associated struts, bracing, shock absorbers, and so on.
Landegeschwindigkeit - Landing speed.
Lange - Length
Langstrecken - Long-range.
Lateral - Pertaining to the side or a location along the lateral axis of an airplane.
Launch - To send off a vehicle under its own power, as in the case of guided aircraft rockets, or to send off an aircraft by means of a catapult.
Leading edge - The forward edge of an airfoil or other body moving through air.
Length - The dimension of an aircraft form nose to tail.
Lift - The total force of air acting on a body such as a wing or an airfoil which is perpendicular to the direction of flight and is exerted, normally, in an upward direction.
Linear - Of or pertaining to a line.
Loaf factor - The ratio of a specified load to the total weight of the aircraft.
Loft bombing - A method of bombing whereby the delivery aircraft approaches the target at a very low altitude, makes a definite pull-up at a given point, releases the bomb at a predetermined point during the pull-up, and tosses the bomb onto the target.
Longeron - A fore-and-aft member of the framing of an airplane fuselage or nacelle, usually continuous across number of points of support.

Glossary

Low-altitude bombing - Horizontal bombing with the height of release between 900 to 8,000 ft (274 and 2,438 m).
Low-monoplane - A monoplane whose wings are located at or near the bottom of the fuselage.
Luftwaffenfuhrrungsstab - Luftwaffe Operations Staff.
Luftwaffengeneralstab - Luftwaffe Air Staff.

M

M-Stoff - Methanol.
Mach number - A number that indicates the relationship of a speed to the speed of sound, named after Austrian physicist Ernst Mach (1838-1916) (pronounced Mock). Mach 1.00 is equal to the speed of sound or 760 mph (1,223 km/h) at sea level (low-level Mach). At 35,000 ft (10.7 km) and above, Mach 1, or the speed of sound, is 660 mph (1,062 km/h) (high-level Mach). Mach number variation with altitude is caused by variation in temperature, not by variation in air density.
Main stage - In multistage rocket, the stage that develops the greatest amount of thrust, with or without booster engines.
Manhattan Project - The US War Department program during World War II that produced the first atomic bomb. The term originated in the project's code name, "Manhattan Engineering District," which was used to conceal the nature of the secret work under way.
Manufacturer's empty weight - Weight of an aircraft, including the structure, power plant(s), furnishings, systems and other items that are considered an integral part of a particular configuration. Essentially, this is a dry-weight, as it includes only those fluids that are contained in closed systems (for example, hydraulic fluids).
Maximum landing weight - The maximum gross weight at which an aircraft is permitted to land, dependent on design or operational limitations.
Maximum takeoff weight - The maximum horsepower developed in an engine at sea level and restricted use to a continuous period of 5 minutes.
Mega - A prefix meaning multiplied by 10x6.
Meter - The basic unit of length of the metric system; 1 meter equals 39.37 inches or 1.094 yards.
MG - Machinengewehr (machine gun). Weapons whose caliber use less than 20 mm.
Missile - Any object thrown, dropped, fired, launched, or otherwise projected with the purpose of striking a target.
Mistel - Mistletow; adapted as a generic term for the lower component of all "pick-a-back" aircraft combinations.
MK - Machinekanone (machine cannon). Weapons whose caliber was 20 mm or greater.
ML - Motor-Luftstrahl (reciprocating engine or turbojet).
Moment - A tendency to cause rotation about a point or axis, as of a control surface about its hinge or of an airplane about its center of gravity.
Monocoque - A type of construction, as of a rocket body, in which all or most of stresses are carried by the skin. A monocoque may incorporate formers but not longitudinal members such as stringers.
Monoplane - An airplane or glider having one pair of wings.

N

NACA - US National Advisory Committee for Aeronautics, the predecessor of NASA.
Nacelle - Enclosed shelter for a power plant or for personnel, usually separate from the fuselage.
NATO - North Atlantic Treaty Organization.
NASA - US National Aeronautics and Space Administration.
Nose cone - The cone-shaped leading edge of an aircraft or rocket vehicle which frequently houses electronic gear such as the radar or weapons systems.
Nose gear - The part of a landing gear wheel located at the forward end of an aircraft.
Nose wheel - The landing gear wheel located under the nose of an aircraft that makes use of a tricycle landing gear.
Nozzle - A duct, pipe, tube, and the like through which turbojet or rocket engine gases are discharged.
Nutzlast - Load capacity.

O

Oblique - Style of German aircraft cannon mounted in the dorsal position on the fuselage and firing in an inclined or slanting position.

OKL - Oberkommando der Luftwaffe (Luftwaffe High Command).
Oleo - A shock-absorbing strut, usually in a landing gear, in which the spring action is dampened by oil.
Oscillation - The act of swinging or moving regularly back and forth. Also known as fluctuation, instability, or variation of an object.
Otto engine - All reciprocating gas/air engines as distinct from Diesel engines. From inventor Nicholaus August Otto.

P

Pak - Panserabwehrkanone (anti-tank cannon).
Parabola - An open curve, all points of which are equidistant from a fixed point called the focus (as in the Horten brothers' Parabola all-wing glider of the mid 1930s Germany).
Pilot - A person who handles the controls of an aircraft from within the craft, and in doing so, guides or controls it in flight.
Piston engine - An engine in which the working fuel is expanded in a cylinder against a reciprocating piston.
Pitch - The angular displacement about an axis parallel to the lateral axis of the airplane, with the result that either the nose or the tail moves up or down.
Planform - The shape or form of an object, such as an airfoil or a wing, as seen from above.
Port - The left-handed side of a ship or airplane as one faces forward.
Power plant - The complete assemblage or installation of an engine or engines with accessories (induction system, cooling system, ignition system, etc.) That generates the motive power for a self-propelled vehicle such as an aircraft.
Power plant unit - A complete aircraft engine package, including accessories and cowling.
Pressurized cabin - The occupied space of an aircraft in which air pressure has been increased above that of the outside ambient atmosphere by mechanical means.
Project - A planned or proposed undertaking of a design to be accomplished or constructed.
Propellant - Any agent used for consumption or combustion in rocket and from which the rocket derives its thrust.
Propeller - A device for propelling an aircraft which has blades on an engine-driven shaft and which, when rotated, produces, by its action on the air, a thrust perpendicular to its plane or rotation.
Propfan - A multi blade turboprop.
Prototype - The first of a series of similar form and design, intended for complete evaluation of suitability and performance.
PSI - Pounds per square inch.
Pulsejet - An engine consisting of a simple duct or tube equipped with light, flap-type check valves: successive explosions propel the exhaust rearward, producing forward motion by reaction (as in the Fieseler Fi V-1 "buzz bomb"). Also known as the "intermittent-duct engine" or "resonant jet engine."
Pylon - An attachment to an aircraft that provides carriage of fuel tanks, engines, and ordnance and other accessories.

Q

Quarter-chord point - The point on the chord of an airfoil section at one quarter of the chord length behind the leading edge.

R

R-Stoff - Rocket fuel known as TONKA-250 and consisting of 57% crude oxide monoxylidene and 43% tri-ethylamine.
Radar - Radio detecting and ranging. A method of using beamed, reflected, and timed radio waves to detect, locate, or track objects such as aircraft.
Ramjet - A type of turbojet engine having no mechanical compressor. The air necessary for combustion is shoved into a specially shaped tube or duct open at both ends and is compressed by the forward motion of the engine; the air passes through a diffuser and is mixed with fuel and burned, the exhaust gases issuing a jet from the rear opening. Also called a Lorin tube.
Reichs - State.
Reichweite - Range.
Reisegeschwindigkeit - Cruise speed.
Remote control - Control of an aircraft or rocket from a distance, especially by means of electronics.
Retractable landing gear - A type of landing gear that may be withdrawn into the fuselage or wings of an airplane while it is in flight, in order to reduce parasite drag.

Reverse thrust - A force frequently used to slow the landing roll of an aircraft achieved in a turbojet engine by deflector vanes incorporated into the engine's exhaust nozzle.

Reynolds number - Parameter used to determine the nature of fluid flow along surfaces and around objects, as in a wind tunnel, and expressed in a ratio named after English physicist Osborne Reynolds (1842-1912).

Rib - A fore-and-aft member which maintains the required contour of the covering material on wings or control surfaces and which may also act as a structural member.

Ritterkreuz - German military award known as the Knights Cross or the Iron Cross, first of five grades of the highest German decoration for bravery, the higher grades being signified by the addition of Oak Leaves, Oak Leaves and Swords, Diamonds, Golden Oak Leaves, and Diamonds and Swords.

RLM - Reichsluftfahrtministerium (State Ministry of Aviation, or German Air Ministry).

Rocket - Any device, typically cylindrical, containing a combustible substance which, when ignited, produces gases that escape through a rear vent and drive the container forward by the principle of reaction.

Rocket airplane - Any aircraft using rocket propulsion for its main or only propulsive power.

Rocket engine - A reaction engine that contains within itself all the substances necessary for its operation and does not require any intake of any outside substances. Also known as a rocket motor.

Rocket fuel - A fuel, either liquid or solid, developed for and used by a rocket.

Rocket propulsion - Propulsion by a rocket engine.

Roll - The rolling or rotating movement of an aircraft about its longitudinal axis.

Rudder - A hinged vertical control on an aircraft used to turn the aircraft right or left.

Rustgewicht - Empty weight.

S

S-Stoff - Rocket fuel used in World War II Germany, consisting of 90%-97% nitric acid and 3%-10% sulphuric acid.

Sailplane - A glider designed for sustained flight utilizing wind currents.

Schnellbomber - Fast bomber.

Semi monocoque - A type of construction in which longitudinal members as well as formers reinforce the aircraft's skin and help carry the stresses.

Service ceiling - The altitude at which the rate of climb is the lowest practical for a service operation.

Short takeoff and landing - (STOL) Describing an aircraft able to clear a 50 ft (15 m) obstacle within 1,500 ft (457 m) of takeoff, or, in landing, to stop within 1,500 ft (457 m) after passing over a 50 ft (15 m) obstacle.

Slipstream - The stream of air discharged aft by a rotating propeller.

Slotted airfoil - An airfoil having one or more air passages (or slots) connecting its upper and lower surfaces to modify the normal force.

Slotted flap - A flap whose leading edge is so shaped that the slot, or slots, between it and the wing improves the flow over its upper surface when the flap is deflected downward.

Snaking - An uncontrolled oscillation in yaw of an aircraft, the amplitude of which remains approximately constant.

Soaring - The act of flying a heavier-than-air craft (a sailplane) without power by utilizing rising air columns (thermals) for motive power.

Solid propellant - Specifically, a rocket propellant in solid form, usually containing both fuel and oxidizer combined and mixed into a solid grain.

Solid rocket - A rocket that uses a solid propellant.

Sonder Kommando - A special detachment.

Span - The length of an airfoil from tip to tip measured in a straight line.

Spannweite - Wingspan.

Spar - A principal spanwise structural member of an airfoil or control surface.

Speed - Rate of motion.

Speed of sound - The speed of sound waves, at sea level 1,116.45 ft per second (340.29 m per second) or 760 mph (1,223 km/h).

Spin - A prolonged stall in which an airplane rotates about its center of gravity while it descends, usually with its nose well down.

Split Flap - A flap set into the lower surface of the wing.

Spoiler - A plate that projects into the airstream to break up, or spoil, the smoothness of the airflow. A spoiler usually projects from the upper surface of an airfoil.

Stabilizer - The fixed airfoil of an airplane used to increase stability; usually the aft fixed horizontal surface to which the elevators are hinged (horizontal stabilizers) and the fixed vertical surface to which the rudder is hinged (vertical stabilizer).

Stall - The abrupt loss of lift when the angle of attack increases to a point at which the flow of air tends to tear away from a wing or airfoil.

Static thrust - In turbojet engines, the maximum thrust in pounds or kilograms at sea level with no forward motion.

Strafing - The firing of cannons by an aircraft at ground targets.

Strategic attack - An attack by an air force on an enemy intended to destroy its ability to wage war.

Streamlined - Having a contour designed to offer the least resistance when moving through air.

Stringer - A slender, lightweight fill-in structural member in an aircraft's fuselage or wings that runs lengthwise and reinforces and gives shape to the skin.

Strut - An external major structural member.

Stuka - Sturzkampfflugzeug (dive bomber) such as the Junker's Ju 87.

Subsonic - Moving through air at a speed less than that of sound in air.

Supersonic - Moving through air at a speed greater than that of sound or air.

Surface boundary layer - The thin layers of air adjacent to the wing's surface.

Sweep back - The backward slant of an airfoil's leading edge such that the wing tip is further aft than the wing root.

SV-Stoff - Rocket fuel used by Germany in World War II, known as Salbei, consisting of 85%-88% nitricacid and 12%-15% sulphuric acid.

T

T-Stoff - Rocket fuel comprising 80% hydrogen peroxide plus oxyquinoline or phosphate as a stabilizer.

Tab - A small auxiliary airfoil usually attached to a movable control surface to bring about a slight change for the purpose of trimming the airplane for varying conditions of power, load, and airspeed.

Tail - The rear part of an aircraft's fuselage.

Tail boom - A cantilever carrying the tail unit of an aircraft on which the fuselage does not perform this function.

tail unit. The combination of stabilizing and controlling surfaces situated at the rear of an airplane.

tailless aircraft. An aircraft whose longitudinal control and stabilizing surfaces are incorporated in the wing surfaces, also known as a flying wing.

Test bed - A base, mount, or frame upon which a piece of equipment, especially an engine, is placed for test operations.

Thrust - Force or pressure directly exerted and usually expressed in pounds or kilograms thrust/horsepower ratio. A formula for converting thrust into horsepower;

$$Hp = \text{thrust pounds or kilograms} \times \text{airspeed feet/sec or meters/sec} \over 500$$

Example: 4,000 lb of thrust equals 4,000 hp at 375 mph and 8,000 hp at 750 mph.
Transonic - The range of speed between subsonic and supersonic, approximately Mach 0.8 to 1.2.
Tricycle landing gear - A three-wheeled landing gear in which the third wheel is placed well forward under the nose of the aircraft and the two wheels that make up the main gear are placed a short distance behind the center of gravity.
Turbojet engine - An engine incorporating a turbine-driven air compressor to take in and compress the air for the combustion of fuel, the gases of combustion being used to rotate the turbine an to create a thrust-producing jet.
Turboprop - A gas turbine engine in which a portion of the net energy is used to drive a propeller.

V

V-1 - A German pulse-jet-powered flying bomb commonly known as the "buzz bomb" with the "V" meaning Vergeltungswaffen or weapon of reprisal. Developed by Wernher von Braun and Walter Dornberger it had a range up to 600 miles, a speed of 350 mph, and a warhead containing 1,600 pounds of high explosives. Approximately 9,000 were fired against Great Britain with about one half that number reaching its target.
V-2 - A German liquid-fueled rocket flying bomb which after launch reached supersonic speed at an altitude of fifty miles with 2,000 pounds of high explosives.
Variable geometry wings - Wings whose wing sweep can be changed or varied in flight.
Vertical and short takeoff landing - (VTOL) Describing an aircraft able to take off and land vertically.
Vortex generator - A movable surface on a wing that may be used as a spoiler to break down the airflow.

W

Washout - The practice of lessening the angle of attack near the wing tip to decrease lift.
Wind tunnel - A tunnel-like chamber through which air is forced, and in which scale models of proposed airplanes are tested, to determine the effects of wind pressure and aerodynamic efficiency.
Wing - A general term applied to the airfoil, designed to develop a major part of the lift of an aircraft.
Winglets - Small, nearly vertical aerodynamic surfaces mounted at the tips of airplane wings. Winglets reduce an aircraft's induced drag by smoothing out the airflow around the wing tip.
Wing loading - The gross weight of an airplane divided by gross wing area.
Wing root - The end of a wing that joins the fuselage.
Wingspan - The distance from one wing tip of an aircraft to the opposite wing tip.

Y

Yaw - The rotational or oscillatory movement of an aircraft about its vertical axis.

Z

Z-Stoff - Water solution of Na (Natrium) and potassium permangenate (Kaliumpermaganat).

Index

A
AAC.1 "Toucan," 249
AD-1, 77
Adder (snake), 51
Aerial Jeep, 70
Aerodynamic Testing Institute, 231
Aerodynamische Versuchsanstalt-Göttingen (AVA-Göttingen), 113, 120, 131
Aerodyne, 282
Aerofoil, 282
Aerophysics Corporation, 70
Aiguillon "Sting" STOL fighter, 179
Air Ministry Museum-Berlin, 158, 163, 164
AJ65 "Avon" turbojet engine, 123
Akaflieg, 28, 118
Aktien Gesellschaft Otto, 144
Albatros B.1, 153
Albatros Flugzeug Werke GmbH, 28, 29, 153
Alt-Lönnewitz, City of, 32
Amerika Bomber, 9, 25, 26, 155, 159, 225, 232, 234, 240, 242, 243, 256, 258, 259, 260, 290, 310, 314, 315, 316
Amundsen, Roald, 103
Antipodal Bomber, 322
Anzani motor engine, 61
Arado Flugzeugwerke GmbH, 7, 28, 29, 30, 46, 147, 150, 238, 241
Ar E.370, 31
Ar E.381, 32, 40, 41, 42, 43, 44, 45
Ar E.580, 46, 192
Ar E.581.4, 15, 44, 45, 46, 47, 281
Ar 8-234, 31, 74
Ar 65 bi-wing dive bomber, 188
Ar 234A, 32
Ar 234B, 10, 26, 28, 32, 37, 38, 98, 100, 130, 139, 168, 192, 250, 313, 337, 338, 339
Ar 234B-2, 14, 29, 30, 31, 32, 33, 35
Ar 234C, 14, 30, 35, 36, 42, 43, 44, 82
Ar 234C-2, 32
Ar 234C-5, 32
Ar 234 V6, 36
Ar 234 V9, 32
Ar 234 V16, 34, 36, 37, 38, 42
Ar 234 V20, 37
Ar 234 V30, 36
Ar P.NJ-1, 29, 39, 40, 41, 42, 107, 109, 150, 192, 202, 282, 338
Ar P.NJ-2, 29, 39, 109, 150
Argentina, 10, 11, 124, 125, 126, 128, 141, 225, 231, 235, 242, 339
Argentine Air Force, 111
Argus-Rohr Pulse-jet Engine, 137, 171, 243, 255
Argus-Schmidt-Schubror As 014 Pulse-jet Engine, 58, 79, 80, 93, 94, 171
Armistice of 11 November 1918, 62, 115, 247
Armstrong-Whitworth Aircraft Ltd., 231
Arrow Wing, 80
Atar E5V turbojet, 129
Atliers Aéronautiques de Colombes - France, 249
Atomic Bomb -United States, 291
Atomic Bomb - Germany, 225, 258
Azores, 316

B
B-9, Berlin, 41, 189
B-17, 6, 19, 21, 24, 43, 48, 49, 57, 84, 85, 124, 141, 169, 239, 284, 294, 301, 311, 326, 327, 330, 332, 333, 334, 335, 336
B-24 "Liberator," 48, 124, 141, 169, 264, 314, 332
B-25, 311
B-29, 24, 69, 97, 99, 101, 120, 124, 126, 311, 314, 335
B-36, 69
B-45C, North American, 38, 338
B-51, 24
B-52, 69
B-58 "Hustler" Convair, 10, 16, 278, 281, 286, 338
Ba P.20, 52, 169
Ba 349A "Natter," 42, 48, 49, 50, 51, 52, 53, 54, 55, 56, 57, 79, 169, 170, 171, 306, 327, 334
Ba 349B "Natter," 10, 26, 54, 132, 192, 333

Ba 349M "Natter," 54
Baade Type 152, East German passenger aircraft project of, 267
Baade, Professor Dr.Ing. Brunolf, 98, 260, 266, 267, 323
Bachem, Dr.Ing. Erich, 7, 42, 48, 50, 51, 53, 54, 55
Bachemwerke GmbH, 48
Backhaus, Dr.Ing., 98
Badetage, 245
Bansemir, Dip.Ing. Wilhelm, 124
Battle of Britain, 155, 188, 228, 237, 294
Battle of the Somme, 59
Baumann, Professor Dr. H. - 60
Bäumer "Sausewind" of, 118, 157
Bäumer, Paul, 115, 118, 119, 124, 156, 157
Bayerische Motoren Werke, 10, 29, 82, 18, 130, 249, 300
Becker, Professor Dr. Adolf, 110
Bell Aircraft Corporation, 10, 16, 111, 290, 298, 309, 310, 311
Bell, Larry, 10, 290, 310
Bell P.59, 24
Bell X-1, 99, 101, 111, 123, 293
Bell X-5, 10, 16, 290, 298, 309, 310, 311, 338
Benz, Dip.Ing. Wilhelm, 14, 42, 52, 159, 162, 164, 169
Bergen, William, 117
Berlin-Adlershof Wind Tunnel- DVL, 305
Berlin to Baghdad Air Race, 292
Betty, Japanese bomber, 135
Betz, Dr.Ing. Albert, 118, 305
Bi 500, 30
Bley Segelflugzeugbauwerke, 326
Blitz Bomber, 301
Blohm und Voss Abteilung Flugzeugbau, 18, 29, 46, 58, 63, 139, 147, 173, 175, 239, 254, 255, 299, 338
Blohm Brothers, 64, 67, 71
Blohm, Rudolf, 63
Blohm, Walther, 59, 63, 64, 71, 115
Blue Max, 28
Blue Maltese Cross, 28
Blume, Walter, 14, 28, 29, 30, 31, 38, 46, 336, 338, 339
BMW 003A, 31, 32, 34, 35, 189, 232, 239, 240, 241, 309
BMW 003B, 126, 182, 187, 188, 189, 240, 241, 258, 265
BMW 003C, 240
BMW 003D, 78, 314
BMW 003R, 34, 36, 130, 150, 234, 240
BMW 018A, 68, 76, 77, 82, 83, 84, 186, 187, 232, 264, 314, 317
BMW 028, 82, 83, 84, 87, 88
BMW 801A, 75
BMW 801D, 65, 68
BMW 801E, 316
BMW BZW-718 liquid bi-fuel rocket drive, 34, 85, 88, 130, 234
BMW IIIa engine, 249
BMW turbojet engines, 74, 83, 119, 157, 240, 294, 300
BMW P.Schnellbomber I, 86, 87
BMW P.Schnellbomber II, 82, 84, 86, 87, 88, 89, 194
BMW P.Strahlbomber I, 82, 84, 85, 86, 194
BMW P.Strahlbomber II, 86
BMW P.3303 gas turbine, 300

BMW-Spandau, 82, 131
Bock, Professor Dr. Günter, 14, 98
Boeing Aircraft Company, 70, 72, 123, 245
Boeing 707, 245
Boelcke-Staffel, 157
Bölkow, Dr.Ing. Ludwig, 316
Bölkow GmbH, 71, 299
Boultan Paul P.111, 338
Brandis/Leipzig Luftwaffe Air Base, 262, 264
Brandner, Dr.Ing. Ferdinand, 265, 320
Bravery Medal of the King of Würettemberg, 62
Bredt, Dr.Ing. Irene, 324
Bristal "Cherub" engine, 274
British Avro 698 "Vulan," 278
British Avro 707A, 338
British Spitfire, 156

Bundes Luftwaffe, 107
Busemann, Dr.Ing. Adolf, 251, 263, 280, 287, 305, 320
Butter, Dip.Ing., 137
Buzz Bomb, 79, 93, 182
Bv Coupled Aircraft Experiments, 69, 70, 139
Bv Pivoting Wing, 339
Bv Swiveling Wing, 77, 338
Bv 40, 41
Bv 138, 66
Bv 141, 60, 61, 63, 64, 65, 66, 67, 75
Bv 238, 66, 67
Bv P.Ae-607, 58, 64, 73, 192
Bv P.111, 62
Bv P.155B, 64, 67
Bv P.180, 62
Bv P.188, 12, 58, 70, 74, 81, 193
Bv P.188.04, 74
Bv P.192 dive bomber, 63, 67
Bv P.194.01, 63, 65, 67
Bv P.196, 58, 63, 68, 74, 75, 80, 193
Bv P.197.01, 58, 64, 76, 193
Bv P.198, 58, 76, 77
Bv P.202, 58, 77, 78, 81, 193, 338
Bv P.208.03, 67 125
Bv P.209, 78, 339
Bv P.209.02, 10, 65, 66, 68, 78, 143, 193
Bv P.211, 65, 68, 78, 79, 161, 193, 239
Bv P.212, 67, 125, 175, 207, 208
Bv P.212.03, 66, 208, 209, 210
Bv P.213, 58, 65, 67, 78, 80, 193, 255
Bv P.214, 189
Bv P.215, 40, 58, 67, 80, 81, 106, 109, 149, 150, 193, 210, 211, 212, 238, 306
Bv P.222, 60, 66, 67
Bv P.237, 63

C
C-Stoff, 19, 54, 57, 128, 130, 272, 283
C-5, Lockheed, 124
C-141 military transport, 117
C-450 "Coleopter," 129, 130, 131
Caesar, Julius, 51
Canadian War Museum-Ottawa, 285
Canards, 73, 184
Cannstatter Wasen, 60, 61
Cessna 172 "Skyhawk," 292
Chef der Technischen Luftrüstung (Chief of Technical Air Equipment), 40, 73, 80,109, 150, 184
Chicago, Illinois, 316
Coleopter, 129, 130, 131
Collins, Arthur, 281
Collins Radio - 281, 282
Columbia, NASA Space Shuttle, 124, 283, 286, 338, 339
Concorde, British-French Supersonic Airplane, 280, 283, 285, 286, 338
Convair, 10, 130, 278, 286, 338
Cordes, Dr.Ing., 323
Còrdoba, City of, 128, 231, 237, 238
Cranfield Institute of Technology, 285
Crescent Wing, 36, 37, 42, 313
Croneiss, Theo, 295
Curtiss F11C-2 Hawk 11, 59, 63, 114

D
da Vinci, Leonardo, 64
Dahlke, Dip.Ing. Otto, 186
Daimler-Benz - Stuttgart, 84, 249, 300
DB 603A piston engine, 102, 105, 108
DB Project A, 86, 89
DB Project B, 86
DB Project C, 86
DB Project D, 86
DB 007 ducted fan turbojet engine, 84, 88, 89, 300
DB 601 piston engine - 73 DB 605D piston aero engine, 131
DB 603LA piston aero engine, 106
DB 603N, 139
DB 606 coupled engines of the He 170, 140, 155
DC-3, Douglas, 248
DDR, 267
De Havilland 108, 338
De Havilland "Vampire," 142
Department "L," 278
Detroit, Michigan, 314
Deutsche Forschungsinstitut für Luftfahrt (DFL), 146
Deutsche Forschungsinstitut für Segelflug (DFS), 11, 92, 137, 139, 148, 178, 179, 188, 274, 275, 277, 278, 285, 288, 293, 294, 325, 333
Deutsche Versuchsanstalt für Luftfahrt (DVL), 14, 30, 55, 146, 148, 149, 181, 182, 187, 189, 232, 237, 241, 251, 263, 305, 330
Deutsches Museum-München, 57, 285

Deutschlandhalle-Berlin, 93
DFS "Kranich," 92
DFS "Mistel," 94
DFS "Seeadler," 93
DFS 194, 271, 277, 278, 279, 289
DFS 228, 94, 95, 96, 148
DFS 230 A-1, 92
DFS-346, 10, 11, 23, 26, 92, 93, 95, 96, 97, 98, 99, 100, 121, 148, 178, 194, 222, 223
DFS 346-P, 96, 97, 98
DFS-486, 100
Dittmar, Heini, 272, 279
Do 15 "Wal," 103, 106
Do 17 "Flying Pencil," 102, 104, 106
Do 18 "Wal," 227
Do 24-K seaplane, 104
Do 26, 104
Do 31 "vertical takeoff and landing aircraft," 107, 108
Do 217, 65, 182, 324, 329
Do 335 "Pfeil," 102, 105, 106, 108, 119, 139
Do P.254, 102, 105, 106, 108
Do P.256, 40, 102, 107, 109, 150, 184, 194, 191
Do P.257, 107
Do RS-1, 106
Do "X" seaplane, 103
Dornberger, Hauptmann Walter, 328, 330
Dornier Metallbauten GmbH, 106, 107
Dornier, Dr.Ing. Claudius, 17, 59, 60, 62, 63, 64, 71, 103, 106, 107, 124, 268, 270
Dornier "Superwal," 103
Dornier "Skyservant" short takeoff and landing aircraft, 107
Dornier Werke, 29, 58, 102, 106, 107, 108, 147, 150, 179, 187, 282, 332
Dresden, City of, 337
Duisburg-Ruhrort GmbH, 30
Dunne, J.W., 235
Duralumin, 79, 247
Dutch Roll, 148, 231, 237

E
Edison, Thomas, 282
Edwards Air Force Base, 77, 101, 286
Egypt, 125, 299
Eiffel Tower, 61
Eiffel, Alexandre G., 61
Einstein, Albert, 241
England Fliers, 144
English Electric "Lighting" interceptor, 123
English Electric P-1, 120, 122, 339
English Electric Ltd., 123
Entwung, 26
Enzian, 25
Eprobungsflugzeug, 26
Eschenauer, Oberst Artur, 228, 229
Espenlaub, Gottlob, 269, 271
Etich, Dip.Ing. Igor, 145
Europa, German ocean liner, 59, 63
EZS, 86

F
F 13, 247
F-84, 69
F-86, 125, 310
F-102 "Delta Dagger" Convair, 10, 278, 281, 286
F-102A Convair, 278
F-104 "Starfighter" Lockheed, 107, 108, 299
F-106 "Delta Dart" Convair, 10, 286
F-111 General Dynamics, 311
F3D "Sky Ray," 338
F7U "Cutlass," Chance-Vought, 41, 42
Fa 224, 93
Fallbobe air-to-air missile, 328
Farman bi-wing, 151
Fatli, Egan, 103
Fernbomber, 241, 316
Feuerspeidenden Berg, 326
Fi 103 (piloted bomb), 32, 44, 79, 93, 136, 137, 154, 182, 255, 261, 332
Fi 156 "Storch," 50, 278
Fieseler, Gerhard, 51, 137
Flessner, SS Heinz, 50, 51
Flitzer, 142, 143
Floating Wing Tests, 69
Flying Flounder, 282, 283, 284
Flying Fortress, 84
Focke, Professor Heinrich Karl, 115
Focke-Wulf Flugzeugbau GmbH, 16, 22, 30, 46, 71, 110, 112, 113, 114, 118, 119, 120, 122, 125, 127, 131, 132, 134, 135, 139, 140, 141, 147, 150, 179, 241, 297, 332, 337, 338, 339
Föhn RZ-73 anti-aircraft rocket missiles, 328
Fouge C. 170R "Magister," 176

Franco, Major Francisco, 106
Franz, Dr.Ing. Anselm, 249, 325
French Bleriot, 61
French "Farman" biplane, 60
French Foreign Legion, 130
Friedrichroda, Gotha facilities at, 147, 231
Fritz X air-to-air missile/bomb, 74, 328, 329
Fritz Z projectile, 330, 331
Führer, 296, 299, 300, 301
Für die Luft, 180
Fw 44 "Stieglitz," 114
Fw 56 "Stösser," 114
Fw 58, 110
Fw 187, 119
Fw 189, 65
Fw 189A-1, 113
Fw 190, 65, 74, 75, 107, 114, 166, 169, 298, 300, 327
Fw 190A-8, 113
Fw 190D, 80
Fw 200 "Condor," 19, 65, 113, 114, 183, 297
Fw 300, 313
Fw 388, 75
Fw P.3x1000A, 134
Fw P.3x1000B, 134
Fw P.3x1000C, 110, 134, 135, 204, 282
Fw Ta P.152, 108, 139

Fw Ta P.183 (Design II), 127, 141, 142, 178
Fw Ta P.183 (Design III), 127, 128
Fw Ta P.183, 8, 10, 15, 17, 23, 110, 111, 113, 114, 116, 117, 120, 122, 123, 124, 125, 126, 127, 128, 142 189, 190, 218, 219, 203, 194, 298, 306, 312, 338, 339
Fw Ta P.283, 110, 132, 133, 134, 195
Fw P."Kamikaze," 84, 110, 135, 137, 138, 139, 140, 194
Fw P."Kamikaze Carrier," 84, 110, 137, 138, 139, 140, 194, 333
Fw P."Triebflügel," 110, 128, 129, 131, 132, 194, 205, 206
Fw P.1, 142, 150
Fw P.2, 40, 107, 109, 110, 143, 150
Fw P.3, 40, 109, 110
Fw P.5, 140, 141, 142
Fw P.6, 140, 141, 142
Fw P.7, 110, 142, 143, 192

G
G 2, 248
G 31, 248
Galland, General der Flieger Adolf, 52, 53, 228, 300, 301
Garber, Paul E., 154, 236
GEKADOS, 324
General Electric Company - Re-entry and Environmental Systems Division, 124
General Dynamics, 311
Georgii, Professor Dr. Walter, 92, 94, 275
Gerhard Fieseler Werke, 50, 51, 137
Gerlach, Dr.Ing., 323
German Academy of Aeronautical Sciences, 110, 111, 135
German Air Ministry, 6, 21, 23, 26, 29, 31, 38, 43, 46, 47, 48, 51, 52, 53, 54, 64, 65, 67, 68, 71, 74, 78, 79, 80, 81 82, 93, 114, 125, 126, 131, 132, 134, 140, 141, 142, 145, 146, 147, 148, 149, 150, 151, 155, 158, 159, 161, 162, 163, 164, 166, 167, 168, 169, 172, 178, 179, 182, 184, 186, 188, 189 233, 235, 236, 237, 238, 239, 241, 242, 243, 249, 255, 258, 260, 261, 264, 275, 276, 284, 285, 286, 289, 290 294, 297, 298, 299, 300, 306 312, 313, 314, 316, 317, 318, 320, 322, 326, 327, 332, 333, 334
German National Prize for Art and Science, 154
German Naval Academy, 226
Germany, 125
Gestapo, 244
Glenn L. Martin Aircraft Company, 123, 124, 298, 338
Go 147B, 145
Go 229, 147
Go 242A-1, 146
Go P.60 series, 147, 149
Go P.60A, 147, 149, 150, 151, 231, 237
Go P.60B, 147, 148, 149, 150, 151, 195, 236
Go P.60C, 40, 106, 109, 147, 149, 150, 151, 195
Goodyear Tire and Rubber, 123
Göring, Generalfeldmarschall Hermann, 19, 21, 22, 23, 25, 28, 29, 120, 155, 158, 159, 225, 228, 235, 236, 241, 242, 243, 244, 258, 260, 293, 294, 295, 296, 299, 300, 301, 314, 316, 317
Gotha, 144, 146, 147, 148, 149, 150, 151, 236
Gotha "G" bombers, 144
Gothaer Waggonfabrik AG, 93, 144, 146, 148, 225, 231, 238, 247
Göthert, Dr.Ing. Rudolf, 12, 146, 147, 148, 149, 150
Göttingen University, 128
Groenhoff, Günther, 269, 270, 271, 272
Grossflugzeug Firma, 251
Günter, 177, 178
Günter Twins, 7, 17, 118, 153, 154, 156, 157, 158, 160, 165

Günter, Siegfried, 11, 13, 14, 39, 99, 100, 154, 156, 157, 158, 159, 160, 172, 173, 174, 175, 176
Günter, Walter, 42, 154, 156, 157, 158, 160, 162, 176
Gurevich, Mikhail I., 10, 15, 262, 339

H
Ha 139, 59
Ha Me 100, 299
Ha Me 200, 299
Hackmack, Hans, 294
Hahn, Otto, 240
Hamburger Flugzeugbau GmbH, 63, 71, 254, 267, 299
Handley-Page B-35/46 "Victor 8," 313, 316
Handley-Page Company, 36, 37, 338
Harth, Friedrich, 291, 292
Hawker Siddeley "Comet," 316
He 50, 188
He 59, 227
He 64, 158, 176
He 70 "Blitz," 22, 154, 156, 158, 162, 176
He 72, 162
He 111, 54, 107, 137, 154, 155, 159, 228, 336
He 112, 157, 162
He 162 "Salamander," 10, 23, 26, 46, 68, 79, 130, 152, 154, 155, 159, 160, 161, 171, 173, 188, 189, 195, 239, 255, 336
He 162A-10, 171
He 162B, 171, 255
He 177, 73, 134, 140, 154, 155, 159, 183, 184, 256, 257
He 211, 178
He 219, 139
He P.176, 156, 158, 159, 162, 163, 164, 167, 195, 261, 276
He P.178A, 164
He P.178B, 42, 153, 155, 158, 163, 164, 165, 167, 261, 337
He P.280, 152, 158, 164, 165, 166, 167, 195, 196, 336, 337
He P.280 V3, 165, 166, 167
He P.343A-1, 13, 82, 83, 152, 159, 167, 168, 176, 195
He P.343B, 83
He P.1073, 79, 161
He P.1077 "Julia," 42, 52, 153, 159, 160, 169, 170, 171, 196, 213, 335
He P.1078, 179
He P.1078A, 11, 14, 153, 159, 172, 173, 196, 306, 337
He P.1078B, 6, 13, 152, 172, 173, 174, 196
He P.1078C, 13, 125, 152, 159, 172, 175, 177, 190, 196
He P.1079A, 11, 13, 152, 159, 172, 175, 176, 177, 178, 184, 196
He P.1080, 152, 159, 171, 178, 179, 196
Heavy water, 241
Heinkel AG-Jenbach, 160
Heinkel AG, 6, 7, 10, 14, 18, 23, 29, 46, 51, 64, 71, 79, 187, 130, 131, 137, 140, 152, 153, 155, 158, 159, 160, 161, 162, 163, 164 165, 166, 167, 168, 169, 170, 171, 172, 173, 175, 177, 179, 239, 241, 255, 261, 276, 277, 278, 296, 297
Heinkel AG-Zuffenhausen, 160
Heinkel AG-Rostock, 160
Heinkel AG-Marienehe, 160
Heinkel AG-Oranienburg, 160
Heinkel, Dr.Ing. Ernst, 6, 11, 13, 14, 17, 18, 60, 61, 115, 130, 153, 154, 155, 157, 158, 159, 160, 165, 166, 167, 175, 179, 225, 230, 249, 251, 261, 276, 278, 284, 293, 296, 297, 299, 300, 336, 337
Heisenburg, Dr.Werner, 241
Hellmuth Hirth Versuchsbau, 181
Henschel Hs 117 "Schmetterling," 182
Henschel Hs 121, 180
Henschel Hs 123, 181, 188
Henschel Hs 126, 181
Henschel Hs 129, 181
Henschel Hs 217 Fohn air-to-air antiaircraft rockets, 49, 51, 54
Henschel Hs 121, 180
Henschel Hs 293 "Egret" missile, 182, 183
Henschel Hs 294 missile, 32, 183
Henschel Hs 295 missile, 183
Henschel Hs 298 missile, 183
Henschel Hs P.75, 184, 185
Henschel Hs P.122, 183, 184, 186, 187, 191
Henschel Hs P.132, 161, 182, 186, 187, 188, 197
Henschel Hs P.135 aircraft project, 187, 189, 191, 197, 283
Henschel Flugzeugwerke, 50, 89, 180, 181, 186, 188, 191, 238, 330
Henschel & Sohon, 180
Henschel, Oscar R., 180, 181, 182, 183, 1
Hertel, Dr.Ing. Heinrich, 9, 250, 252, 253, 257, 258, 259, 260, 261, 265, 267, 316, 336, 338
Herz Mountains, 258
HeS 3B turbojet engine, 163, 249
HeS 8A turbojet engine, 165, 166
HeS 011 turbojet engine, 34, 37, 40, 47, 73, 77, 102, 105, 107, 108, 109, 112, 113, 119, 126, 127, 134, 140, 141, 151, 155, 157, 159, 167, 172, 175, 176, 183, 191, 225, 232, 238, 239, 240, 242, 243, 251, 257, 259, 260, 290, 291, 300, 302, 306, 311, 313, 318 337
HeS 021 turboprop, 300

Hess, Rudolph, 19
HFB - Hansa Düse airplane, 267
Hill, G.T.R., 145
Himmler, Reichsführer SS Heinrich, 42, 50, 53, 54, 170
Hirschgeweih radar, 298
Hirth, Hellmuth, 60, 153
Hitler, Adolf, 19, 20, 21, 29, 53, 63, 154, 158, 162, 180, 181, 243, 244, 251, 258, 278, 284, 296, 299, 300, 301, 336
Hitler Youth, 161
Ho 1, 226, 227
Ho 2, 226, 227
Ho 3, 240
Ho 3B, 73
Ho 3G, 226
Ho 4A, 226
Ho 5B, 227, 236
Ho 5C, 226
Ho 7, 227, 239
Ho 9, Concept of, 228
Ho 9A, General concept, 148, 197, 231, 232, 235, 236, 237, 238, 239
Ho 9A V1, 227, 237
Ho 9A V2, 146, 147, 175, 225, 228, 232, 236, 241, 250, 338
Ho 9A V3, 237, 239, 240
Ho 9B, 224, 230, 237, 338
Ho 229, 148, 149, 174
Ho 229 V3, 8, 10, 12, 18, 25, 26, 146, 147, 225, 229, 230, 231, 339
Ho 229 V4, 236
Ho 229 V5, 236
Ho 229 V6, 231, 236
Ho 229 V7, 236
Ho 229 V8, 236
Ho P.9B, 197, 238
Ho P.9C, 197
Ho P.10A, 197, 224, 231, 232, 239
Ho P.10B, 7, 197, 225, 232, 239, 282
Ho P.10C, 198, 225, 233, 235, 239, 282, 338
Ho P.13A, 225, 232, 240
Ho P.13B, 7, 198, 225, 232, 233, 234, 240, 338
Ho P.18A, 198, 225, 232, 241, 242, 243, 256, 257, 258, 314, 315
Ho P.18B, 25, 26, 155, 159, 198, 225, 240, 242, 243, 259, 260, 316, 317
Ho P.18C, 198
Hochgeschwindikeit (high speed), 10, 306
Horten Brothers, 6, 7, 10, 17, 18, 20, 25, 73, 80, 82, 145, 146, 148, 155, 159, 174, 175, 225, 229, 230, 231, 232, 233, 234, 235, 238, 239, 241, 242, 243, 250, 273, 274, 277, 280, 314, 316, 317 338
Horten, Dr. Max, 226
Horten, Dr. Reimar, 11, 25, 224, 225, 230, 231, 233, 234, 237, 238, 239, 240, 241, 242, 258, 259, 260, 272, 315, 338
Horten Flugzeugbau GmbH, 54, 160, 256, 317
Horten, Frau Elizabeth, 226
Horten, Walter, 25, 52, 224, 228, 231, 236, 237, 241, 242, 258, 337, 238, 239, 272, 314
Horten, Wolfram, 226, 227
Huber, Dr.Ing., 86
Hugo Junkers Society, 253
HWK, bi-fuel liquid rocket drive of, 40, 44, 54, 57, 99, 150, 261, 272, 298, 326
HWK R1, 162, 163, 203, 277, 278, 279
HWK R2, 279
HWK 501, solid fuel booster rocket of, 33, 264
HWK 509A-1, 11, 32, 36, 42, 43, 55, 80, 85, 93, 128, 132, 133, 141, 145, 151, 230, 251, 254, 260, 272, 273, 275, 283, 287, 289, 293, 299, 313
HWK 509A-2, 41
HWK 509C-1, 54, 55, 170, 243, 260, 261, 262

I
IAe-33, 126
IAe-37, 233, 239, 338
IAe-48, 238
Ilyushin Il 11-2, 293
Ilyushin Il 16, 313
Ilyushin Il 26, 338
Ilyushin Il 28, 338
Ilyushin, Sergei, 293
Imperial College, 123
Imperial War Museum, 285
Inspection of fighters Command, 228, 229
Instituto Aerotècnico (IAe), 231, 237, 239, 338
Inter Avia magazine, 146

J
Jägdgruppe (JG) 26, 52, 226, 228
Jakobs, Hans, 94, 271
Japan, 63, 120, 154
JATO (Jet Assistance Take-Off), 257
Jensen, Dip.Ing. Adolf, 164
Jeschonnek, Generaloberst Hans, 23
Joliot-Curie - Frederic, 241
Ju G 38, 247, 248

Ju J-1000, 250
Ju 1, 246
Ju 10, 247
Ju 52, 244, 249, 251
Ju 52/3m, 248
Ju 86, 181
Ju 87, 59, 67, 71, 74, 185, 186, 188, 189, 251
Ju 88, 107, 139, 249, 251
Ju 90, 250
Ju 188, 251
Ju 287, 14, 250, 251, 252, 253, 256, 262, 264, 267, 337, 338
Ju 287 V1, 199, 251, 255, 262, 264
Ju 287 V2, 251, 262, 264
Ju 287 V3, 7, 10, 74, 82, 83, 243, 256, 262, 263, 264, 265, 266
Ju 287 V4, 264, 265
Ju 290, 313, 314
Ju 352, 264
Ju 388, 264
Ju P.009, 243, 254
Ju P.126 "Elli," 79, 170, 198, 242, 243, 252, 255
Ju P.127 "Walli," 10, 42, 52, 169
Ju P.128, 125, 175, 189, 190, 198, 243, 251, 252, 261, 306
Ju P.130, 199, 243, 245, 256, 257
Ju P.140, 9, 198, 234, 243, 258, 259, 315
Ju P.248, 9, 191, 199, 250, 251, 252, 254, 260, 262, 274
Ju P.131, 7
Ju R-1, 247
Jülge, Hauptmann, 264
Julia, Heinkel target-defense interceptor, 14, 42, 52, 169, 213, 335
Jumo 004A, 249, 250
Jumo 004B, 18, 31, 32, 34, 35, 36, 90, 126, 142, 166, 167, 189, 224, 228, 232, 236, 237, 241, 250, 263, 264, 265, 279, 286, 287, 298, 300, 301, 302, 325
Jumo 004C, 189
Jumo 004D, 250, 289
Jumo 004H, 250
Jumo 012, 90, 251
Jumo 022K turboprop, 90, 91, 250, 320
Junkers Flugzeugbau, 10, 46, 51, 64, 98, 131, 144, 160, 238, 241, 242, 243, 244, 245, 247, 249, 252, 253, 254, 255, 258, 259, 260, 261, 262, 263, 265, 298, 315, 336, 338
Junkers L-1, 249
Junkers L-2, 249
Junkers L-5, 249
Junkers Luftverker, 247
Junkers Mortoren Werke (Jumo), 29, 31, 119, 181, 244, 249, 252, 253, 254, 320
Junkers, Professor Hugo, 26, 180, 235, 243, 244, 245, 246, 248, 249, 250, 254, 256, 268

K
Kaiser Wilhelm Institute, 118, 241
Kamikaze pilots, Japanese, 137
Kamikaze, theory of ,135, 137, 139
Kamikaze aircraft, German thinking on, 137, 139, 333
Kappus, Dipl.Ing., 325
Käther, Dipl.Ing., 122
Kawasaki of Japan, 63
Keller, Richard, 256
Kern County (California) Airport, 18
Kesselring, Generalleutnant Albert, 20, 23, 106
KG40, 183
Kirchheim, Village of, 55
Klages, Dip.Ing. Paul, 124
Klemm Flugzeugbau, 63
Klemm Kl 35, 54
Klöckner, Eric, 93
Knemeyer, Oberst Siegfried, 26, 73, 74, 241, 242, 258, 314, 315
Kobe, Japan, 62
Koch, Erich, 181
Komet, 93, 162, 272, 283, 284
Konzentrationslagers (KZ), 53
Koppenberg, Dr.Ing. Heinrich, 244, 251
Korea, 6
Korean War, 338
Kosin, Dip.Ing. Rüdinger, 31, 36, 37, 42, 258
Kracht, Dr.Ing. Felix, 7, 11, 23, 93, 95, 98
Kraemer, Fritz, 271
Kramer, Dr.Ing. Max, 330
Kronach Lorin Duct-Ramjet Engine, 288, 289
Kuibyshyev, USSR, 320
Kummersdorf-West, 330
Kunzel, Dr.Ing., 98, 260
Kupper, Dr.Ing. August, 145

L
Lake Constance, 62, 106, 282
Lake Zell, Austria, 299
Laminar-flow wings, 32
Lancaster, British Heavy Bomber, 24
Langenhagen, City of, 22

Langley Field Virginia, 281
Legion "Condor," 188
Lehmann, Dr.Ing., 98, 260
Leist, Professor Dr. Karl, 249
Lerche-1, Heinkel AG VTOL, 130
Lerche-2, Heinkel AG VTOL, 131, 132
Li Delta-I, 271, 274, 275
Li Delta II, 274
Li Delta III, 272, 274, 275
Li Delta IVc, 275, 277
Li DM-1, 9, 10, 200, 224, 230, 238, 239, 274, 275, 276, 277, 281, 285, 286, 288, 338
Li DM-2, 275
Li "Glieter Bomber," 214, 215
Li Storch-5, 269
Li X-112 "Flying Flounder," 282, 283, 284
Li P.11, 199, 279, 281, 284, 286, 287
Li P.12, 9, 199, 279, 287, 288, 338
Li P.13A, 9, 199, 216, 217, 231, 238, 280, 281, 288, 289
Li P.13B, 199, 288, 289
Li P.15, 199, 289
Li P.20, 280
Li Project "X," 271, 273, 277, 280
Lilienthal, Otto, 118, 152
Lilienthal Prize Award, 225, 230
Lilienthal Society, 70
Lindbergh, Charles, 154
Lippisch, Dr. Alexander Martin, 6, 7, 9, 10, 16, 17, 18, 24, 93, 106, 115, 134, 145, 149, 162, 172, 175, 187, 224, 225, 232, 233, 234, 236, 237, 238, 239, 240, 251, 261, 268, 269, 270, 271, 272, 273, 274, 275, 276, 277, 278, 279, 280, 281, 283, 284, 285, 286, 287, 288, 289, 291, 293, 316, 317, 336, 337, 338, 339
Lippisch Research Corporation, 282
London to Isle of Man Air Race, 292
Lorenz, Dr.E., 276
Lorin Duct-Ramjet Engine, 8, 286, 287, 288, 320, 325
Lucht, Robert, 28
Luftfahrt Forschungs Anstalt (LFA), 132, 179, 280, 285, 287, 320, 322
Lufthansa, 59, 244, 294
Luftwaffe Sonder Kommando #9, 229
Luftwaffe Quarter Master's Office, 228, 235, 250
Luftwaffe , 21, 23, 28, 29, 38, 40, 48, 51, 59, 108, 114, 154, 159, 188, 189, 280, 294, 297, 301, 316, 317, 326, 327, 329, 330, 331
Lulko turbojet engines, 266
Lutz, Otto, 320
Lycoming Jet Engine Division - AVCO Corporation, 249, 325
LZ-127 "Graf Zeppelin" airship, 114

M
Mader, Professor Dr. Otto, 251
Magazine for Flight Sciences, 30
Manchurian War, 63
Martin Aircraft Company, 117, 124, 125
Martin-Bell "Dyna Soar" space shuttle, 124
Martin-Marietta Aerospace Corporation, 124, 298
Mathias, Dip.Ing. Gotthold, 124
MBB, 72, 299
Me Bf 108 "Taifun," 290, 292, 293, 299
Me Bf 109, 32, 94, 139, 155, 291, 292, 298, 327
Me Bf 110, 145, 146, 250
Me Bf 109E, 291
Me Bf 109G, 291
Me Bf 109H, 67
Me Bf 109K, 80
Me Bf 163, 160, 278
Me 20, 294, 295
Me 155, 67
Me 163B, 9, 10, 18, 19, 23, 24, 26, 162, 230, 237, 260, 261, 264,272, 273, 274, 278, 280, 283, 284, 285, 286, 288
Me 163B V1, 284
Me 209 V-1, 291, 292, 293, 297
Me 262A-O, 300
Me 262A-1A, 301
Me 262-C1A, 299
Me 263, 291
Me 328, 80, 93, 94, 137, 171
Me 410, 107, 139
Me 423 "Giant," 22
Me P.262, 10, 26, 68,75, 76, 159, 160, 166, 167, 200, 250, 260, 290, 292, 293, 294, 296, 297, 298, 299, 300, 302, 303, 304, 305, 309, 317, 325, 336, 337
Me P.262 HG-2, 10, 200, 290, 298, 300, 301, 302, 303, 316
Me P.262 HG-3, 306
Me P.264 V1, 82, 310, 311, 312, 313, 314, 315, 316
Me P.1065, 10, 293, 294
Me P.1099, 302, 303
Me P.1101, 10, 11, 16, 25, 125, 155, 159, 169, 172, 200, 290, 298, 304, 305, 306, 308, 309, 310, 311, 312, 317, 337, 338
Me P.1103, 170
Me P.1104, 170, 306

Me P.1106, 189, 190, 200, 219, 290, 298, 307, 312, 317
Me P.1106R, 313
Me P.1107.2, 313, 314
Me P.1107A, 200, 290, 310, 313, 314, 315, 316
Me P.1107B, 201, 290, 316
Me P.1108, 201, 283, 290, 316, 317
Me P.1110, 125, 317
Me P.1111, 125, 201, 315, 317, 318, 319, 338
Me P.1111/1112, 317
Me P.1112, 201, 291, 298, 317, 318, 319, 338, 339
Mercedes D 111a engine, 249
Messerschmitt, AG, 7, 22, 23, 46, 51, 71, 145, 160, 241, 242, 243, 254, 258, 278, 280, 285, 286, 290, 291, 293, 294, 295, 296, 297, 298, 299, 300, 301, 303, 304, 305, 306, 307, 308, 309, 312, 313, 314, 316, 317, 318, 338
Messerschmitt Flugzeugbau GmbH, 292
Messerschmitt "Kabineroller" KR 200 auto, 295
Messerschmitt post-war prefabricated houses, 295
Messerschmitt, Professor Willy, 7, 10, 17, 28, 115, 125, 155, 156, 230, 251, 273, 278, 279, 280, 284, 290, 291, 292, 293, 294, 295, 296, 297, 298, 299, 300, 301, 304, 305, 306, 317, 336
Messerschmitt Zick-Zack sewing machine, 295
Messerschmitt-Bölkow-Blohm (MBB), 254, 299
Meyer, Dr.Ing. Eberhard, 94
MG 151 20 mm cannon, 32, 254
MiG-15, 6, 14, 15, 114, 124, 128, 178, 338, 339
MiG-19, 128, 339
MiG-I-270, 191, 262
MiG turbojet fighters, 15, 125
Mikoyan, Artem I., 10, 15, 262, 339
Milch, Generalfeldmarschall Erhard, 20, 21, 31, 155, 159, 162, 166, 225, 244, 251, 284, 293, 294, 295, 297, 317
Miniatur Jäger Program, 79, 80, 255
Mistel (mistletoe), 139
Mitchell Field, New York, 69
Mittelhüber, Dr.Ing. Ludwig, 120, 122, 141, 142
Mittelklasswagen P.511 auto, 295
Mittleholzer, Walter, 106
MK 108 30 mm cannon, 40, 73, 79, 81, 109, 127, 239, 242, 259, 261, 284, 289, 302, 309, 318
MK 108 20 mm cannon, 43
MK 103 30 mm cannon, 133
MK 103 20 mm cannon, 289
MK 103 30 mm cannon, 76, 131, 238, 241, 287, 298, 317
Mödel, Dip.Ing. Rolf, 96, 188
Morel, August, test pilot, 130
Multhopp, Folker, 118
Multhopp, Frau, 118
Multhopp T-Tail, 76, 99, 117, 121, 124, 125, 127, 128, 263, 290, 312, 313, 314, 316, 338, 339
Multhopp, Hans, 6, 7, 8, 15, 17, 18, 23, 110, 111, 112, 113, 114, 116, 117, 119, 120, 121, 122, 123, 124, 125, 126, 127, 128, 132, 133, 141, 142, 178, 336, 338, 339
Multhopp, Heiko, 118
Multhopp, Ralf, 118
Muroc Air Force Base, 101
Myhra, Olaf, 6

N
N-1M flying wing, 146, 235
Nachschub Amt (Supply Office), 23
Nacht Jäger, 40
NASA AD-1 experimental aircraft, 72, 77, 339
NASA Dryden Flight Research Center, 77
National Advisory Committee for Aeronautics (NACA), 120, 281, 286
National Aeronautics and Space Administration, 71, 72, 77, 81, 154, 338
National Air and Space Museum-Washington, 30, 31, 38, 57, 147, 224, 225, 236, 237, 285
National Educational Television, 282
National Socialist German Worker's Party, 68, 243, 244, 245, 295, 299, 333
Naumann, Dip.Ing., 122
Naval Air Material Center, 281
Nazism, 23, 153, 299, 333, 336
Nene turbojet engines, 266
Never Satisfied, 21
New York City, 225, 314
Nickolaus, Friedrich, 180
NK-12M Soviet version of the Jumo 022, 91, 251
Norsk Hydro Hydrogen Electrolysis, 241
North African War Front, 300
North American Aircraft, 68, 339
North Dakota, 6
North Pole, 227
Northrop All-Wing Aircraft, 257
Northrop Corporation, 231
Northrop, John (Jack), 146, 233, 235
Norway, Vemonk, 241
Nuclear-powered Aircraft, 71
Nürmberg War Crimes Tribunal, 299

O

Oberammergau, Messerschmitt Research Center, at, 78, 290, 299, 304, 306, 307, 308, 309, 317, 319, 338
Oberkommando der Luftwaffe (OKL), 126, 134, 189, 191
Oblique wing, 71, 81
Oestrich, Dr.Ing. Hermann, 82
OHKA - Japanese piloted suicide bomb, 135, 136
OKB-2, 14
OKB-4, 14
Opel Rak.2, 269
Operation Paperclip, 69, 70, 281, 298, 310
Operation Seahorse, 236
Oranienburg Luftwaffe Air Station, 98, 224, 225, 230, 236, 241
Oranienburg (Heinkel AG), 160
Orbital Bomber, 320
Order Pour le Mérite, 22, 28, 30, 118, 156, 157

P

P-47 "Thunderbolt," 6, 37, 46
P-51, 6, 46
P-80 Lockheed, 76
P5M-2, Martin Seaplane, 124
Pabst, Dr.Ing. Otto, 122, 131, 132, 141
Pan Am, 245
Paris to Rome Air Race, 292
Parseval, Dr. August, 284
Payen Pa-49, 338
PC-1000 RS armor-piercing rocket, 189
Peenemünde, 272, 326
Peenemünde-West, 279
Peròn, Juan, 10, 231, 111, 112, 113, 122, 124, 125, 126, 127, 128, 339
Peschke Flugzeugbau Werke, 229
Pilotus sailplane, 292
Piper Aircraft, 124
Pittsburgh, Pennsylvania, 316
Pobjoy "R" engine, 275
Podberesje, USSR, 14, 99, 253, 260
Pogo, 130
Pohlmann, Dip.Ing. Hermann "Hans," 71, 186
Poland, 278, 336
Polish Matter, 336
Porro, General Jelice, 225
Prandtl, Professor Dr.Ing. Ludwig, 111, 112, 118, 124, 234, 249
Priem am Chiensee, 274
Prone flying position, 187, 188, 189
Pulqui Dos, 10, 17, 112, 122, 125, 126, 127, 128, 339

Q

R

R4M air-to-air missiles, 333, 334
Raketenbramnkammer, 322
Ramenskoje, USSR, 265, 266
Rasche, Theo, 115
Rebeski, Dip. Ing. Hans, 31
Rechlin Test Establishment, 255
Red Devil, 158
Red Army, 122, 182, 183, 188, 189, 191, 249, 251, 252, 262, 287
Reflex-swept wing, 74
Regner, Dip.Ing. Hans, 162
Reichenburg, piloted Fi 103, 136, 137, 154
Reichsverteidigungs, 251
Reichwehr (German Defense Ministry), 154
Reissner, Professor, 246
Reitsch, Hanna, 93, 137, 227, 332, 333
Republic Aircraft Corporation, 18
Reynolds Number, 274
Rhein-Flug, 282
Rhön-Rossiten Gesellschaft (RRG), 92, 272, 274
Rhön Wasserkuppe, 274
Rhön Wasserkuppe, annual summer sailplane competitions of, 73, 226
Richard Vogt Way, 72
Rocket Propulsion Establishment, Westcott, England, 284
Rockwell International, 68
Rohrbach "Bero," 118
Rohrbach "Rofix," 115, 118, 119
Rohrbach "Roland," 115
Rohrbach "Romar" flying boat, 114, 118
Rohrbach, Dr. Ing. Adolf, 106, 114
Rohrbach Metallflugzeugbau, 119
Rolls-Royce "Avon" turbojet, 142
Rolls-Royce Limited, 123, 156, 158, 162
Rolls-Royce "Kestral V-12" engine, 156, 158
Rolls-Royce "Merlin" engine, 156
Roma, Italian battleship, 329
Roosevelt, President F.D., 241
Royal Aircraft Establishment (RAE), 112, 113, 121, 122, 123, 339
Royal Air Force, 285, 297
Royal Air Station - Braunschweig, 147
Ruhrstahl, AG, 327
Russia, 114, 157, 159, 189, 296, 324
Russian "Neuer," 302, 303
Russians, 124, 159

S

SA, 295
SAAB, 15
SAAB 29A, 10, 15, 338
SAAB 210, 338
SAAB J-35, 189, 190
Samolot 5-2, 100
Sänger, Dr.Ing. Eugen, 6, 18, 179, 320, 321, 322, 323, 324, 325
Sänger Orbital Bomber, 18, 201, 321, 322, 323, 324
Sänger-Bredt, Dr.Ing. Irene, 6, 320, 322, 324
Saur, Karl, 23
Sausewind, 157, 158
Sawjolovo, USSR, 98
Saxony Air Races of 1928, 292
SC-1000 "Hermann" armor-piercing rocket, 189
Schainer, George, 71
Schaudt, Herr, 62
Scheidhauer, Heinz, 236
Schelp, Hellmut, 249, 300
Schlageter, Albert Leo, 291
Schlageter Squadron, 228, 291
Schmidding, solid fuel rocket boasters of, 54, 170, 334
Schneider, Dip.Ing. Horst, 90, 91, 323, 324
Schnellbomber, 89, 316, 317
Schnellstjäger, 78
Schulze, Ferdinand, 118
Schutzstaffeln (SS), 55, 68, 69, 280
Schwabenland, German steamship, 59
Schwalbe "Swallow," 299
Science Museum, 285
Scientific Society for Aviation, 30
SD-1400 air-to-air missile, 328
SE-1801, 265
SE-1802, 265
Secret of Flight, The, 282
Segelflugzeugbau Harth und Messerschmitt, 292
Segelsdorf, Village of, 37
Seidemann, Hans, 158
SFECMAS Ars. 5.5 pilotless target aircraft, 322
Shakespeare, William, 51
Shenstone, Beverly, 271
Siebel Flugzeugbauwerke KG, 93, 100, 254
Siebert, Leutnant Lothar, 52, 53, 55
SM01 engine, 179
SNCASE, 148
SNECMA, 129, 131, 253, 267
Society of German Engineers, 70
Soldenhoff, Alexander, 145
Sombold So P.344, 201, 327, 328, 331, 333
Sombold, Heinz G., 326, 327, 328, 329, 333,
Soviet Army, troops of, 36, 37, 41
Soviet Union, 122, 125
Soviet 022K turboprop, 91, 251
Soviets, 99, 124, 188, 191, 324
Space Vehicle 5 (SV-5), 124
Spain, 107, 188, 299
Speer, Albert, 20, 23, 78, 251, 299
Spiegel (radar scanner), 150
Spielzeug, 237
SS, 130, 170
Stamer, Fritz, 162
Stampa, Dip.Ing., 122
Stanley, Robert, 310
Stevens Institute - Hoboken, New Jersey, 249, 300
STOL, 179
Strahlbomber, 74
Stuka, dive bomber, 59, 114, 188, 189
Stumff, Generaloberst Hans, 23
Sturmvögel "Stormbird," 299
Stuttgart Oberrealschule, 60
Swiss N-20, 179

T

T-55, 250
T-Stoff, 19, 54, 57, 128, 130, 272, 283
T40-A-6 Allison turboprop, 129
Tailless aircraft, 235
Tank, Dr.Ing. Kurt, 10, 11, 16, 17, 19, 21, 84, 106, 110, 113, 114, 115, 118, 119, 120, 124, 125, 126, 127, 128, 132, 141, 142, 284, 336, 339

Taube, 145
Technical High School of München, 106
Technical College of Berlin, 115
Technical University of Berlin, 267
Technical University of Hannover, 28
Technical University of München, 292
Technisches Amt (Technical Office of the German Air Ministry), 23, 275, 276, 300
Tempelhof Field-Berlin, 246
Thalau, Dip.Ing. Karl, 124
Thick cantilever wing, 246
Third Reich, 70, 78, 79, 106, 124, 144, 160, 291, 296, 299, 301
Trauter, Dip.Ing. Barnard, 99
Treaty of Versailles, 17, 21, 117, 144, 153, 243, 244, 248, 328
Trimble, George Jr., 117
Troisiéme Congress of 1958, 323
Tu 20 "Bear" Soviet long-range bomber, 91
Tu 114 Soviet long-range passenger aircraft, 91
Twain, Mark, 275
Type 131, Soviet aircraft project, 266
Type 140, Soviet aircraft project, 266, 267
Type 150, Soviet aircraft project, 338, 267
Type 151, Soviet aircraft project, 338

U
U.S. Army Technical Intelligence (USATI), 281
U.S. 8th and 15th Army Air Force, 48, 297
U.S. 9th Army, 36, 37
Udet's Folly, 134, 154
Udet, Generaloberst Ernst, 19, 20, 21, 23, 28, 59, 63, 64, 65, 75, 93, 94, 119, 158, 162, 225, 227, 241, 245, 251, 279, 284, 292, 293, 294, 296, 297, 314
Uhu, 145
United States of America, 297
University of Bonn, 226
University of Göttingen, 157, 231
University of Heidelberg, 280
University of Stuttgart, 58
Ural Mountains, 23, 106, 189, 250, 257
US Air Force Museum, 311
USAAF T2-1011, 57
USSAF, 257
USSR, 7, 11, 12, 14, 39, 99, 122, 159, 183, 189, 191, 251, 257, 265, 266, 267, 338, 280, 296, 299

V
V-1 Flying bomb, 50, 137, 255
V2 rocket, 157, 322, 330, 323, 324
V-E Day, 122
V-type tail-rudder control system, 313, 316, 318
Valiant MK.1, 337
Vampyr glider, 28
Vemonk, Norway, 258
Vereignite Flugtechische Werke (VFW), 160
Verein für Raumschifahrt (VFR), 273
VFW Fokker 614, 88
Vickers Supermarine 508, 176, 177, 178
Victor 8 Handley-Page Bomber, 36, 37, 313, 331, 338
Vogt Design, 65
Vogt, Dr.Ing. Richard, 58, 59, 60, 61, 62, 63, 64, 65, 66, 67, 68, 69, 70, 71, 72, 73, 74, 75, 76, 77, 78, 79, 80, 81, 106, 139, 150, 173, 251, 338, 339
Voigt, Dip.Ing. Woldemar, 7, 10, 11, 16, 290, 292, 293, 298, 300, 301, 302, 304, 305, 306, 309, 310, 312, 313, 317, 318, 336, 337, 338, 339
Volksjäger "People's Fighter," 10, 23, 47, 65, 69, 78, 142, 154, 171, 224, 239, 255, 336
von Blomberg, Wernher, 21
von Braun, Wernher, 6, 11, 51, 157, 162, 330, 323
von Gronau, Wolfgang, 106
von Hindenburg, Generalfieldmarshall, 250
von Holst, Dr.Ing. Erich, 128, 130, 131
von Ohain, Dr.Ing. Joachim Pabst, 153, 154, 157, 163, 164
von Opel, Fritz, 162, 270, 271, 273, 274, 277
von Parseval, Professor Dr. August, 113, 117, 118
von Richthofen, Manfred, 28
von Richthofen Squadron, 22
von Richthofen Zirkus, 296
von Zborowski, Dip.Ing. Hellmuth, 130, 131
von Zeppelin, Ferdinand Graf, 106, 117, 332
VTOL, 128, 129, 130, 131, 132

W
Wagner, Dr.Ing. Herbert, 89, 181, 182, 249, 330
Wahpeton, North Dakota, 6
Walter, Hellmuth, 11, 132, 162, 179, 261, 275, 277, 278, 279, 280
Walter KG rocket drive, 156, 162
Walter Raketentriebwerken, 261, 276, 279
War Museum - Canberra, Australia, 285
Warnemünde, 62
Warsitz, Erich, 153, 156
Washington, DC, 314
Weber, Professor Moritz, 118
Wehrmacht, 21, 134, 262
Wendel, Fritz, 292
Werde Gesellschaft - Freilassing, 168
Wever, Generalleutnant Walther, 19, 22, 23, 104, 106, 249, 250
White Hells of Pitz Palo-the Alps, 20
Whittle, Sir Frank, 154
Wiegmeyer, 271
Winter, Martin, 112, 122, 123, 339
Wirtschafts Amt (Contracts Office), 23
Witt, Major Hans-Hugo, 226
Wocke, Dip.Ing. Hans, 6, 7, 14, 250, 251, 252, 253, 256, 257, 258, 263, 267, 336, 338
Wolff, Dip.Ing. Herbert, 124
Woods, Robert, 309
World War I, 17, 21, 23, 28, 30, 59, 117, 153, 246, 247, 248, 292, 296
World War II, 18, 24, 70, 76, 107, 111, 148, 181, 185, 189, 249, 283, 314, 324
World Soaring Championship 1952, 231
Wright Brothers, 29, 235
Wright Field, Dayton, Ohio, 298
Wright, Orville, 28, 246, 268, 269
Wright-Patterson Air Force Base, 69, 70, 148, 236, 274, 281, 309, 310
Wright, Wilbur, 246
Wulf, Georg, 115
Wurster, Dr.Ing., 293

X
X-1, Bell, 111, 112, 123
X-4 Ruhrstahl, AG missile, 327, 330, 331
X-4, Northrop, 318, 338
X-5, Bell, 309
XB-46, Convair, 36, 38, 338
XB-48, Martin, 37, 38, 338
XB-51, Martin, 117, 124
XF-84H, 17, 18
XF-86, 338
XF-92A "Cutlass" Convair, 277, 278, 286, 338
XF2Y-1 "Sea Dart," 338
XF7U-1 "Cutlass" Chance-Vought, 38
XFV-1, Lockheed, 129, 130
XFY-1, Convair, 129, 130
XP-55, 184, 185
XP-87, Curtiss, 37
XP6M, Martin Seaplane, 124
XT40, Allision Turboprop, 17

Y
Yaeger, Charles "Chuck," 99, 100, 101, 112, 293
Yak-1, 293
Yak-3, 293
Yak-7, 293
Yak-15, 10
Yakvlev, Aleksandr, 10, 293
YF-102 "Delta Dagger," 338

Z
ZAGI, 99, 324
Zeise, Leutnant Wolfgang, 98, 99, 100
Zeppelin Abteilung Flugzeugbau GmbH, 332
Zeppelin Airship, 62
Zeppelin Flugzeugwerke, 62, 270, 271,
Zeppelin Project "Rammer," 201, 220, 221, 332, 333, 334, 335
Zerstörer "Destroyer," 175
ZIAM, 324
Ziller, Leutnant Erwin, 228, 236
Zindel, Dip.Ing. Ernst, 248, 250, 252, 253, 257, 258, 336, 338,
ZTL "Zweikreis-Turbine Luftstrahl," 88
Zübert, Hans, 51, 54

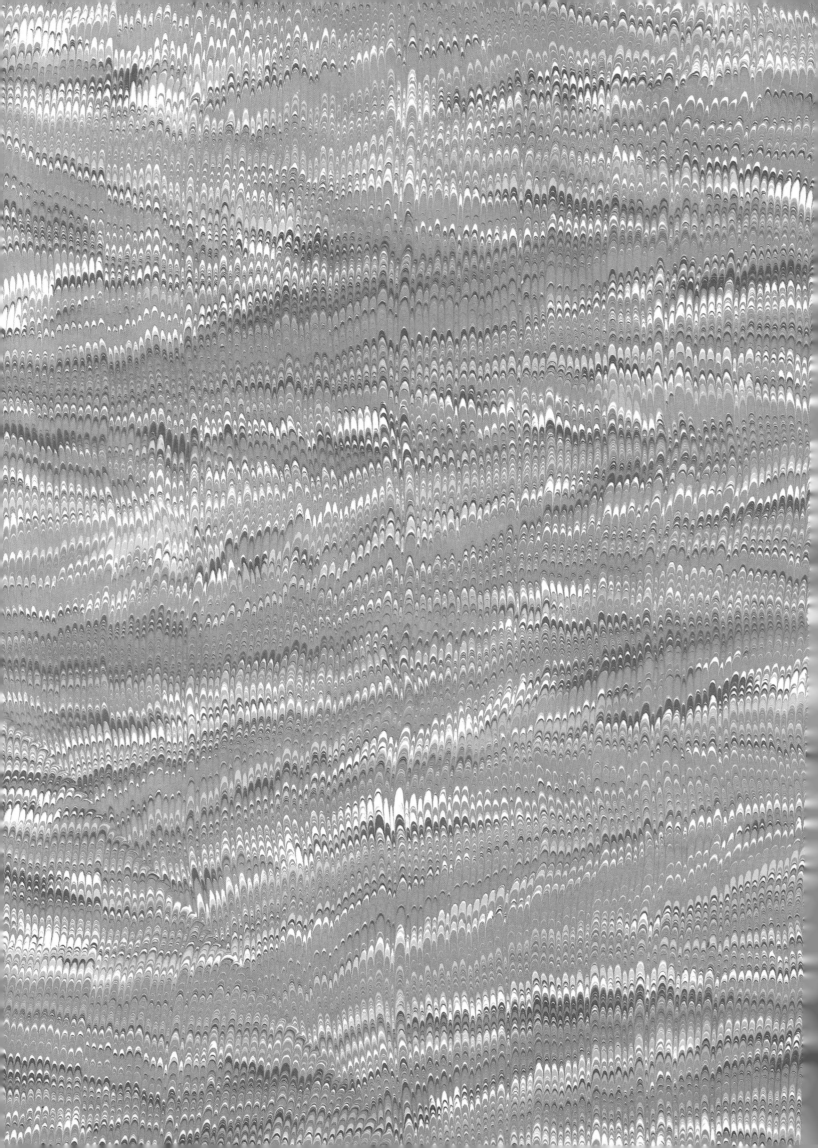